Bernd Gischel

EPLAN Electric P8
Reference Handbook

2nd Edition

Hanser Publishers, Munich   Hanser Publications, Cincinnati

The Author: Dipl.-Ing. Bernd Gischel, Lünen, Germany
Translated by think global GmbH, Berlin, Germany

Distributed in the USA and in Canada by
Hanser Publications
6915 Valley Avenue, Cincinnati, Ohio 45244-3029, USA
Fax: (513) 527-8801
Phone: (513) 527-8977
www.hanserpublications.com

Distributed in all other countries by
Carl Hanser Verlag
Postfach 86 04 20, 81631 München, Germany
Fax: +49 (89) 98 48 09
www.hanser.de

The use of general descriptive names, trademarks, etc., in this publication, even if the former are not especially identified, is not to be taken as a sign that such names, as understood by the Trade Marks and Merchandise Marks Act, may accordingly be used freely by anyone.

While the advice and information in this book are believed to be true and accurate at the date of going to press, neither the authors nor the editors nor the publisher can accept any legal responsibility for any errors or omissions that may be made. The publisher makes no warranty, express or implied, with respect to the material contained herein.

Library of Congress Cataloging-in-Publication Data

Gischel, Bernd.
  EPLAN electric P8 reference handbook / Bernd Gischel ; translated by Think Global GmbH. – 2nd ed.
    p. cm.
  ISBN-13: 978-1-56990-503-6 (softcover)
  ISBN-10: 1-56990-503-7 (softcover)
  ISBN-13: 978-3-446-42674-0 (softcover) 1. Electrical engineering–Data processing. 2. Electrical engineering–Computer programs. 3. EPLAN electric P8. I. Title.
  TK153.G5413 2011
  621.30285–dc22
                                    2011007238

Bibliografische Information Der Deutschen Bibliothek:
Die Deutsche Bibliothek verzeichnet diese Publikation in der Deutschen Nationalbibliografie; detaillierte bibliografische Daten sind im Internet über <http://dnb.d-nb.de> abrufbar.

ISBN 978-3-446-42674-0

All rights reserved. No part of this book may be reproduced or transmitted in any form or by any means, electronic or mechanical, including photocopying or by any information storage and retrieval system, without permission in writing from the publisher.

© Carl Hanser Verlag, Munich 2011
Production Management: Stefanie König
Coverconcept: Marc Müller-Bremer, www.rebranding.de, München
Coverdesign: Stephan Rönigk
Printed and bound by Kösel, Krugzell
Printed in Germany

Content

Foreword and acknowledgements 11
Important notes 12

1 Install EPLAN Electric P8 15
1.1 Hardware 15
1.2 Installation 16
1.3 Important notes for Version 1.9.x users 23
1.3.1 Parallel operation with Version 1.9.x 23

2 The basics of the system 25
2.1 Five principles that apply to working with EPLAN Electric P8 25
2.2 Directory structure, storage locations 27
2.3 Settings – General 30
2.3.1 Settings – Project 30
2.3.2 Settings – User 41
2.3.3 Settings – Workstation 47
2.3.4 Settings – Company 48
2.4 EPLAN and multiple starts? 48
2.5 Properties 49
2.5.1 Project properties 50
2.5.2 Page properties 51
2.5.3 Symbol properties (components) 52
2.5.4 Form properties 53
2.5.5 Plot frame properties 53
2.6 Buttons and popup menus 53
2.6.1 Device dialog buttons 54
2.6.2 Buttons in dialogs (configuring) 55
2.6.3 Buttons in dialogs such as filter schemes 55

2.6.4	Small blue triangle button	56
2.6.5	Property arrangements (components)	56
2.6.6	Format properties	57
2.6.7	Buttons (small black triangles)	58
2.6.8	Dialogs for schemes	59
2.7	Master data	60
2.8	Operation	61
2.8.1	Using the keyboard	61
2.8.2	Using the mouse	62
2.9	The user interface – more useful information	62
2.9.1	Using workspaces	62
2.9.2	Dialog display	63

3 Projects 67

3.1	Project types	68
3.1.1	Project types in EPLAN	69
3.1.2	Project types in EPLAN	69
3.1.3	Project templates and basic projects	70
3.2	Create a new project	72
3.2.1	A new project (from a basic project)	72
3.2.2	A new project (with project creation wizard)	75

4 The graphical editor (GED) 83

4.1	Page navigator	83
4.1.1	Page types	85
4.1.2	The Popup menu in the page navigator	86
4.1.3	Page navigator filter	101
4.2	General functions	101
4.2.1	The title bar	102
4.2.2	The status bar	102
4.3	Coordinate systems	103
4.3.1	Graphical coordinate system	103
4.3.2	Logical coordinate system	104
4.4	Grid	104
4.5	Increments, coordinate input	106
4.5.1	Increment	106
4.5.2	Coordinate input	107
4.5.3	Relative coordinate input	108
4.5.4	Move base point	108
4.6	Graphical editing functions	109
4.6.1	Graphical objects - lines, circles, rectangles	110

4.6.2	Trim, chamfer, stretch and more	111
4.6.3	Group and ungroup	117
4.6.4	Copy, move, delete,	118
4.6.5	Dimensioning	119
4.7	**Texts**	**122**
4.7.1	Normal (free) texts	123
4.7.2	Path function texts	123
4.7.3	Special texts	124
4.7.4	The Properties – Text dialog	125
4.8	**Components (symbols)**	**129**
4.8.1	Insert components (symbols)	129
4.8.2	Properties (components) dialog – [device] tab	131
4.8.3	The Display tab	132
4.8.4	The Symbol/function data tab	135
4.8.5	The Parts tab	137
4.9	**Cross-references**	**137**
4.9.1	Contact image on component	138
4.9.2	Contact image in path	140
4.9.3	Special feature: pair cross-reference	142
4.9.4	Device selection settings	144
4.10	**Macros**	**145**
4.10.1	Types of macros	145
4.10.2	Macros with value sets	149

5 Navigators 157

5.1	**Overview of possible navigators**	**158**
5.1.1	Additional navigators and modules	160
5.2	**Device navigator**	**162**
5.2.1	Devices / Assign main function	164
5.2.2	Devices / Cross-reference DTs	166
5.2.3	Devices / Synchronize function texts	167
5.2.4	Devices / Interconnect devices	168
5.2.5	Devices / Number (offline)	175
5.2.6	Device navigator / New	178
5.2.7	Device navigator / New device	179
5.2.8	Device navigator / Place / [Representation type]	180
5.3	**Terminal strips**	**181**
5.3.1	Terminal strips / Edit	182
5.3.2	Terminal strips / Correct	185
5.3.3	Terminal strips / Number terminals	186
5.3.4	Terminal strip navigator / New	190
5.3.5	Terminal strip navigator / New functions	190

5.3.6	Terminal strip navigator / New terminals (devices)	192
5.3.7	Terminal strip navigator / Place / [Representation type]	194
5.4	Plugs (Sockets)	195
5.4.1	Plugs / Edit	195
5.4.2	Plugs / Correct	195
5.4.3	Plugs / Number pins	196
5.4.4	Plug navigator / New...	196
5.4.5	Plug navigator / New functions	197
5.4.6	Plug navigator / Generate plug definition	197
5.4.7	Plug navigator / Generate pin	198
5.4.8	Plug navigator / Place / [Representation type]	199
5.5	Cables	199
5.5.1	Cables / Edit	200
5.5.2	Cables / Number	201
5.5.3	Cables / Generate cables automatically	204
5.5.4	Cables / Automatic cable selection	206
5.5.5	Cables / Assign conductors	207
5.5.6	Cable navigator / New...	209
5.5.7	Cable navigator / Place / [Representation type]	209
5.5.8	Cable navigator / Number cable DT	210
5.5.9	Cable navigator / Number DT	210
5.6	PLC	211
5.6.1	PLC / Write back connection point designations	213
5.6.2	PLC – Set data types	215
5.6.3	PLC / Addresses / assignment lists	216
5.6.4	PLC / Address	218
5.6.5	PLC navigator / New...	220
5.6.6	PLC navigator / New functions	220
5.6.7	PLC navigator / New device...	222
5.6.8	PLC navigator / View	222
5.7	Parts / Devices / Bill of materials	223
5.7.1	Popup menu functions	225
5.8	Assign function (navigators)	229

6 Reports ... 231

6.1	What are reports?	232
6.2	Report types	233
6.3	Types of graphical reports	233
6.3.1	Report types (forms)	233
6.3.2	Special connection diagrams	247
6.4	Settings (output options)	248

6.4.1	The Display / output project setting	248
6.4.2	The Parts project setting	250
6.4.3	The Output to pages project setting	251
6.5	Generate reports	256
6.5.1	The Reports dialog	257
6.5.2	Generate reports without templates	259
6.5.3	Popup menus in the Reports tab	265
6.5.4	Generate reports with templates	266
6.6	Other functions	272
6.6.1	Generate project reports	272
6.6.2	Update	273
6.6.3	Settings for automatic updates	273
6.7	Labeling	274
6.7.1	Settings	274
6.8	Edit properties externally	280
6.8.1	Export data	282
6.8.2	Import	288

7 Management tasks in EPLAN ... 291

7.1	Structure identifier management	291
7.1.1	Tabs in the structure identifier management	294
7.1.2	The graphical buttons	297
7.1.3	Sort menu	298
7.1.4	Extras menu	299
7.2	Message management	301
7.2.1	The visual appearance of message management	302
7.2.2	Project checks	303
7.2.3	Message classes and message categories	304
7.2.4	Filters in the message management	309
7.2.5	Various ways of editing messages	311
7.3	Layer management	313
7.3.1	Standard layers	313
7.3.2	Export and import layers	314
7.3.3	Create and delete your own layers	315
7.3.4	Uses of layers	316
7.4	Parts management	318
7.4.1	Structure of the parts management	321
7.4.2	Tabs in the parts management	321
7.5	Revision control	329
7.5.1	General	330
7.5.2	Generate new revision	330
7.5.3	Execute changes	331

7.5.4	Complete page(s)	335
7.5.5	Generate reports	336
7.5.6	Complete a project	337

8 Export, import, print 339

8.1	Export and import of DXF / DWG	340
8.1.1	Exporting DXF and DWG files	340
8.1.2	Import of DXF and DWG files	344
8.2	Image files	347
8.2.1	Export of image files	348
8.2.2	Insertion of image files (import)	350
8.3	Print	353
8.3.1	The Print dialog and its options	354
8.3.2	Important export/print setting	356
8.4	Export and import of projects	357
8.4.1	Export of projects	357
8.4.2	Import of projects	357
8.5	Print attached documents	358
8.6	Import PDF comments	359
8.6.1	Import of commented PDF documents	359
8.6.2	Delete PDF comments	362
8.7	Generate PDF documents	364
8.7.1	Export of PDF files	365

9 Data backup 373

9.1	Zipping and unzipping of projects	373
9.1.1	Zip projects	373
9.1.2	Unzip projects	374
9.2	Backup and restoring of projects	375
9.2.1	Back up projects	375
9.2.2	Restore projects	381
9.3	Other important settings	384
9.3.1	Default settings for project backup (global user setting)	384
9.3.2	Compress project (remove unnecessary data)	385
9.3.3	Automated processing of a project	387
9.4	Backup and restoring of master data	390
9.4.1	Back up master data	390
9.4.2	Restore master data	392
9.4.3	Default settings for the backup of master data (user)	394
9.5	Send project by e-mail directly	394

10 Master data editors 397

10.1 Preparatory measures 400
10.2 Overview of forms 400
10.2.1 First option - manual overview 400
10.2.2 Second option - automatic overview 402
10.3 Forms 404
10.3.1 Create new form (from copy) 407
10.3.2 Edit existing form 412
10.3.3 Create new form 413
10.4 Plot frame 414
10.4.1 Create new plot frame (from copy) 414
10.4.2 Edit existing plot frame 416
10.4.3 Create new plot frame 417

11 Old EPLAN data (EPLAN 5...) 419

11.1 Import options 420

12 Extensions 421

12.1 EPLAN Pro Panel 421
12.1.1 EPLAN Pro Panel – what is it? 421
12.1.2 General 422
12.1.3 Toolbars in EPLAN Pro Panel 424
12.1.4 Menus in EPLAN Pro Panel 437
12.1.5 Navigators in Pro Panel 447
12.1.6 Settings for EPLAN Pro Panel 452
12.1.7 A practical example with EPLAN Pro Panel 454
12.2 EPLAN Data Portal 468
12.2.1 What are the advantages of the EPLAN Data Portal? .. 468
12.2.2 Before the first start 469
12.2.3 How the Portal works 472
12.3 Project options 476
12.3.1 What are project options? 477
12.3.2 Terminology in the Project options module 477
12.3.3 Creating options and sections 478
12.3.4 Generate options overview report 485

13 FAQs 487

13.1 General 487
13.2 Parts 517

13.3	Terminals, plugs	522
13.4	Cables	526
13.5	PLC	529
13.6	Properties, layers	529
13.6.1	Master data	534
13.7	Data exchange	540
13.8	Reports	541

Index 545

Foreword and acknowledgements

Dear users,

EPLAN Electric P8 is a CAE software with multiple innovations and innumerable project editing options, and it is constantly being developed further.

The new Version 2.0 continues to improve upon the previous versions, incorporating a number (*over 1000 new features / improvements / function expansions*, etc.) of user requirements and requests that have arisen during the practical use of EPLAN.

This book, in its second completely revised and expanded edition, is meant to demonstrate the range of functions in EPLAN Electric P8 Version 2.0. The book is meant to make it easier to start using the software and smooth over the initial bumps. As in previous editions of this book, practical examples demonstrate what is and could be possible.

Of course, this edition, like its predecessors, cannot and will not be able to describe all of the software's functions or provide examples for every conceivable function. EPLAN Electric P8 becomes increasingly comprehensive with every new version, and it offers a variety of functions that cannot be completely covered in a book of this length. If we were to describe everything this software can do, then our readers would have to read a 4000-page tome, and I don't think this is what they want to do. And EPLAN Electric P8 Version 2.0 is teeming with new features, function expansions and improvements. Therefore, I had to leave aside some of the functions in this edition.

As the reader will again see, this version has several paths to the same goal. A few solutions are presented and discussed; in this way, you will discover other solutions yourself and ask yourself why no one has ever tried it this or that way before.

The book should and will help to show solutions or propose approaches to solutions, thereby simplifying everyday practical work. It can also make the decision-making process a bit simpler, without taking away the reader's power to make decisions.

The book once again is addressed to everyone who uses EPLAN Electric P8 for electrical engineering constructions - the daily EPLAN Electric P8 user and the sporadic P8 user, as well as engineers, electrical engineers, pupils, and students.

I would like to express my thanks to Sieglinde Schärl and her team at the Carl Hanser Verlag for the opportunity to write and publish this book. I would also like to sincerely thank my family, especially my wife Susanne. They have always been, and continue to be, very patient with me.

At this point, I would also like to thank all of the readers who have made this book a success! All of your feedback, whether criticism or praise, has been a strong motivator for me to write this edition!

And finally, I would again like to thank EPLAN Software & Service GmbH & Co. KG for their consistent and very friendly support and collaboration in compiling some of the information for this new second edition of the EPLAN Electric Manual.

■ Important notes

All of the examples and explanations assume local installation and local operation of EPLAN. Furthermore, the book assumes that the user has all of the user rights in EPLAN and is logged in as the local administrator.

It is possible that, depending on the user's license and module package, a functionality or function described in the book will not be available or executable in the way in which it is explained and illustrated. Therefore, you should always check to see which licensed add-ons you have (see image) (HELP / ABOUT / LICENSED ADD-ONS TAB).

For this book, EPLAN Electric P8 Professional Edition 2.0.5 Build 4602 was used with the following licensed add-ons (see image).

NOTE for users of previous versions: Certain parts of the functions described here exist in EPLAN Electric P8 Versions 1.7.x, 1.8.x, and 1.9.x, although the use and range of their functionality may differ in Version 2.0.

NOTE for users of the EPLAN Electric P8 Education Version: The example project that goes with this book is also available in an education format and can be downloaded under
http://www.eplan-efficient-engineering.com/handbook.

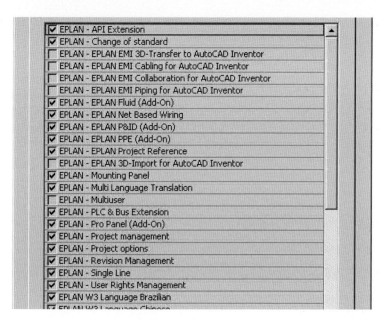

Help / Info / Licensed add-ons

Fig. 1 Help / Info / Licensed add-ons tab

Additional notes on reading this book: Some settings were used, such as those for filters or schemes, that differ from a standard installation of EPLAN. All of this additional data is normally stored in the example project. Some custom, non-standard shortcut keys were also used.

Text boxes are used to visually highlight notes, tips, etc. They have the following meaning:

NOTE:

TIP:

TO DO:

Further information

Furthermore, symbols in the margins denote the following:

You will find examples, questions and answers here.

1 Install EPLAN Electric P8

Since installation requires few steps and can only be performed by the system administrator, this chapter provides a basic description of the installation process. EPLAN is usually already installed on the workstation.

Installation of EPLAN generally requires administrator rights. At the very least, the system administrator also defines an EPLAN administrator who will later manage the EPLAN users (also known as rights management). If rights management (user management) is not used, then EPLAN can be started by all users without requiring passwords etc.

User management (an add-on that must be purchased separately and is not always present with every license) is not described in this book. Brief general information is provided as necessary at the appropriate points.

1.1 Hardware

For Version 2.0.4, EPLAN recommends a CPU comparable to a 3-GHz Intel Pentium D or a 2.4-GHz Intel Core 2 Duo with 4 GB of RAM.

EPLAN has no special requirements for the graphics card or other hardware components. A standard computer as used for (e.g.) Office applications is sufficient. And this also applies to the graphics card: the more available memory, the smoother EPLAN runs. However, if you are going to use an EPLAN Pro Panel module, then the graphics card must at least be able to handle OpenGL Version 2.1. This should not be a problem for the latest ATI or Nvidia graphics cards.

A single-screen solution can no longer be recommended for EPLAN due to the many additional modular dialogs that can be displayed, such as the various navigators. A two-screen solution is clearly preferable, and a three-screen solution with each screen having a resolution of 1680 × 1050 pixels is the ideal system.

Of course, EPLAN still functions with a normal single-screen solution. However, this should have a resolution of at least 1280 × 1024 pixels.

■ 1.2 Installation

As far as installation is concerned, EPLAN is a normal Windows program. Apart from a few entries, the installation wizard performs most of the work. There are only a few entries in the Windows registry, which is commendable and is not always the case today.

The installation is usually started from the installation CD provided. The Setup.exe file in the root directory of the CD is started to begin installation. The same applies when you perform the installation after downloading the installation package from the EPLAN homepage (the downloaded ZIP file unpacks the installation data into the same directories that would be on the installation CD).

Fig. 1.1 Installation directory

The first installation dialog is displayed, and a number of installation settings must be defined here.

- **Program variant** (usually)   Electric P8
- **Program language** (usually)   English (USA)
- **Installation language** (usually) English (USA)

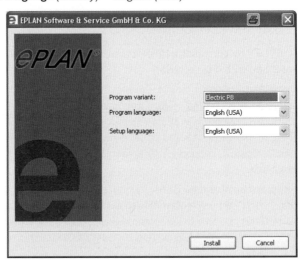

Fig. 1.2 Starting the installation in the installation dialog

NOTE: Running EPLAN Electric P8 requires the .NET Framework system components. If the .NET Framework is not installed, or is not installed in the required version, it must be installed before you can proceed with the installation of EPLAN.

After setting all the required information and clicking the INSTALL button, the Windows Installer prepares the required components and displays a second installation dialog with general information.

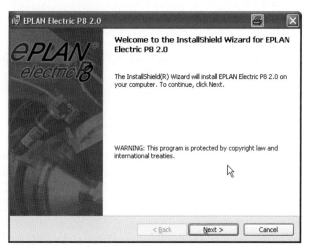

Fig. 1.3 Installation dialog – Welcome

Installation continues after confirmation by clicking the NEXT button, and the copyright information dialog is displayed. This must be confirmed via the radio buttons in the lower area of the dialog. If the copyright notice is not accepted, EPLAN will not continue with the installation.

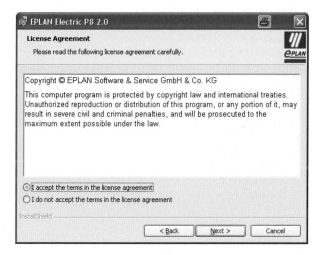

Fig. 1.4 Installation dialog – Licensing agreement

After clicking the NEXT button, the **User information** dialog is displayed. Enter the appropriate information here and select one of the options in the radio buttons in the lower area of the dialog.

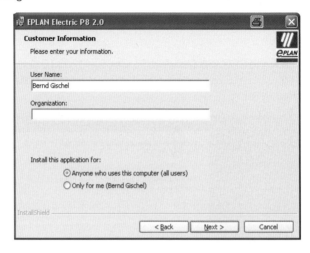

Fig. 1.5 Installation dialog – User information

After clicking NEXT, the **Target directories, settings** dialog is displayed. This is where you set the *program directory*, the *system master data directory*, the *company code*, and the directories for *user, workstation and company settings* . You must also define the *units of measurement* for the system.

You must also define the directory for the original *EPLAN master data*. This ensures that later synchronization of your own master data always takes place with the original EPLAN master data.

EPLAN recommends installing every new P8 system (new version number) in its own program directory. All other settings (data etc.) should always be made in the same directory to avoid duplication of data, which can result in "data chaos" (multiple data directories containing a great deal of different data).

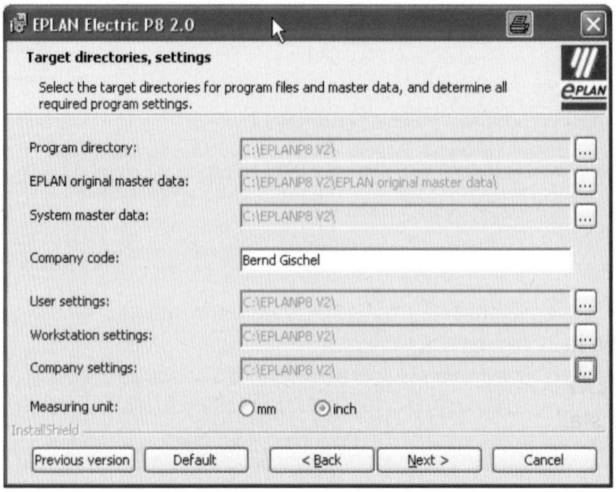

Fig. 1.6 Installation dialog – Target directories, settings

Example directories of an installation of Version 2.0.5:

Program directory \\EPLAN\Electric P8 V2\2.0.5

Data directory   \\EPLAN\Electric P8 V2\

User settings   \\EPLAN\Electric P8 V2\

Company settings \\EPLAN\Electric P8 V2\

In the example, the program directory changes for the new version, but all other directories remain the same. EPLAN later installs and updates all the data during installation.

TIP: EPLAN does not replace your own system master data. If you would like to work with EPLAN's new system master data at a later date, then you must synchronize the system master data!

By design, EPLAN does not overwrite user-related master data, because the user may have modified the original system master data and saved this under the original name (as assigned by EPLAN).

During installation, EPLAN does not recognize whether this data has been changed and would therefore simply replace it. The user does not usually want this to happen.

Once all settings have been made, clicking NEXT continues the installation. The next dialog asks what is to be installed. The *Complete* option can be selected here without further consideration.

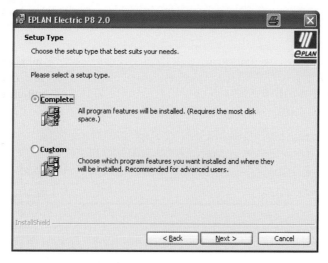

Fig. 1.7 Installation dialog – Setup type

The *customized setup* option in the *User-defined* setup type allows particular program components to be included or excluded from the installation.

After clicking NEXT and then confirming by clicking INSTALL in the subsequent dialog, EPLAN begins installing EPLAN Electric P8.

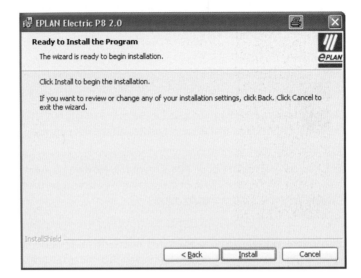

Fig. 1.8 Installation dialog – Ready to install

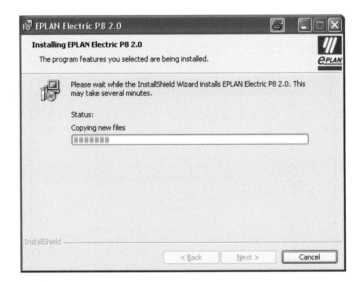

Fig. 1.9 Installation dialog – Installing EPLAN Electric P8 2.0

The installation may take a while depending on the computer's speed. Once installation is complete, EPLAN displays the completion dialog, and EPLAN Electric P8 is now installed.

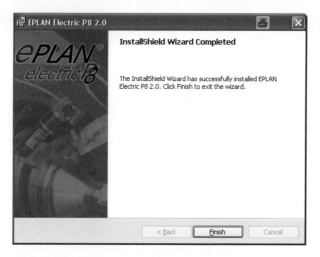

Fig. 1.10 Installation dialog – Installation complete

EPLAN Electric P8 can now be selected in the Start menu.

If a license has not yet been installed, a one-time dialog requesting the validation code (license number) for the corresponding dongle (hardware protection) is displayed before the actual program starts.

EPLAN starts with the **Select scope of menu** dialog. Here you can choose between the options *Beginner* (Beginners – only the basic menus allowing graphical drawing of a project and/or working with macros), *Advanced* (Advanced – allows more extensive display options such as (e.g.) minimum text size or display of empty text boxes), or *Expert* (Expert – all menus and functions available), and then confirm your selection by clicking OK. The Beginners, Advanced, and Expert options are hard-coded into EPLAN and cannot be changed or extended.

 NOTE: The Select scope of menu dialog is only displayed when EPLAN is used without rights management.

If a previous version is being used, then there is a one-time option to import the settings (user, workstation and company) from this version. By clicking the CANCEL button, none of the previous version's settings will be imported.

Fig. 1.11 Starting up EPLAN for the first time

Fig. 1.12 Import settings dialog

EPLAN opens the *Default* workspace (a standard view) and displays the **Tip of the day**.

After installing a previous version or an additional module, it is then possible that EPLAN will want to perform a *master data synchronization* (of original EPLAN data with customer data). If none of the original EPLAN system master data has been changed, then you can run this synchronization without further consideration.

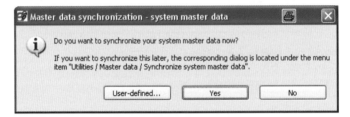

Fig. 1.13 Master data synchronization

In general, it is always advantageous to synchronize your own system master data with the new EPLAN original master data. This ensures that all of the new features, improvements and function expansions are automatically updated for your own system master data.

Installation is now finally complete and you can begin working with EPLAN!

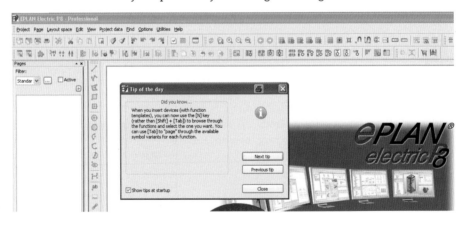

Fig. 1.14 Tip of the day

1.3 Important notes for Version 1.9.x users

You should use the export function to export the schemes, filters, etc., that you have created in Version 1.9.x so that you will be able to import them later into Version 2.0 as necessary.

1.3.1 Parallel operation with Version 1.9.x

Because EPLAN has again made some far-reaching changes to the databases in Version 2.0, which are no longer compatible with the previous versions, we recommend that you install Version 2.0 and its master data in a separate directory.

This applies above all to changes in the parts database. If you open the parts database with Version 2.0 and reformat it for Version 2.0, you will no longer be able to write to the file when you open it with Version 1.9.x! However, you will still be able to read the file with Version 1.9.x. Please keep this in mind!

2 The basics of the system

This chapter provides a brief explanation of some important principles, functions, and working methods of EPLAN, and uses a number of examples to illustrate selected facts and system settings.

Important points are the directory structure, settings, multiple starts, notes on the properties of projects or pages, notes on particular dialog properties, handling of schemes, forms, and plot frame overviews, symbol structures and an overview of my personal shortcut keys.

■ 2.1 Five principles that apply to working with EPLAN Electric P8

TO DO: 1st principle
Errors during project editing are allowed on principle in EPLAN Electric P8!

A basic principle is: Mistakes are allowed when working with EPLAN Electric P8. This basic principle, errors are allowed, will be illustrated with an example.

A contactor may have two auxiliary contacts in the schematic, with both of them initially having the same connection point designation. Things that are not physically possible are initially „allowed" by EPLAN while working on a project. The user is thus not slowed down by „irritating" errors or editing messages.

Of course, this type of error appears in the message management, but only as a message entry. This entry has no further consequences at the moment. When the project editing has progressed far enough, or is finished, then EPLAN can perform certain checks on the project. Erroneous entries such as those described above will then be listed in the project, if they do not already exist in the message management.

Naturally, this error must be fixed in order to have a correct practical reference. However, this is not **compulsory**!

This means that EPLAN allows the user to decide whether a project is error free or not and which priority a message (error, warning, or note) should have. Of course, it is possible to avoid even such errors. With the *Prevent errors* check option, the above approach would not be possible. But this is a freely adjustable setting.

TO DO: 2nd principle
As a matter of principle, one always edits in EPLAN Electric P8 whatever is selected!

This will be clarified with a small example. If I select three texts on a page and start the translation function, then exactly these three texts will be translated.

If I select this page in the page navigator, then the translation function will translate all texts on the entire page, depending on their settings.

TO DO: 3rd principle
As a matter of principle, EPLAN Electric P8 stores data and their references online!

EPLAN is an online system. All references or device data are kept constantly (i.e. online) up to date. The only limitations (if you want to call them this) relate to connection data. For performance reasons, these are only updated as desired via a few specific actions, the rest is performed completely independently by EPLAN.

A typical example of this is the editing of a page followed by a page change. If required, here you need to manually start the updating of the connections. There is of course a setting that allows EPLAN to do this type of connection updating automatically. However, this can negatively affect the performance of a project.

In my opinion, constant (online) updating of connections is not really necessary because relevant actions such as graphical project reports or automated procedures – such as device numbering – automatically update the connections before the actual action is performed.

TO DO: 4th principle
As a matter of principle, EPLAN Electric P8 can be operated using a graphical approach.

This means that the devices (symbols) can first be placed in the schematic and the devices (symbols) can then be subsequently assigned the parts, including the associated function definitions.

This is not compulsory, and you have a completely free choice when editing a project.

TO DO: 5th principle
As a matter of principle, EPLAN Electric P8 can also be operated using an object-oriented approach.

This means that (e.g.) external motor or other component lists can be read into the system as device lists and the project can be started from this end.

NOTE: Of course, principles 4 and 5 can be combined with each other. There are absolutely no limitations when working with EPLAN Electric P8.

■ 2.2 Directory structure, storage locations

This can be described in few words. EPLAN can use any desired directory structure. EPLAN allows the user a free choice here. Data such as project or master data can therefore be easily integrated into an existing company data storage structure.

EPLAN recommends installing/running the program files locally, with only the data stored in the network. I can only agree with this recommendation. Any further structuring or organization of the remaining master data is the responsibility of the user.

EPLAN creates the following directories by default: The program directory with the following sub-directories:

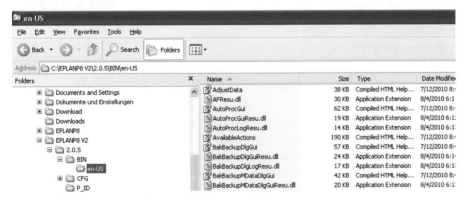

Fig. 2.1 Program directory with subdirectory

- BIN – contains the program modules and, in additional subdirectories, the language files (e.g., en-US for English).
- CFG – contains configuration files for the user, company, workstation, and projects.
- P_ID – of interest only to the PPE users.

EPLAN also creates certain default directories for particular system data and other data during installation, according to a similar scheme.

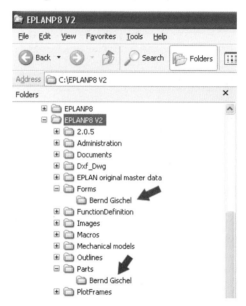

Fig. 2.2 Default customer directories

For example, a main directory called *Parts* is created, followed below by the directory with the *Company code*. EPLAN locates this code from the information entered during the installation – in this case, for example, *Bernd Gischel*. The following *directories* are initially recommended as default directories and filled with the corresponding data during the installation:

- *Parts* – contains the parts databases and the configuration files for importing and exporting parts.
- *Images* – contains all images.
- *Documents* – contains documents like PDF documents or Excel tables.
- *Dxf_Dwg* – contains drawings in DXF or DWG format.
- *Forms* – contains all forms (system master data).
- *FunctionDefinition* – contains the files for the function definitions.
- *Macros* – contains all macros such as window macros (*.ema) or page macros (*.emp).
- *Mechanical models* – contains mechanical data.
- *PlotFrames* – contains all plot frames (system master data).
- *PPE* – contains files for the PPE module.
- *Projects* – default directory for projects.
- *Revisions* – contains the revision database.
- *Schemes* – contains preconfigured or user-created schemes, for example personal filter or sorting settings.
- *Scripts* – contains the corresponding script files *.cs.

- *Symbols* – contains all symbol libraries (system master data).
- *SymbolMacros* – contains all symbol macros.
- *Translation* – contains the dictionaries (translation databases).
- *Administration* – as standard, contains the project database into which all projects are read from the project management and other management databases such as the rights management database.
- *Templates* – contains templates and basic projects, as well as exchange files for exporting project data (labels).
- *Xml* – contains XML files.

These are standard EPLAN defaults. The EPLAN user can, of course, change these directory structures to suit his needs. For practical reasons, this should be done after the first program start and before starting the project management for the first time.

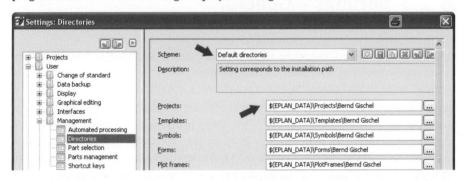

Fig. 2.3 Directory settings

These directories can be viewed and modified under OPTIONS / SETTINGS / USER / MANAGEMENT / DIRECTORIES. A useful feature is that separate schemes can be created and configured for different directory settings here.

 TIP: It is also possible to start EPLAN Electric P8 with an appropriate command line parameter and thus automatically set a particular directory structure. Example of a call: E:\EPLANP8 V2\2.0.5\BIN\W3u.exe /PathsScheme:OwnSchemeName

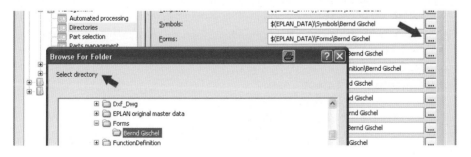

Fig. 2.4 Select other directories

Clicking the button causes EPLAN to display the BROWSE FOR FOLDER dialog. You can select a different directory here, or use the CREATE FOLDER button to create one at the desired storage location.

> **NOTE**: These changed settings, however, will not be adopted until the next, that is, new, program start of EPLAN!

2.3 Settings – General

To put it simply, a project contains all the relevant data that it uses and requires. This means that a project is independent of the general system master data, because EPLAN stores all required and used data within the project automatically upon first-time use.

The advantage of this method of „storage" should not be underestimated because even after several years, this type of project can still be opened and edited with exactly the original master data such as plot frames, symbols, etc. This is also true when (e.g.) a company plot frame in the graphics has changed in the meantime (for example the company logo has been revised).

Of course, this behavior can be configured in EPLAN via settings (parameters) – if desired. EPLAN has four basic setting areas for this.

The settings are accessed via the OPTIONS / SETTINGS menu item. Storing this variety of project-related data naturally increases the data volume of the project.

 An empty project has an average size of approx. 4 to 5 MB.

This section provides a brief overview of the properties. Many of the properties mentioned here are explained in more detail in other chapters because they relate to the chapter topics.

2.3.1 Settings – Project

All project properties (and only these) are defined under SETTINGS / PROJECTS [PROJECT NAME]. The settings in the Projects node are available only if at least one project is open in the **page navigator**.

In the project settings, you can (e.g.) define whether the project master data should be automatically synchronized with the system master data when the project is opened. You should always carefully consider the use of this, and many other project settings (i.e., "automation" when a project is opened), because changing these settings can have wide-reaching effects on the project to be opened.

EPLAN has no "undo" button at this point and, if no data backups exist, the old data is irretrievably lost!

For example, some important project properties are located under OPTIONS / SETTINGS / PROJECTS / PROJECT NAME / MANAGEMENT / PAGES and under the entry for the *Default plot frame*. This entry for the global plot frame applies to all pages if no other plot frame has been assigned via the page properties. Symbol libraries are located under OPTIONS / SETTINGS / PROJECTS / PROJECT NAME / MANAGEMENT / SYMBOL LIBRARIES. After selection and entry in the settings, these are also subsequently stored with the current state in the project.

There are numerous settings – too many to include in the scope of this book. For this reason, only specific, important project properties are listed and explained.

2.3.1.1 Setting Projects [project name] / Reports

This area affects the output of graphical reports (refer to chapter 6 for further explanations).

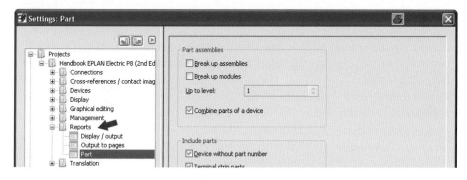

Fig. 2.5 Settings for reports

- **Display / output** – settings that affect the output or display of data in reports (forms)
- **Part** – settings that affect the output of parts data and their behavior in reports
- **Output to pages** – project-wide settings for reports (forms) for the different report types in a project

2.3.1.2 Setting Projects [project name] / Devices

In the area of devices, different settings such as numbering schemes or syntax checks for devices are defined.

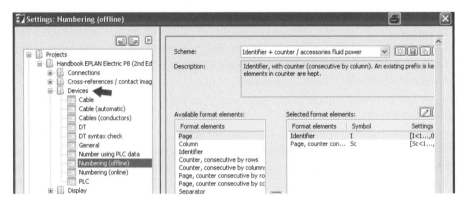

Fig. 2.6 Settings for devices

- **General** – among other settings, contains the parameter for *conversion to uppercase* for devices. If the parameter is changed, automatic conversion of uppercase to lowercase letters or vice versa does not occur. It is only possible to work with either uppercase letters (parameter activated) or with upper and lowercase (parameter not activated). Generally, EPLAN initially always uses uppercase letters for identifiers. This setting can also be used to define the devices that automatically receive a preceding sign on insertion.
- **DT** – this is where settings for preceding signs or conversion to upper case are defined.
- **DT syntax check** – these settings define the *special characters* possible for the structure identifiers and for devices.
- **Cable** – default settings for inserting cable definition lines (cables), shields, and their connection definition points (i.e., which connection definition point graphical symbol is the first choice).
- **Cables (conductors)** – these settings allow for default settings; for example, defining the symbols to be assigned by EPLAN to a connection definition during insertion.
- **Cable (automatic)** – contains the settings for automatic cable generation, cable selection, and cable numbering. All settings are defined via filter schemes. You can use predefined schemes or define your own schemes.
- **Number using PLC data** – default values of a scheme for numbering connected devices with PLC data. The PLC numbering can be used to number the following devices (among others): Terminals, pins, and general devices.
- **Numbering (offline)** – settings for a scheme for numbering devices offline (subsequently). Offline numbering is used to subsequently give a schematic a different DT layout. For example, the devices were first numbered with the default [identifier counter]. However, the devices must now be numbered according to the scheme [page identifier column]. The scheme is set as the default here for this. However, this does not have to be used. The scheme for offline numbering can be changed at any later time.

- **Numbering (online)** – contains the defaults for a numbering scheme for online assignment of devices. Such schemes are applied immediately when creating a schematic or inserting symbols, macros or copy operations. This setting also defines the identifier set used for the project or how identifiers are to be handled when inserting symbols or macros.
- **PLC** – the settings for outputting PLC assignment lists (among others) and the format of the output.

2.3.1.3 Setting Projects [project name] / Display

This setting is used to change the display of project structures in the navigators as well as in the page navigator. The Display node also contains the display or format of numbers, time and date.

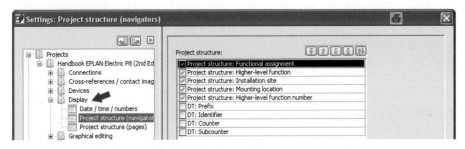

Fig. 2.7 Setting of project structures and formats display

- **Date / time / numbers** – settings for selecting the format of the date, time and numbers display to suit the operating system or a project-specific format (for example to set the date or time format to match typical foreign displays).
- **Project structure (navigators)** – these settings allow you to change the sequence of the display of structure identifiers and/or of the DT in the navigators (except for the page navigator). However, this relates only to the display in the various navigators. There are no changes to devices.
- **Project structure (pages)** – these settings are used to affect the display sequence especially in the page navigator.

 NOTE: This setting changes only the representation or display of the project structure or devices in the navigators. The page structure and the device tag structure are not changed!

2.3.1.4 Setting Projects [project name] / Graphical editing

Here are the general settings for graphical editing (selected).

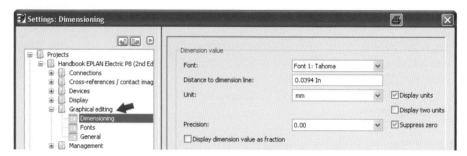

Fig. 2.8 Setting for graphical editing

- **General** – among other settings, the default values for representing connection junctions are set here, i.e. whether the display should be target-oriented, a point, or shown as drawn. However, the setting depends on the SETTINGS / USER / GRAPHICAL EDITING / CONNECTION SYMBOLS user setting and especially the *Connection symbols with prompt* parameter.

  With target specification – this default means that (e.g.) a T-node is drawn as target wiring. This T-node is also first placed and can then subsequently be changed to point wiring or a different target. This parameter has the effect that the representation remains as target wiring, regardless of how the T-node is internally set.

  Point wiring – this default allows T-nodes (junctions) to be entered with the visual display of point wiring. Internally, this type of junction can also be set for target wiring.

  As drawn – this means that (e.g.) a T-node is generally placed as target wiring. Changing this while placing (inserting) in the schematic is not possible. After the T-node has been placed, the T-node [direction] dialog can be called up by double-clicking and the targets can be changed. It is possible to change the display setting of this T-node from target wiring to point wiring. This representation is then also shown in the schematic. This setting thus allows a mixed representation of target and point wiring.

  With these parameters, it is generally possible to switch between the point wiring and target wiring representations. This means that if all T-nodes (junctions) have been drawn with the point representation, the target wiring parameter allows the representation to be changed to target wiring.

  Insertion of T-nodes (junctions) is controlled at a higher level by the SETTINGS / USER / GRAPHICAL EDITING / CONNECTION SYMBOLS parameter using the *Connection symbols with prompt* parameter.

  If this query parameter is not switched on, the T-node (junction) is first placed and can be changed later. If it is switched on, a dialog for setting the target tracking or point/target wiring is displayed before the T-node (junction) is placed. You should note this!

  Aside from the settings for the connection symbols, there are a number of additional general settings contained in this node.

2.3 Settings – General

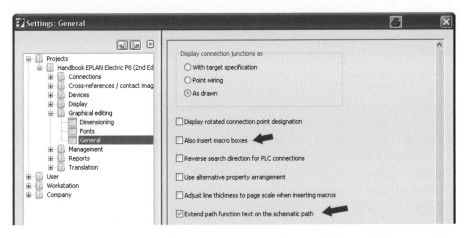

Fig. 2.9 Additional general settings

- **Also insert macro boxes** – if this setting is active, the macro boxes, if they are present on the macro, will be inserted as well.
- **Extend path function text on the schematic path** – if this setting is placed, the insertion point of a path function text does not have to be directly in the path of the device in order to be adopted. It is sufficient if the path function text is in the path (of the plot frame).
- **Dimensioning** – format setting for entering dimensions, such as font or dimension line termination. Changing this setting has no effect on existing dimensions, and only affects dimensions that are subsequently inserted.
- **Fonts** – defines the fonts that may be used in the project (up to ten fonts). These settings have priority over the font settings in the company settings (*Company / Graphical editing / Fonts*).

2.3.1.5 Setting Projects [project name] / Cross-references / contact image

These settings offer many parameters defining the representation of references for interruption points, motor overload switches, contactors, general areas (for example the representation of references between different page and report types and the symbols to be used), and the display itself (the options for the overview cross-references between the types multi-line, single-line, and the overviews are set here).

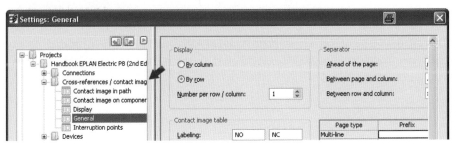

Fig. 2.10 Cross-references / contact image settings

The settings at the devices always have priority.

However, if no change to the cross-reference representation is selected on the devices, the default settings are taken from these parameters (depending on the symbol properties).

- **Interruption points** – default setting for representing cross-references: by column or row, with or without page names and much more
- **General** – settings for displaying general cross-references, how and which separators are placed between pages and columns or if the cross-reference is enclosed in brackets
- **Display** – setting defining between which types (multi-line, single-line or overview) of device cross-references should be displayed
- **Contact image on component** – default setting for displaying cross-references at components, such as (e.g.) motor overload switches, circuit breakers
- **Contact image in path** – default setting for displaying (e.g.) cross-references of the arrangement (contact image) of contactor coils

2.3.1.6 Setting Projects [project name] / Connections

The default values for connection properties are defined in this node.

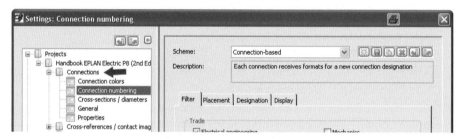

Fig. 2.11 Connection settings

- **General** – setting for determining the source and target
- **Properties** – default settings for certain connection properties such as the unit (mm^2), etc.
- **Cross-sections / diameters** – default setting for cross-sections and diameter values for the project
- **Connection colors** – editable and expandable table with color codes and color names
- **Connection numbering** – defines the scheme used for numbering connections. Pre-defined schemes or your own scheme can be used.

2.3.1.7 Setting Projects [project name] / Management

This includes default settings defining how EPLAN should handle (e.g.) master data or particular modules in the project.

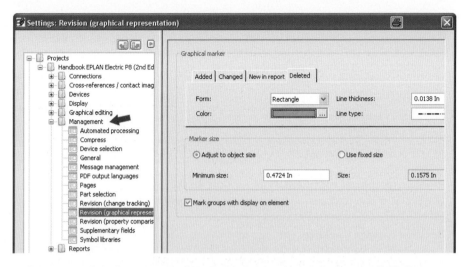

Fig. 2.12 Settings for managing the project

- **3D import** – default setting of a function definition during the import of 3D data
- **General** – this setting contains three very important parameters. The *Synchronize project master data when opening* parameter means that when the project is opened, the project master data, e.g. a terminal diagram form or a symbol library, is synchronized with the system master data and possibly updated. The *Synchronize plot frames* setting synchronizes plot frames immediately after the project start. The last setting, *Synchronize stored parts when opening*, means that parts are synchronized with the master data automatically.

 NOTE: However, forms are only synchronized when in the Projects [project name] / Reports / Output to page setting, the Synchronize option for the corresponding forms has also been set.

Before finally synchronizing the master data, EPLAN displays a security message that must be confirmed by clicking the YES button. Only then is the project master data updated, and only when it is older than the system master data.

- **Part selection** – preselection of parts (for example, filter schemes can be used to (e.g.) only display particular types of contactor in the parts selection), and also which parts database EPLAN should use
- **Automated processing** – a definable active script. This script is the default setting for the Automated processing function.
- **Device selection** – in addition to part selection, EPLAN also has device selection. Whereas the part selection is independent of the function definitions used in the schematic, device selection occurs based only on the function definitions belonging to the devices. A preselection of how EPLAN should handle the device selection is defined here: For example, are existing function data to be used and if so, which ones, etc.?
- **Compress** – default scheme used for compressing the project.

- **Message management** – an important setting that defines the range and type of project checks for the current project. Project checks are not essential and do not necessarily have to be performed. However, they uncover possible editing or data entry errors during project editing and list the problem areas in the message management. This setting defines the project checks and the scope of each check for this project. The number of project check messages can also be limited here.
- **PDF output languages** – default setting for outputting languages in the PDF document to be generated.
- **Revision (property comparison of projects)** – settings for the revision markers of connections.
- **Revision (graphical representation)** – setting for the revision control of projects: What color or line thickness should be assigned to revision markers for changed, deleted or added objects or changes in the reports?
- **Revision (change tracking)** – settings for revision control with change tracking, for example: What watermark is to be displayed on an incomplete page? Should a project be stored as a PDF upon completion? As well as many other settings.
- **Pages** – defines the handling of pages. Are letters allowed as subpages, what form should the path numbering take, pagewise, projectwise, or according to the structure identifier? The most important parameter in these settings is the definition or (subsequent) changing of the global plot frame for the project. A separate plot frame can be defined for every page in the page properties independently of the global plot frame setting. The page settings always have priority over the global project settings.
- **Symbol libraries** – defines the symbol libraries used in this project. New entries (addition) of symbol libraries are automatically stored in the project after the setting is saved.
- **Supplementary field** – EPLAN provides the user with 100 freely definable supplementary fields for each of the project, page, and function areas. The *Part reference* tab has five freely definable supplementary fields. Freely definable means that the user of the property can provide their own name for the project user supplementary field 1. This name is then displayed in the corresponding areas instead of the previous name.

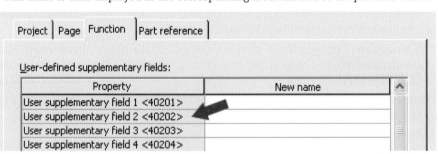

Fig. 2.13 User-defined supplementary fields tabs

In the MANAGEMENT / SUPPLEMENTARY FIELDS setting, the user-defined supplementary fields do not have names of their own. The properties with the standard names defined by EPLAN would thus be displayed on a function.

2.3 Settings – General

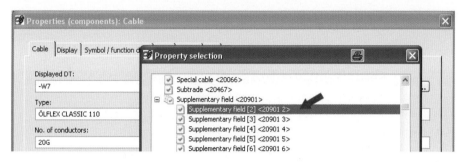

Fig. 2.14 User-defined supplementary fields on a function (here: cables)

Now, you can assign separate names to these user-defined supplementary fields so that they can be assigned by the terms better.

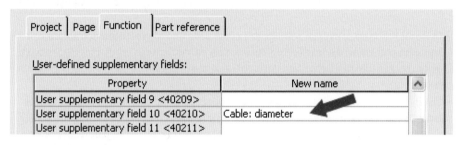

Fig. 2.15 Assignment of a separate name for the user supplementary field 10 <40210>

Using a function (here: a cable), this user supplementary field 10 <40210> can then be added with its new name.

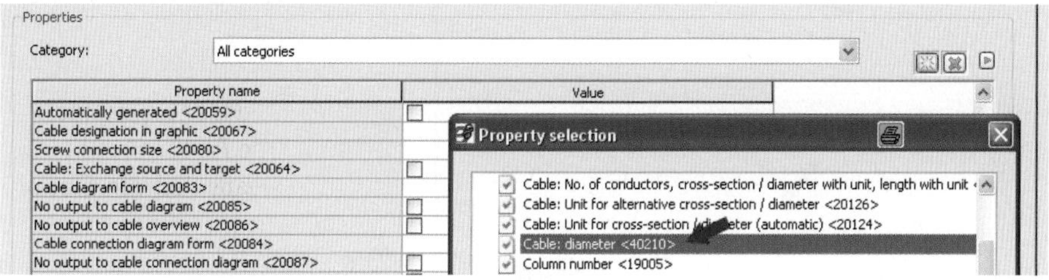

Fig. 2.16 Property 40210 with the „new" name

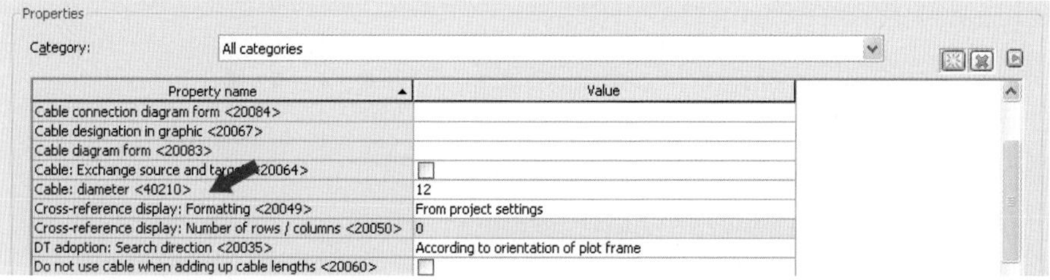

Fig. 2.17 The user-defined supplementary field integrated with the properties

The advantage of such assignment of separate names for user-defined supplementary fields is self-evident: clear defaults for entries with very specific properties, which may have to be evaluated correctly later on. This helps avoid incorrect entries and tedious extra work to fix values that have been assigned to the wrong properties.

2.3.1.8 Setting Projects [project name] / Translation

This setting defines the databases to be used for translation or the scope of translation that EPLAN should use.

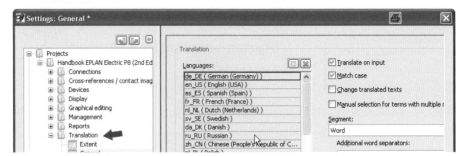

Fig. 2.18 Translation settings

- **General** – project language settings are made here and also settings defining the handling of texts when translating during data entry. Settings for displaying translation languages (the displayed language) are also made here. The source language is also defined here.

- The definition of a **source language** is crucial, because EPLAN always displays this language in the various dialog boxes **first**. This helps to prevent operating errors, e.g., entering German in the wrong "language line" by mistake.

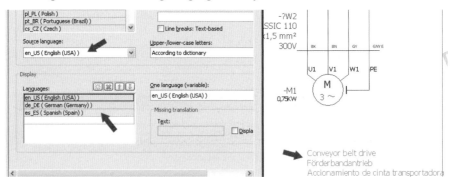

Fig. 2.19 Translation sequence „German English Spanish" setting

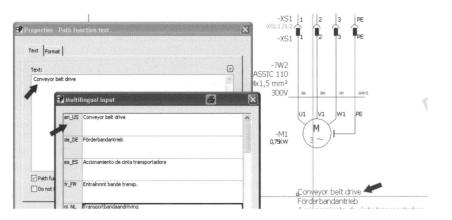

Fig. 2.20 Display in graphical editing

If in the settings the sequence of the *displayed languages* is changed, the *source language* will still be shown first in the dialogs!

Fig. 2.21 Changing translation sequence „Spanish English German"

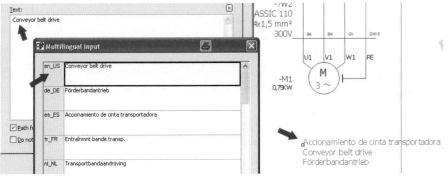

Fig. 2.22 Display in graphical editing

The sequence of the displayed languages has been changed correctly, but the sequence of the languages in the dialogs remains as is. This way, translations can no longer be entered by mistake.

- **Extent** – this item allows the user to define the range of the properties to be translated for particular texts in areas such as forms, projects, or components. Properties set to active are translated, and when the check box is deselected, they are not translated. This allows specific properties to be excluded from translation. The properties of the areas are defined by EPLAN and cannot be extended.
- **Translatable pages** – this setting is similar to the range setting. However, it generally relates to a page type. If page types are excluded from a translation, then they are not translated by a subsequent automatic translation run.

2.3.2 Settings – User

User settings relate to personal settings such as the user interface structure, which dialogs are opened on which screen, the directory structure for certain personal data, or the different representations of settings. All these settings depend on the login name entered at the Windows login and are stored in the directory \\Basic\...\USER...

Fig. 2.23 User settings

CFG in a workstation-specific manner (by computer). If you place these settings on (e.g.) a network drive, then you can call up your own EPLAN configuration from any computer in the network.

2.3.2.1 Setting User – Display

This user settings area contains the personal user interface settings or default values for a user-related workspace.

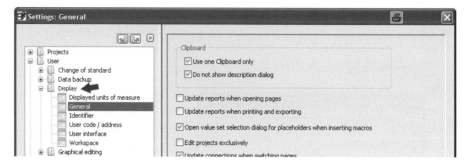

Fig. 2.24 General display settings

A number of important settings are explained below.

- **General** – settings defining whether multiple clipboards or associated description dialogs should be used.
- This settings area also contains other important settings:

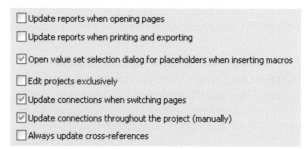

Fig. 2.25 Additional general settings

- **Update reports when opening pages** – if this parameter is active, then reports that are already generated in the project and which exist as graphical output are automatically updated by EPLAN when a page of this type is opened.
  This may be intentional, but it may also be the case that these reports should not be graphically changed by EPLAN, e.g. for revision reasons or for tracking changes. This parameter should then be switched off by deselecting the option check box.
  This parameter should also be switched off for performance reasons if project editing becomes very slow (EPLAN must constantly update the data in the background and then update the report as soon as it is opened – this requires a fast computer). This also applies to the *Update connections when switching pages* setting. The following applies to the Update connections throughout the project (manually) setting: Once set, the connections will be updated throughout the entire project. If the setting is inacti-

ve, only connections will be updated that have been selected (one or several pages or selected devices, etc.).
- **Update reports when printing and exporting** – this setting is useful when activated. This way, the reports are always up-to-date when the project is printed or exported.
- **Edit projects exclusively** – if this parameter is active then the project can only be edited exclusively by a single user. Setting this parameter allows faster editing of projects in EPLAN.

 NOTE: If EPLAN is operated in multi-user mode, this parameter must not be active, otherwise multiple users cannot access the project at the same time because multi-user operation is locked out.

- **Displayed units of measure** – here the units of length and weight are defined.
- **Workspace** – the preferred user workspace is defined here. The workspace can be changed at any time during project editing. Workspaces will be handled in more detail later in this chapter.
- **User code / address** – entries such as code, name, phone, or the e-mail address of the user are made here. Some of these entries are used for the page or project properties, and others are used for system messages, for example when a conflict with another user occurs during project editing. These entries are also useful for forms, plot frames, etc.
- **Identifier** – default settings defining how structure identifiers that are newly assigned during project editing are to be handled in the Structure identifier management dialog.

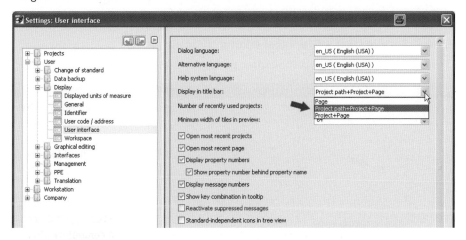

Fig. 2.26 Extended interface settings

- **User interface** – general settings for the dialog language, or language settings for the help files. Further user interface settings are possible here: Should the last projects be reopened when the program starts, or should the property number be displayed with the properties in EPLAN (my recommendation)? The *Reactivate suppressed messages* setting is interesting. If this is set, then dialogs whose displays have been deactivated are once more displayed. Apart from this, there is also the *Display in title bar* setting. Here you can define which information is to appear in the title bar: only the current page, only the project name and the current page or the project name including project path and the currently open page.

2.3.2.2 Setting User – Data backup

This user settings item defines default values for backing up or not backing up master data.

Fig. 2.27 Data backup settings

These are just default values and can be subsequently changed during the actual data backup.

- **Filter / [data type]** – here data can be included in or excluded from the data backup and/or can always be supplied as default for data backups.
- **Default settings / Projects** – particular details such as e-mail message split size, or the backup drive as default settings for project backups.

2.3.2.3 Setting User – Graphical editing

These settings define further user interface settings such as colors or grid sizes.

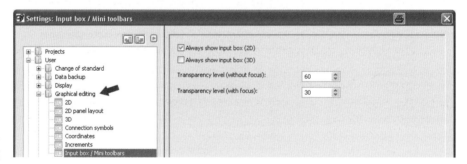

Fig. 2.28 Settings for graphical editing

- **2D** – color settings, background color, grid settings, cursor settings, and the behavior when scrolling.

The following settings, especially the grid sizes, have proven useful in practice. Levels of 4; 2; 1; 0.5, and 0.25 have proven to be very practical.

- **2D panel layout** – settings for part placements, their mounting options and settings to apply the dimensions from the parts master data or via manual input.
- **3D** – settings for the 3D area, grids, color settings or how terminal strips are to be handled for placement purposes.
- **Coordinates** – editable coordinate tables for electrical engineering and fluid power (only integer values such as 3 or 4 are possible) and graphics areas (real numbers such as 3.475 or 1.222 are possible).
- **Increments** – editable increment table for the grid area (only integer values such as 6 or 12) and units area (real numbers such as 33.445 or 0.234 are possible).
- **Connection symbols** – an important setting causing the connection dialog to be displayed before placing a connection junction (T-node) or to disable this. This settings area also contains several parameters for switching property dialogs on and off when inserting shields or fluid distributors.

2.3.2.4 Setting User – Change of standard

An important area for projects where a change of standard has taken place. These entries remain empty until a change of standard has occurred!

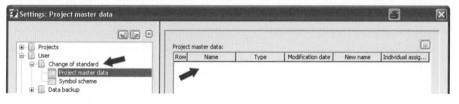

Fig. 2.29 Change of standard

- **Project master data** – contains assignments of project master data. This area is only filled with data when a change of standard has occurred.
- **Symbol scheme** – contains symbol assignments. The same applies as with master data. If no change of standard has been performed, then this area is empty, too.

2.3.2.5 Setting User – Interfaces

EPLAN offers a number of different interfaces for importing and exporting data. The interface settings are defined in this area.

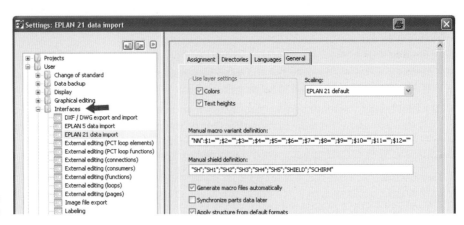

Fig. 2.30 Various interface settings

These settings are adopted as default values if no other settings are used during the direct output.

- **Labeling** – the default settings for the most frequently used schemes are defined here, in order to transfer data to the labeling system.
- **Image file export** – default settings for outputting image files in BMP and TIFF format.
- **DXF / DWG export and import** – default settings for importing and exporting DXF/DWG files.
- **EPLAN 5 data import** – settings for importing EPLAN 5.x data .
- **EPLAN 21 data import** – settings for importing EPLAN 21 projects such as symbol library assignments, directory settings, languages, and general import parameters.
- **External editing [type]** – defines the most frequently used schemes for external editing. These settings affect a number of different areas such as functions, pages, and also connections.
- **PDF export** – settings for the internal PDF output of pages or projects.

2.3.2.6 Setting User – Management

In this area, EPLAN allows user related settings for management tasks.

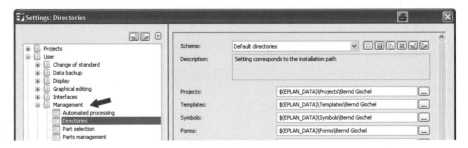

Fig. 2.31 Management settings

- **Part selection** – defines the part selection, where the parts are selected and the interface to be used (API, internal, ODBC, or COM interface). This can also be an external PPC system.

- **Parts management** – this defines the parts database and various settings affecting the parts and how they are displayed.
- **Automated processing** – editor entry allowing scripts to be directly edited. EPLAN provides a parameter allowing the generated system messages to be displayed after a script has run. This parameter can be switched off.
- **Data Portal** – settings for the user, the portal and the connection settings to the portal.
- **Shortcut keys** – user related settings defining the shortcut keys for EPLAN commands. The shortcut keys can be individually assigned or deleted here, or the keyboard assignments can be restored to the default settings.
- **Directories** – a complete overview of the installed directories for the various master data and databases in EPLAN. You can also use your own schemes to change all the directories in a single step.

2.3.2.7 Setting User – Translation

General definitions as to how languages and their translations are to be handled.

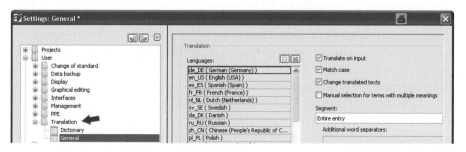

Fig. 2.32 Translations, foreign languages

- **General** – definition of how project-independent texts are to be translated or, also, the choice of source language to be used. The source language is the language for which translations will be generated subsequently.
- **Dictionary** – defines the dictionary the user wishes to work with. Simultaneous display of the foreign languages available in the database. The parameter also allows default settings for importing new translations and defines the behavior when texts are entered (AutoComplete or AutoCorrect).

2.3.3 Settings – Workstation

These are general settings for the current workstation.

- **Display / General** – this allows for operational optimization to be set if you are using a terminal server.

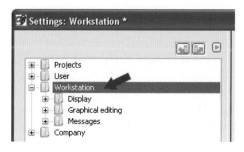

Fig. 2.33 Workstation settings

- **Graphical editing / Print** – various global options revolving around the topic of printing, such as print margins or print size.
- **Messages / System** – the path to, and the maximum size of, the system messages file are defined here.

2.3.4 Settings – Company

The company settings contain a number of parameters to make (e.g.) comparisons of projects from external suppliers easier.

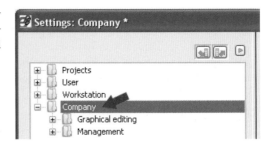

Fig. 2.34 Company settings

- **Graphical editing / Fonts** – default values of up to ten company fonts. These ten different fonts allow for subsequent switching from one font to another in the entire project without any problems, provided that one of the global fonts has been set on the objects in question. However, priority is always given to the project-specific fonts (setting under *Projects / [project name] / Graphical editing / Fonts*)!
- **Management / Property comparison of projects** – settings determining the properties to be used for the purposes of project comparison. You can create your own schemes for such comparisons.
- **Management / Project management database** – the storage location of the company-related project database can be set here.
- **Management / Add-ons** – allows automated control of P8 when started according to manually specified defaults in an Install.xml file, the location of which is defined here.
- **Management / Change tracking (numbering of the revision index)** – settings (controllable via your own schemes) for the index of revisions.

■ 2.4 EPLAN and multiple starts?

EPLAN can be started as often as desired. You may ask why, since you can generally work on several projects at the same time in EPLAN. Generally, it is not necessary to start EPLAN multiple times, but it is possible.

The practical benefit of EPLAN multiple starts is that (e.g.) you can have the parts management and dictionary open at the same time, allowing missing entries to be entered in parallel to working on a project, without interrupting the project work.

Of course, when editing is finished, for example in the parts management, this data should, and must, be synchronized with the project and a decision must be made as to where these parts are to be stored. After calling up for the first time, this can then always occur automatically. Here too, you must carefully consider whether automation at this point is actually useful. EPLAN lets the user decide.

■ 2.5 Properties

In addition to the actual graphics of symbols, forms, or plot frames, EPLAN also outputs logical information. This logical information must be specified for the symbols, forms, or plot frames so that it is visible. This is done by assigning properties.

Every property has a property name and a property number that is only valid for this single property. You can use the property number to gain an *approximate* idea of the area that the property belongs to (project, page, etc.). The property value assigned to each property is unique to the property.

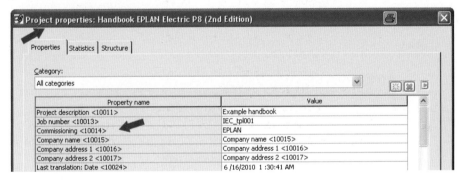

Fig. 2.35 Project properties

The *Project name <10000>* property contains the *<Project name>* value (the name of the project).

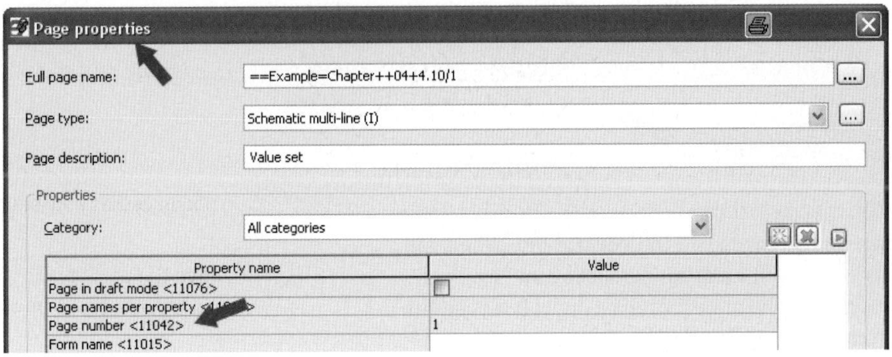

Fig. 2.36 Page properties

The *Page number <11042>* property contains the *<n>* value (n = the actual page number).

EPLAN generally distinguishes between several different types of properties. There are *Project properties* (relevant project information such as the project name; 10000), *Page properties* (properties of a page in the project itself, e.g. the current page number; 11042), *Symbol properties* (property of a symbol such as the displayed device tag; 20010), *Form properties* (properties for constructing a form and reporting data such as terminal and pin designations; 20030), *Plot frame properties* (e.g. the search direction for transfering the device tag, whether the device tag should be automatically adopted from the first device to the left; 12103), and numerous function properties.

2.5.1 Project properties

Project properties can also be described as global properties. These properties can be used everywhere as Special text – Project property, for example, in graphical editing via the INSERT / SPECIAL TEXT / PROJECT PROPERTIES menu item. This is then followed by the **Special text – Project properties** dialog.

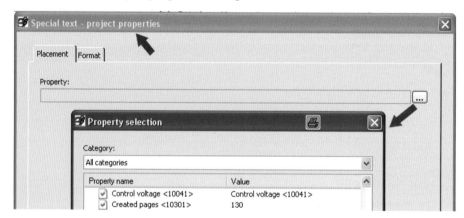

Fig. 2.37 Select and insert project property

Fig. 2.38 Adopted project property

Clicking the ... button opens the **Property selection** dialog. Here you can select the desired project property and accept it by clicking OK. The project property is loaded into the Property field of the **Special text – Project properties** dialog.

The special text can be formatted after this (font, width, italics, etc.) on the **Format** tab. When you have finished entering data, confirm this by clicking OK. The project property now hangs on the cursor and can be placed.

Fig. 2.39 Place property

Project properties can be added to the project via the graphical button ⊠ in the project properties. For this purpose, the project properties are opened for editing via the PROJECT / PROPERTIES menu. Clicking the ⊠ button now opens the **Property selection** dialog.

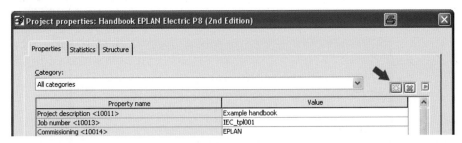

Fig. 2.40 Add project property

In the **Property selection** dialog, the usual Windows functions can be used to select one or more properties and these are then loaded into the project properties of the project by clicking the OK button.

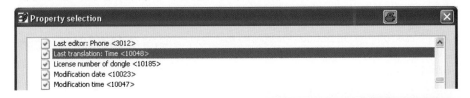

Fig. 2.41 Property selection

Using the popup menu (right mouse button) or the graphical button, it is possible to sort the display of the project properties. To do this, you must select the CONFIGURE command in the popup menu. The **Property arrangement** dialog then opens. The graphical buttons can be used here to arrange the properties as you wish.

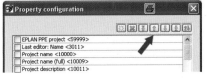

Fig. 2.42 Property configuration

2.5.2 Page properties

EPLAN **Page properties** are those properties belonging to a page. Page properties can also be used everywhere. The INSERT / SPECIAL TEXTS / PAGE PROPERTIES menu item in graphical editing opens the **Special text – Page properties** dialog. In the *Property* field, the button can be used to open the **Property selection** dialog.

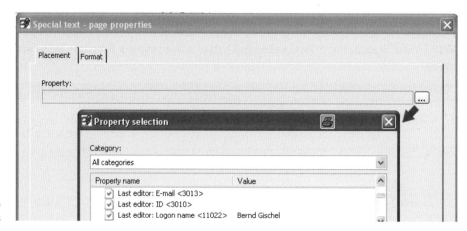

Fig. 2.43 Add page properties

The desired property can then be selected, and clicking the OK button loads it into the *Property* field of the **Special text – Page property** dialog.

After formatting the text (if necessary, on the *Format* tab), the **Special text – Page properties** dialog is closed via the OK button. The special text now hangs on the cursor and can be placed as desired.

Fig. 2.44 Place page property

2.5.3 Symbol properties (components)

Symbol properties are properties that are assigned especially to symbols (components). There are symbol properties that can be directly reached, such as connection point designations or device tags.

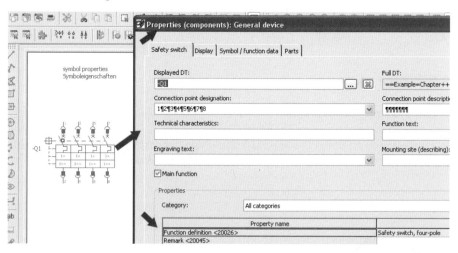

Fig. 2.45 Properties of a symbol (component)

New properties are added in the same way as other properties (project properties or page properties). Click the ▨ graphical button on the PROPERTY (DEVICE TYPE) tab for this. EPLAN then opens the **Property selection** dialog. In the Property selection dialog, you can now select the desired property or several properties to adopt it or them in the symbol properties via the OK button. The property has been added.

It is also possible to adjust the sequence in which properties are displayed in a symbol. To do this, you either right click and use the popup menu or click the ▬ button. Select the CONFIGURE entry in the menu that opens up. EPLAN opens the **Property arrangement** dialog. Here you can use the familiar graphical buttons to move or sort the properties.

2.5.4 Form properties

EPLAN defines **Form properties** as properties of a form. Forms (reports) are distinguished by report type. There are properties shared by all form types, but there are also properties that apply only to specific report types. Form properties can only be edited in the form editor.

2.5.5 Plot frame properties

Plot frame properties are properties assigned to the structure and reports of normal page types, such as the Schematic multi-line page type. An example of this is the Path areas property and its size. Plot frame properties can only be edited in the plot frame editor.

2.6 Buttons and popup menus

EPLAN has many dialogs. This is not unusual in such an extensive program. At this point, I would like to draw special attention to a number of dialog elements that occur repeatedly in many dialogs. These dialogs are similar in many different places in EPLAN. It is therefore important to know how they are constructed, operated and used.

Graphical buttons play an important role because they provide easy access to most of the functions and procedures in EPLAN.

2.6.1 Device dialog buttons

Fig. 2.46 Various buttons

Apart from a few small details, all EPLAN device dialogs are very similar. This means that they always have the same structure, regardless of whether it is a device dialog for a motor overload switch or a transformer or any other component.

EPLAN uses different buttons in different tabs to simplify the operation. What follows is a list of the most commonly used buttons and what they mean.

- Add new property to property arrangement
- Remove property from property arrangement
- Move up property by one place
- Move down property by one place
- Undock property (release)
- Dock property (link to upper area)
- Save
- Delete
- Import
- Export

2.6.2 Buttons in dialogs (configuring)

The **Configure** menu item is used to arrange or re-sort properties in dialogs.

Fig. 2.47 Buttons in the property configuration dialog

These can be page properties but also settings in other dialogs such as the UTILITIES / GENERATE REPORTS / SETTINGS / OUTPUT TO PAGES dialog. The number of buttons can vary here. In principle, the meaning/function of the buttons remains the same.

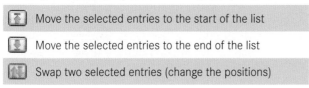

Move the selected entries to the start of the list

Move the selected entries to the end of the list

Swap two selected entries (change the positions)

In contrast to the graphical buttons in the device dialog, multiple entries can be selected here. The functions provided by the other graphical buttons have already been described.

2.6.3 Buttons in dialogs such as filter schemes

Fig. 2.48 Buttons in various settings dialogs

In addition to the already familiar buttons, dialogs such as **Filter, Sorting** or other settings dialogs have additional graphical buttons.

Additional buttons in the Filter or Sorting dialogs:

Edit entry

Copy entry

2.6.4 Small blue triangle button

As soon as the button [▸] appears in any dialog, the right mouse button can be used to call up a popup menu. Clicking on the button itself also displays the same popup menu of the right mouse button.

Fig. 2.49 Blue button

Fig. 2.50 Further function options

Depending on the dialog, the popup menu "behind" the blue button can have different functions. It is very useful in dialogs that open files or directories. Here you can quickly set the default directories, without having to click through all the way to the default directory.

2.6.5 Property arrangements (components)

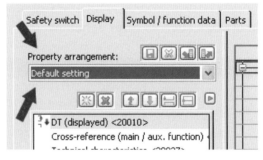

Fig. 2.51 Default setting

EPLAN generally offers for property arrangements a *Default setting* as well. The default setting is defined in the symbol structure and cannot be changed here. To change these default settings, you will have to edit the symbols. But I strongly advise against it!

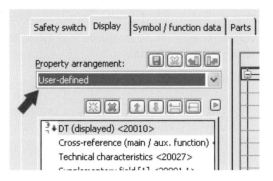

Fig. 2.52 User-defined arrangement

But you can change these property arrangements. Once they have been changed, they appear in the property arrangement, initially, under the designation *User-defined*.

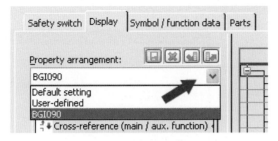

Fig. 2.53 Undo

To undo this user-defined property arrangement, simply select Default setting from the *Property arrangement* selection list. This returns all manually changed settings on a symbol back to the default values defined for the symbol.

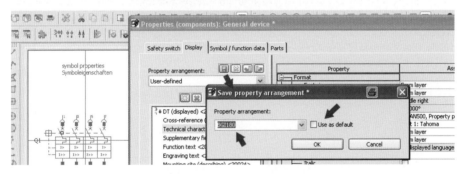

Fig. 2.54 Save your own property arrangement

If a property has been changed, then EPLAN no longer regards it as a default value and the display in the *Property arrangement* field changes to *User-defined*. This user-defined arrangement can also be saved via the button and then called up or set at similar symbols. You can assign a descriptive name to the property arrangement. After you click on the OK button, this property arrangement is saved, entered into the selection field and is then available to all symbol variants.

Accordingly, the property arrangement of a safety switch cannot be transferred to the property arrangements of a terminal, but only to the same „types" of devices.

The *Use as default* option allows for EPLAN, starting immediately (at the next insertion of such a symbol), to adopt your own property arrangement, that is, to use it as a default setting.

2.6.6 Format properties

Different formatting, for example of texts, can be easily implemented in EPLAN. Despite any possible differences in the dialogs, the basic principle always remains very similar. A normal text dialog is used as an example here. To insert a text, the INSERT / GRAPHIC / TEXT menu is used to open the **Properties** text dialog.

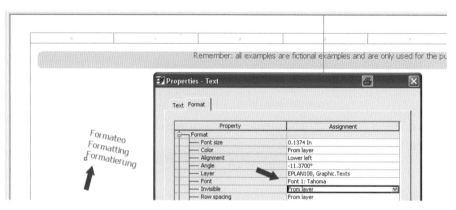

Fig. 2.55 Formatting options for texts (as an example)

The desired text can now be entered into the **Enter text** field of the *Text* tab in the opened dialog. To format the text according to your personal wishes, you now switch to the **Format** tab. The user has the full range of possibilities for formatting the text just entered. All selection fields can be freely edited or you can select default values. Some selection fields, like *Alignment* and *Layer*, are specified by the system and cannot be changed; the existing defaults must be used.

Fig. 2.56 Examples of different formats

This example also shows how different the arrangement and appearance of text can be if it is formatted using the default setting *Alignment* (the small black dots are the handles of the text).

2.6.7 Buttons (small black triangles)

In a number of dialogs, the buttons have a small black triangle. Additional menus are "hidden" behind these buttons.

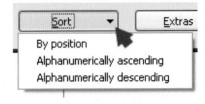

Fig. 2.57 Parts management

Fig. 2.58 Structure identifier management

2.6.8 Dialogs for schemes

Dialogs for schemes allow you to export schemes created and/or import schemes.

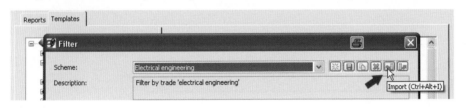

Fig. 2.59 Import and export options of schemes

 TIP: User-created schemes should be regularly exported, preferably as soon as they have been created. These schemes are then always available, even for new versions, other projects or other workstations.

In many areas, EPLAN offers a special identifier for the scheme type. This way, during an import only those scheme types are displayed that can actually be imported into the current scheme.

CBnu	Scheme for cable numbering
LB	Scheme for labeling
WS	Scheme for workspaces

■ 2.7 Master data

EPLAN distinguishes between two types of master data: the **system master data** and the **project master data**.

System master data are stored in the directories with the associated user directory that were set during installation. In addition to the user-specific system master data, the original EPLAN system master data are also installed in the EPLAN original master data directory (depending on the directory selected during the installation).

It should be noted that on installation, only the original EPLAN system master data are overwritten or updated in the \EPLAN directories. In case of a new installation or installation of an update, the user-specific system master data are not overwritten or updated. Generally, though, it is possible that the user's own system master data can be synchronized with the original EPLAN system master data. To do this, in the UTILITIES / MASTER DATA menu, call up the SYNCHRONIZE SYSTEM MASTER DATA menu item. In the following dialog, you can then update your own system master data accordingly.

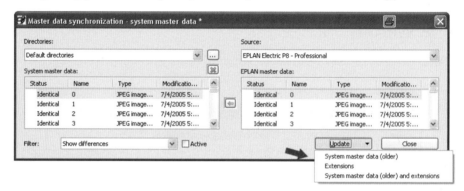

Fig. 2.60 Master data synchronization

If, for example, the user has modified the supplied table of contents form F06_001.f06, EPLAN would not update this form in the user-specific master data directories \\Forms\Company code with an updated, possibly improved, version but will only overwrite this form in the \\Forms\EPLAN directory.

Among other data, SYSTEM MASTER DATA contain symbol libraries with the associated symbols, function definitions, forms, plot frames, project templates, macros, and more.

Project master data are the other type of master data. Project master data, after initial use, are moved from the system master data and stored in the project; after being stored, they are independent of the system master data.

2.8 Operation

EPLAN can be completely operated with the mouse. However, to increase your own working speed, EPLAN is very flexible in allowing functions accessed with the mouse also to be accessed via definable keyboard shortcuts. Normally functions can be assigned to a keyboard shortcut if they are also accessible in the main menus, such as Page, Project etc.

To translate text in a Text dialog, you can select the appropriate command with the mouse via UTILITIES / TRANSLATION. This function can be assigned a keyboard shortcut because it is a normal menu command. For me, that would be the key combination CTRL+L.

2.8.1 Using the keyboard

EPLAN allows the user a great deal of freedom in assigning keyboard shortcuts to functions and commands.

NOTE: One limitation here is that certain shortcuts are only permitted in combination with the CTRL and/or ALT keys.

Of course, important standard Windows shortcuts such as CTRL+ C (copy) or CTRL + V (paste) should **not** be reassigned if at all possible. This can be done, but it is not recommended!

Here are a number of recommended keyboard shortcuts that have proven useful in practice.

Copy page from / to	CTRL + SHIFT + K
Edit / Copy format	ALT + K
Edit / Assign format	ALT + Z
Edit / Delete placement	DEL (NUMERIC KEYPAD)
Find / Synchronize selection	CTRL + SHIFT + Y
Project data / Device navigator	CTRL + SHIFT + B
Project data / Update connections	F11
Utilities / Parts / Management	ALT + V
Utilities / Translation / Dictionary	ALT + Ü

To assign your own shortcut to functions, in the OPTIONS menu, call the SHORTCUT KEYS menu item. The **Shortcut keys** dialog is displayed.

Fig. 2.61 Assign shortcut key to a function

In this dialog, you select the desired action and assign it a shortcut key using the CREATE button. The shortcut key is applied – and fully "operational" – as soon as you click on OK.

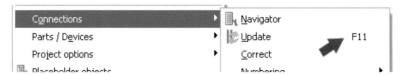

Fig. 2.62 Applied: Key combination

2.8.2 Using the mouse

EPLAN can be completely operated using the mouse. Admittedly, many menu items are easier to reach with the mouse than with cryptic key combinations that one usually cannot remember.

2.9 The user interface – more useful information

EPLAN provides a great deal of support to the user when editing projects, but also in general operation of the program.

A very nice and practical feature is the ability to freely configure the EPLAN user interface. The so-called workspaces themselves are defined depending on the logged-in EPLAN user. This makes it possible to always call up your own user interface configuration on the same computer when EPLAN starts.

2.9.1 Using workspaces

Put simply, a **workspace** is the dialog and toolbar layout you wish to have when performing certain operations in EPLAN. EPLAN allows the creation of separate workspaces for particular areas that contain exactly these desired toolbars, views or dialogs.

The Workspace function is accessed from the graphical editing menu via VIEW / WORKSPACE. When you select this function, EPLAN opens the **Workspace** dialog and you can use the selection field to select one.

2.9.1.1 Create and select a workspace

A workspace is customized by placing desired dialogs or toolbars to suit your personal preferences. To permanently save a new workspace, you open the **Workspace** dialog via the VIEW / WORKSPACE menu.

A new workspace is created via the [⊞] button. EPLAN then opens the **New scheme** dialog. A name should be entered into the empty *Name* field and a sensible description of the workspace to be created should be entered into the *Description* field.

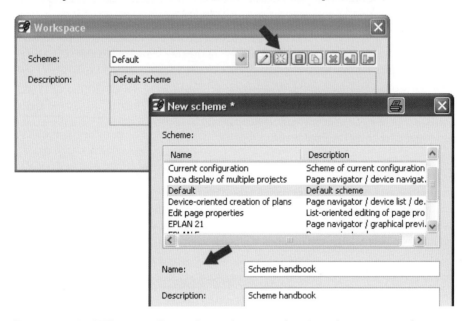

Fig. 2.63 Create a workspace

Clicking on the OK button will save the workspace and set it as the current workspace.

2.9.2 Dialog display

EPLAN provides a lot of data and additional information. EPLAN has a number of different ways of dialog representations in the navigators to make it easier to gain an overview of the information.

2.9.2.1 Tree view

The so-called **tree view** is one type of representation. All the information is displayed in a tree with small symbols, similar to the Windows Explorer.

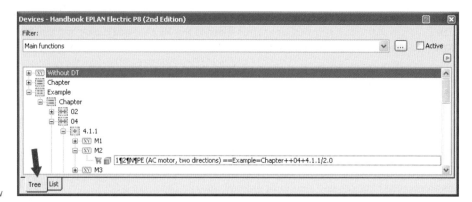

Fig. 2.64 Tree view

This representation type provides a clear overview, making it relatively easy to find objects. In this representation, EPLAN always shows all devices (depending on the function definition). No changes to the form and content of the tree view are possible. EPLAN provides no options for this here.

Only the display sequence of the devices can be configured under OPTIONS / SETTINGS / PROJECTS [PROJECT NAME] / DISPLAY / PROJECT STRUCTURE (NAVIGATORS) and/or (PAGES), where you can define how devices should be shown in the tree view, for example, by identifier or by the page prefix.

2.9.2.2 List view

The **list view** is another representation type. This representation type offers many more customization options to suit your working habits. As with the tree view, EPLAN shows all devices here as well (again depending on the navigator that was selected). The cable navigator shows only cables, the terminal navigator shows only devices with the terminal function type, etc.

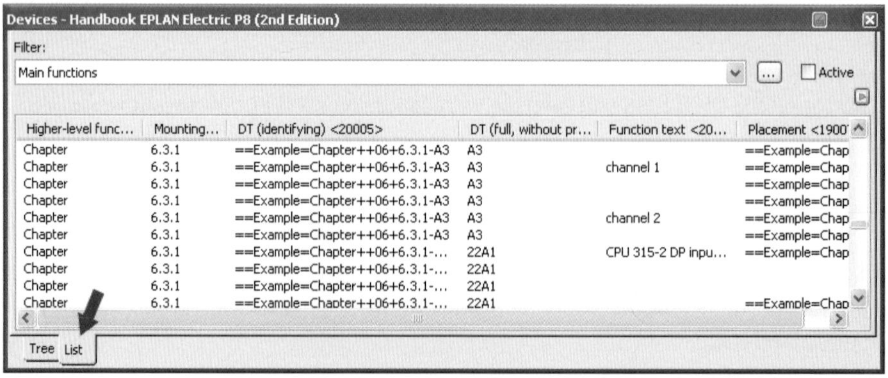

Fig. 2.65 List view

Unlike the tree view, it is possible to add further information to the list view. The popup menu is called up via the popup menu (right mouse button) or via the ⬛ button, and the CONFIGURE COLUMNS function is selected. EPLAN opens the **Column configuration** dialog. The desired columns can now be selected or deselected. When the OK button is clicked, the settings are saved and the list view becomes personally customized.

2.9.2.3 Combined tree and list view (part)

Apart from the previous representations, the *Part master data* navigator also allows for a combined display of list and tree view, which is known as the **combination**.

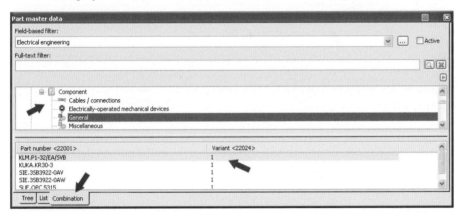

Fig. 2.66 Combined display

In the upper area, a tree view is displayed, which allows for some preselection (e.g., Select cable node), while in the lower area, only the corresponding parts (from the preselection) are displayed neatly in list view with further specific information.

2.9.2.4 Edit in table

Edit in table is another method of conveniently displaying and editing properties. Edit in table is usually accessed from the tree view by right clicking to access the popup menu or via the ⬛ button.

Unlike the previous representations (tree or list view), only the devices and their functions that were previously selected in the navigator are displayed.

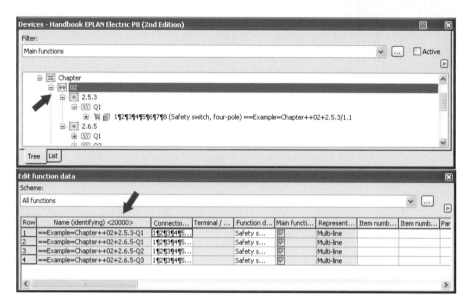

Fig. 2.67 Edit function data in table

The displayed devices can now be edited in a table in the **Edit function data** dialog. This means that Copy and Paste or other editing functions are no problem when editing in tables.

3 Projects

A project must first be created before you can edit schematics in EPLAN.

 NOTE: In EPLAN the storage location of projects and the names of the individual projects can be freely chosen. The only limitations are with the naming conventions of the Windows system being used.

Via the definition of page and device structures, a project contains all the necessary properties such as the *device structure* and also contains all the master data such as forms, parts data, function definitions, symbol libraries, etc., that are necessary for editing projects.

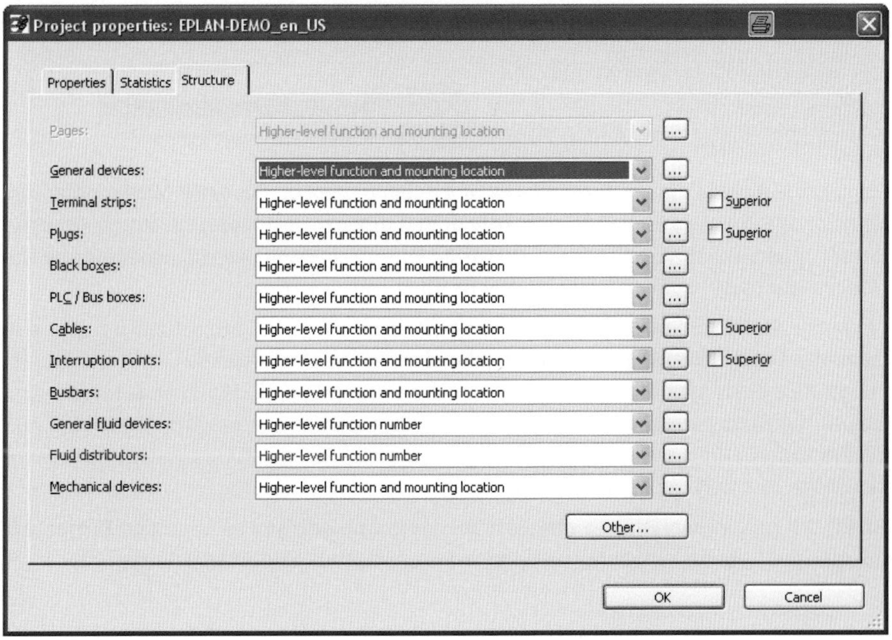

Fig. 3.1 Project properties - Structure tab

This data is all **completely** stored in the project. Data added later, such as forms, are also stored in the project.

This ensures that this project can later be edited with exactly the same data used when the project was created, or which was generated at the beginning of project editing, or which was later stored in the project.

■ 3.1 Project types

EPLAN basically distinguishes between two **project types**: a *schematic project* (this is the usual project in practice) and a *macro project*.

A *macro project* is used for creating macros, automatic generation of window macros, and for managing macros. Logical functions such as (e.g.) cross-references or connection information are not supported in a macro project nor displayed.

If necessary, a *schematic project* can easily be modified by changing the <10902 Project type> property from schematic project to macro project.

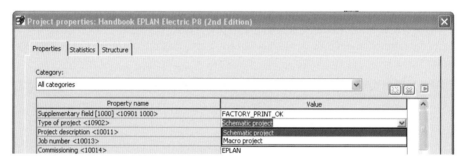

Fig. 3.2 Project properties - Properties tab

Other project types such as form project or symbol project do not exist. Master data (e.g., forms or symbols) are always temporarily edited directly in the project, stored after the form is closed, and then possibly synchronized with the current project (depending on the settings and usage) automatically (also depending on the settings).

Of course, the system master data can be synchronized with the project master data also manually. The synchronization can also take place in the other direction, with the project master data being synchronized with the system master data. If EPLAN finds inconsistencies in the master data at this point, then a corresponding message is generated and EPLAN cancels the synchronization because generally incompatible master data cannot be used to overwrite existing master data.

An EPLAN project consists of a directory **Projectname.edb** and an operation **Projectname.elk** (the *.elk file extension represents the normal editable schematic project).

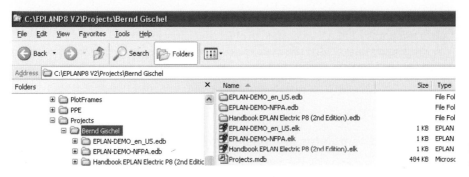

Fig. 3.3 Project directory with projects

Via the *Projectname.elk* operation, it is possible also to start the project from the respective file manager in use (e.g., Explorer) with a double click. You can also drag the *Projectname.elk* into the page navigator while holding down the left mouse button. EPLAN then opens the project.

3.1.1 Project types in EPLAN

Normal schematic projects in EPLAN are subdivided into different **project types**.

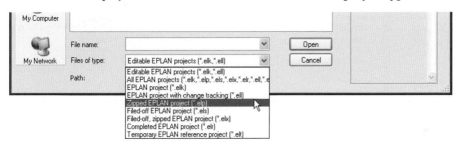

Fig. 3.4 Open project dialog with various file types

Project types define the project by its functional meaning, for example, as a normal project or a revision project. EPLAN distinguishes between the following project types:

3.1.2 Project types in EPLAN

The most important project for the user is the „normal" EPLAN project (*.elk corresponding to the schematic project). Aside from this project type, there are also the following project types in EPLAN.

*.elk: a normal editable project (the EPLAN schematic project)

*.ell: Project with change tracking

*.elp: a zipped project

*.els: a filed-off project

*.elx: a filed-off and zipped project

*.elr: a completed project

*.elt: a temporary EPLAN reference project (comparison)

Central project editing occurs in a normal project. All other types of projects (e.g., a completed project, an archived project, a basic project, or a project template) are derived from the normal EPLAN project (schematic project).

3.1.3 Project templates and basic projects

EPLAN allows the user to quickly and precisely create new projects based on existing *basic projects* and *project templates*.

A project template contains preconfigured values. However, when creating a template for a new project, the project and page structure can still be changed **once**. You should note this.

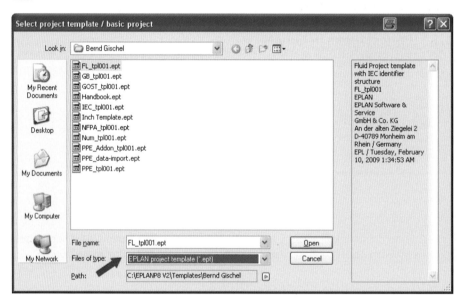

Fig. 3.5 Open project template

In contrast to a basic project, a new project created from a project template (via Project - New (Wizard)) usually has no pages because the later page structure has yet to be defined! Note: But project templates can also contain pages.

A *project template* has the ***.ept** file extension (and *.epb – old project templates from previous program versions) and cannot be directly opened or modified. You can, however, easily overwrite an existing project template that has incorrect properties with a new project template.

Project templates can be generated from existing projects in the graphical editor via the **Project / Organize / Create project template** menu.

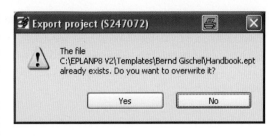

Fig. 3.6 Overwrite project template

To do this, first select the project that is later to become a project template with your mouse in the page navigator and then start the following EPLAN process for creating a project template via the PROJECT / ORGANIZE / CREATE PROJECT TEMPLATE menu.

The **Create project template** dialog is then displayed, where you must define the storage location and the name of the new project template. It is a good idea to create a *project templates* directory below the root directory, possibly with different folders per customer.

Basic projects are (e.g.) prefinished projects with corresponding customer values, such as a predefined page structure, sample pages, graphical report templates, various master data, and much more.

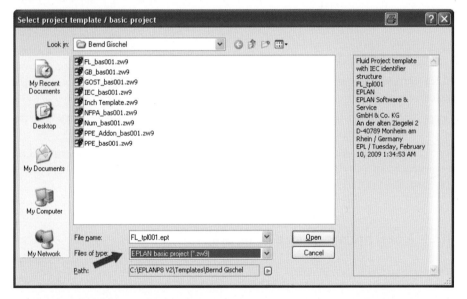

Fig. 3.7 Open basic project

In a project created from a basic project, the project and page structures are fixed and can no longer be modified.

A basic project has a ***.zw9** file extension and, like project templates, cannot be directly opened.

Basic projects, like project templates, can be generated from existing projects in the graphical editor via the PROJECT / ORGANIZE / CREATE BASIC PROJECT menu.

EPLAN then executes a number of functions to create the basic project. This is then followed by a **Create basic project** dialog where the directory and the project name of the basic project to be created are defined.

Here too it is a good idea to create a **basic projects** folder below the root directory, possibly with various customer folders.

Once basic projects or project templates have been created, they can no longer be subsequently (i.e., directly) changed! However, they can be overwritten with new or modified data!

3.2 Create a new project

To get quickly started with a project, new projects can be generated in various ways in EPLAN. There are two ways of doing this directly from the *project editing*.

A new project based on an existing basic project or project template can be created "at the push of a button" via the PROJECT / NEW menu.

EPLAN comes with several basic projects and project templates. The project template *IEC_tpl001.ept* contains the IEC identifier structure, while, for example, the supplied project template *Num_tpl001.ept* contains a sequential numbering structure.

Selecting PROJECT / NEW (WIZARD) generates a new project in several steps, each of which requires the entry of certain data (EPLAN provides detailed information at the bottom of each dialog).

Templates to be used for creating a new project are the basic projects (*.zw9) and the project templates (*.ept and *.epb).

The third way of creating a new project is via the optional project management. A new project based on an existing basic project or project template can be created here just as fast as via the PROJECT menu.

3.2.1 A new project (from a basic project)

We select the menu item PROJECT / NEW or use our own keyboard shortcut for this.

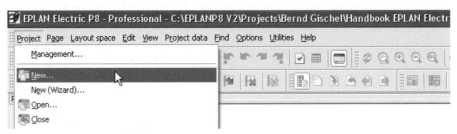

Fig. 3.8 Project / New menu

The **Create project** dialog is displayed. In this dialog, you must and/or can define the following settings:

- Project name (you need not worry about file extensions; these will be added by EPLAN later on automatically)
- Storage location
- Creation date
- Creator

After defining the above entries, you must now select a template. This can be a project template or a basic project.

For this purpose, click on the More button to open the **Select project template / basic project** dialog; in this example, a basic project, **Handbook.zw9**, serves as the basis for creating the new project. Note: You can select the corresponding project type in the *File type* field.

Fig. 3.9 Create project dialog

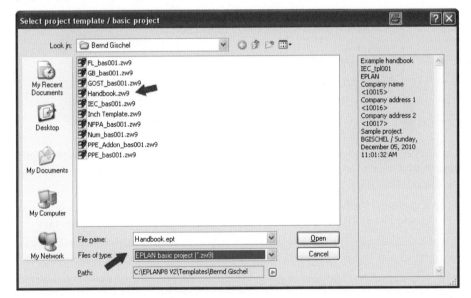

Fig. 3.10 Select project template / basic project dialog

An existing basic project is selected and adopted via the **Open** button in the SELECT PROJECT TEMPLATE / BASIC PROJECT dialog.

Then, EPLAN returns to the **Create project** dialog and imports the selected basic project (or the project template) into the *Template* field.

Now, you only have to confirm this dialog by clicking on the OK button, and EPLAN will generate the new project in the specified directory.

The new project is created from the selected basic project. This may take a while depending on the hardware and the storage location (server, local).

Fig. 3.11 Create project dialog including template

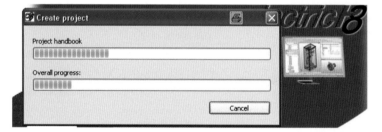

Fig. 3.12 The new project is generated

Once EPLAN has successfully generated the data for the new project, the **Project properties** dialog is displayed. You do not necessarily need to edit these at the moment and can do this at a later stage in the project editing.

EPLAN opens the **Project properties** dialog with the *Properties* tab. The project properties can now be adjusted or completely changed in the *Properties*, *Structure* etc. tabs.

There is a limitation here. The *structure of the pages*, in the *Structure* tab, cannot be changed (grayed out). It is fixed because the page structure was defined in the basic project.

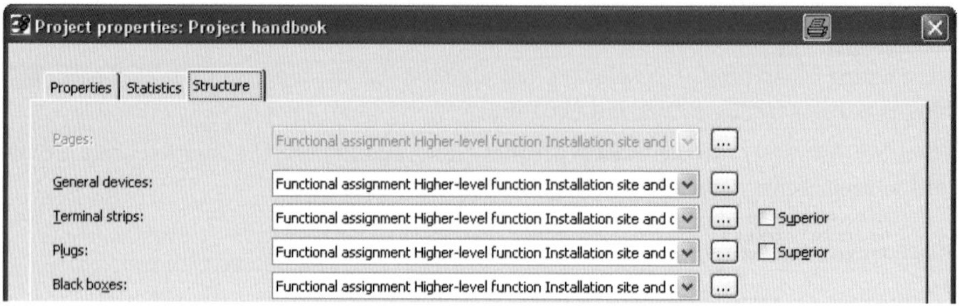

Fig. 3.13 Project properties - Structure tab

After clicking the OK button, the project properties are saved and the project is opened in the page navigator instantly (it can also be opened via the F12 key or you can display the overview via the PAGE / NAVIGATOR menu item).

The project can now be edited.

3.2.2 A new project (with project creation wizard)

We open the PROJECT / NEW (WIZARD) menu, or once more use our own keyboard shortcut.

Fig. 3.14 Project / New (Wizard) menu

The wizard opens the **Create project** dialog. The dialog contains several tabs (*Project, PPE, Structure, Numbering,* and *Properties*), and you must enter data into at least the *Project* tab (the PPE tab is only visible when you have a license for this).

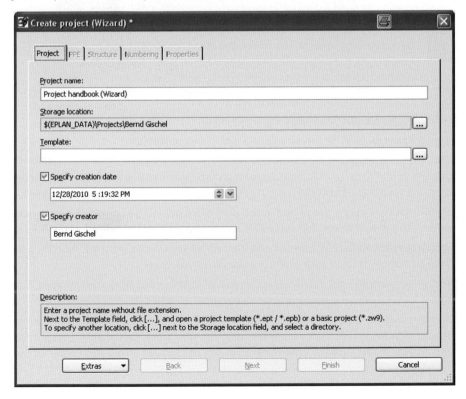

Fig. 3.15 Create project using the wizard

The lower area of the **Create project** dialog contains five buttons (EXTRAS, BACK, NEXT, FINISH, and CANCEL), which have the following meanings.

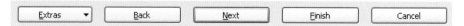

The EXTRAS button allows you to see a summary of the project in the browser. However, this only functions after the first *Project* tab has been successfully completed.

The BACK button returns you to the previous tab. Here too, this only functions when all necessary information has been entered in the *Project* tab and EPLAN has created the project or the NEXT button allows you to switch to the next tab.

Once all necessary entries have been made on the current tab, the NEXT button becomes enabled and you can switch to the next tab.

The FINISH button has a special function. It allows the project creation to be finished without making any further manual entries.

A condition is that a new *Project name*, a *Template* (template project or basic project), and *Storage location* have been defined in the *Project* tab of the new project.

The CANCEL button is easy to explain. It cancels and exits the **New project** wizard at any point.

Back to the tabs. The *Project* tab represents the main core of the project wizard.

The information queried in this tab is essential for creating a new project.

 TIP: All the other tabs can be edited, but this is not compulsory for creating a project via the wizard, and they can be edited later. (**Exceptions** are the *Structure* tab and the *Pages* selection field, which cannot be edited later.)

The *Project* tab has the following compulsory fields:

Project name input field - the new project name is entered here.

Note: The project name is not checked to see if it already exists until a template (project template or basic project) has been selected and the NEXT button has been clicked. The action can then still be cancelled.

EPLAN always suggests the project name *New name*, possible also with a consecutive number. This can, of course, be changed to any desired value.

Template selection field - via the selection button you select a template, with the usual range of templates and basic projects available for selection.

Storage location selection field - specifies the storage location of the new project. The storage location can be selected to any value via the button.

After entering these three pieces of information, EPLAN actually requires no further information. If the FINISH button is clicked, then the wizard can immediately generate the new project. In this case (clicking the FINISH button), EPLAN would just adopt the settings on the other tabs for the new project, generate the project, and close the project wizard.

You can, but do not have to, fill in the information under *Specify creation date* and *Specify creator*.

 NOTE: These fields cannot be changed later on. You must therefore make sure to enter the data correctly!

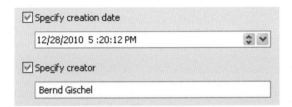

Fig. 3.16 Information on creation date and creator

When the NEXT button is clicked, EPLAN generates the basic elements of the new project based on the selected template. The **Import project** dialog is displayed. The process can be cancelled here if desired. Depending on the amount of data, the import may take a while.

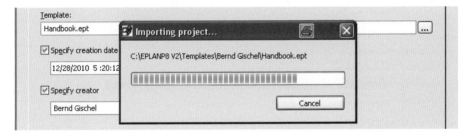

Fig. 3.17 The wizard creates the new project

EPLAN then remains at the *Project* (or *PPE*) tab. You can still select (e.g.) a different template project or storage location by clicking the BACK button once more.

In the *Structure* tab you define the subsequent page structure (after the final creation step, i.e. when the project is saved, this can no longer be changed) and the structure of the individual device groups for the project.

 NOTE: The structure for the pages, once saved, cannot be changed anymore after it has been adopted and/or saved! But this applies only to the structure settings of pages. All other structure settings, like the general devices, etc., can be adjusted later on!

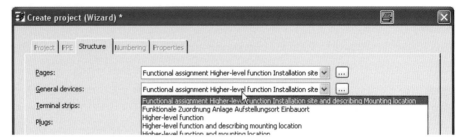

Fig. 3.18 Define project structure and its components

You can use the ▭ button at the end of each device group to select a different finished structure (a scheme) via the SCHEME selection field or create a new structure (a new scheme) via the ▭ button.

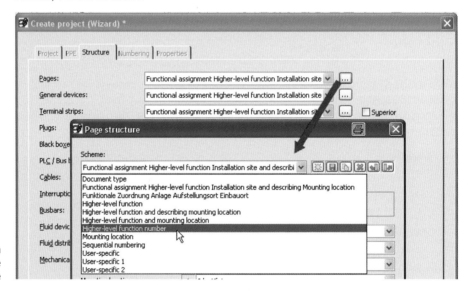

Fig. 3.19 Selection of a different structure scheme

The **Superior** option allows the user to define reports for terminal strips, plugs, cables, and interruption points independently of the structure scheme that is set. Here EPLAN evaluates the presence or absence of the "-" character in front of a device tag with a different result.

The *Higher-level function and mounting location* scheme was set for interruption points. The *Superior* option was also set. If an interruption point in the schematic is designated by –**L12**, EPLAN assigns the device tag the higher-level function and mounting location in the schematic (if present) based on the defined *Higher-level function and mounting location* scheme.

Input: -L12

Result: = *Higher-level function* + *Mounting location*-L12

The *Higher-level function and mounting location* scheme for interruption points also has the *Superior* option set. If an interruption point in the schematic is designated by **L12** (without the preceding sign "-") in the schematic, the *Superior* option means that EPLAN does **not** assign the device tag the higher-level function and mounting location in the schematic (if present).

Input: L12

Result: L12

The OTHER... button in the **Structure** tab opens the **Extended project structures** dialog, where you can define the *separators* and the nesting of devices for the new project.

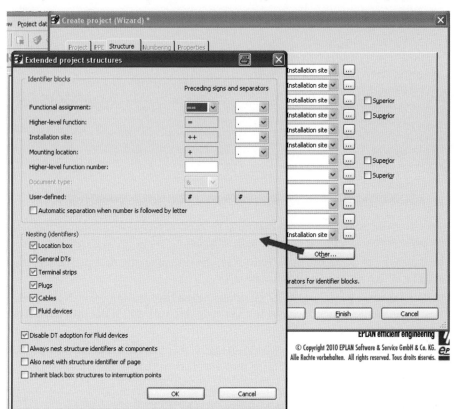

Fig. 3.20 Settings for further project structures

Here you can check existing and enter new separators. The possible changes, however, depend on the page scheme set. Certain preconditions also exist at this point for the *nesting of devices*.

 NOTE: Apart from the page structure, which is the most important property, all other settings can be changed at a later date.

Here, it is a good idea to select at least the *Higher-level function and mounting location* scheme. When such a scheme is set (as is also true of other structure identifiers), EPLAN does not necessarily expect a higher-level function or mounting location. Therefore, you can also use this scheme to create schematics with consecutively numbered pages.

If you later discover that one of the two identifiers is to be used after all, the pages can still easily be changed to the desired higher-level function or mounting location structure.

After setting and saving all the desired settings, click the NEXT button. The **Numbering** tab is displayed.

Fig. 3.21 Options and defaults for numbering

Here you make general project settings, such as *Standard plot frame* (default setting for a new page) or *Path numbering* (page- or path-oriented).

The standard plot frame can be selected from the system master data pool and is then stored by EPLAN in the project automatically.

The path numbering can be globally set as page-based here (default value *Page-oriented*; every page then (e.g.) begins with path 1 and ends with path 10), or the path numbering should function across all pages.

Across pages (default value *Project-oriented*) means that the paths are numbered across all pages, e.g., they begin with path 1 on page 1 and end with path 30 on page 3.

 TIP: All these settings can be modified in any manner of your choice later on in the project settings!

In addition, you can use ▣ at the drop-down field in this tab to specify through predefined *DT numbering* schemes how new devices are to be numbered online when they are inserted.

You can use the ▣ button to select a scheme from the list, or you can create a new scheme via the NEW button, enter the desired values, and adopt this new scheme for the current project.

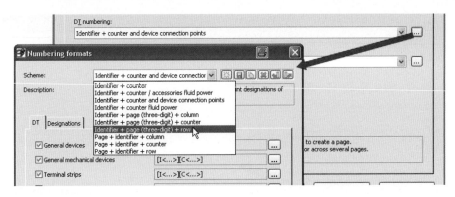

Fig. 3.22 Settings for DT numbering

The same applies to the setting in the *PLC numbering* selection field on the *Numbering* tab. Here too you can select existing schemes from the selection list. You can also use the button to select a scheme (from the selection list). They can be adopted or changed at this point.

You use the NEW button to create new schemes and later assign them to the project.

 TIP: All schemes created in EPLAN can be exported or imported. This makes it easy to exchange these settings between different EPLAN workstations. You should use a unified directory structure to make exported schemes easier to find.

This completes all fields on the *Numbering* tab and clicking the NEXT button now brings us to the *Properties* tab. The *Properties* tab contains general, non-essential descriptive information about the project.

For example, to fill the plot frame with information, such as customer name or the name of the person responsible for the project, from the project properties, these project properties have to be entered.

This is not essential at this point because this is truly only descriptive information about the project which can be entered at any later date.

After clicking the FINISH button, the project is generated and can then be opened in the page navigator (the page overview). You can access the page navigator via the F12 key or the PAGE / NAVIGATOR menu.

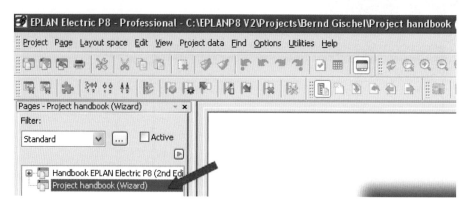

Fig. 3.23 The newly created project in the page navigator

You can now proceed with editing the project.

4 The graphical editor (GED)

The main area for working with EPLAN Electric P8 is the graphical editor, or GED for short. Here, in the graphical editor, all functions necessary for project editing are available to the user. It is, as it were, the main hub of EPLAN Electric P8.

■ 4.1 Page navigator

In addition to the actual data such as parts data, a project also contains function templates consisting of individual pages, e.g., schematics, construction drawings, or terminal diagrams that can also be divided into particular structures such as higher-level functions or mounting locations and installation sites.

Fig. 4.1 Open the page navigator

To provide a certain level of clarity and allow easy editing of the page properties, EPLAN has the **page navigator**. The **page navigator** itself is opened via the PAGE / NAVIGATOR menu.

As usual in EPLAN, the **page navigator** offers a choice between tree structure or list view viewpoints.

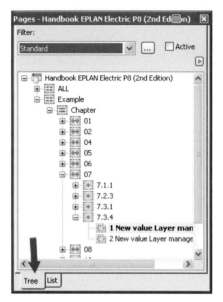

Fig. 4.2 Tree view

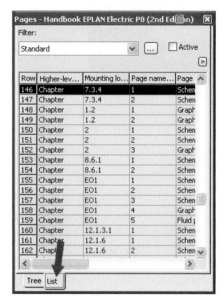

Fig. 4.3 List view

The tree view is ideal for providing a visual overview of all pages in a project. On the basis of the small preceding symbols, the identifiers and the pages are distinguished graphically.

The display sequence can be defined via the parameter in the OPTIONS / SETTINGS / PROJECTS [PROJECT NAME] / DISPLAY / PROJECT STRUCTURE (PAGES) menu.

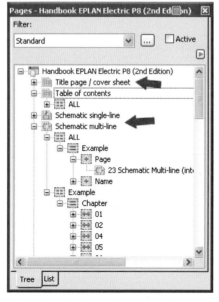

Fig. 4.4 Structured according to page type

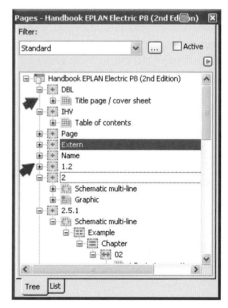

Fig. 4.5 Structured according to mounting location

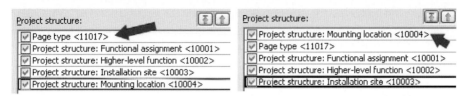

Fig. 4.6 Project setting according to page type

Fig. 4.7 Project setting according to mounting location

However, this only affects the display in the **page navigator**. The sorting of the identifiers is defined in the structure identifier management, i.e. whether (e.g.) higher-level function = A1 is sorted before or after higher-level function = A2.

4.1.1 Page types

EPLAN offers a range of different page types for specific purposes. These pages have small symbols in EPLAN to make them easier to be recognized at a glance.

EPLAN basically distinguishes between logical and graphical pages.

Examples of logical pages are, for example, *Schematic multi-line*, *Schematic single-line* or also *P&I diagram*. These page types are examined for logical information by EPLAN and evaluated correspondingly (cross-references, etc.).

The *Graphical* page type or the model view, however, is a purely graphical (non-logical) page that initially contains no further logical information.

To refine things more, EPLAN also makes a distinction between pages that can be edited (interactive pages) and pages that are generated (automatic pages). An interactive page, for example, is the *Panel layout* page type; the *Terminal diagram* page type is an example of a generated (automatic) page.

All page types in EPLAN are differentiated on the basis of small symbols in the page navigator, as described above and shown in Fig. 4.8.

Fig. 4.8 Page types

The exception is that there is no available page type in the *layout space*. The layout space is the basis of the 3D representation of enclosures or other components used for the panel layout; here it provides a view of the 3D data and their further processing.

4.1.2 The Popup menu in the page navigator

Page editing, but not the graphical editing of the page, is done via the PAGE menu or via the popup menu in the page navigator.

The *popup menu* functions are accessed using the right mouse button or simply the button ▶ in the **page navigator**.

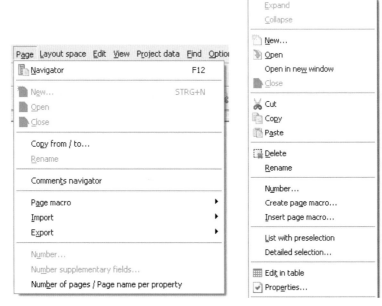

Fig. 4.9 Page menu   Fig. 4.10 Popup menu

4.1.2.1 Create new pages

Generally, the pages of a project are created from scratch (leaving aside copying for the time being). This can be done using the CTRL + N shortcut key or using the PAGE / NEW menu. The procedure coming from the popup menu is the same, except that in this case the NEW menu entry is selected.

The **New page** dialog is then displayed.

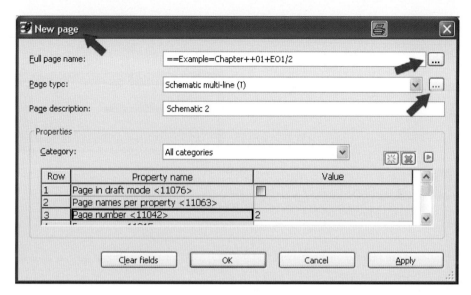

Fig. 4.11 New page dialog

There are now two ways of integrating the new page into an existing page structure. You can define the full page name in the *Full page name* field. The existing entry can be adapted manually, or a completely new value with a prefix for the structure identifier can be entered. Alternatively, the button ⊟ is used to call up the *Full page name* dialog.

EPLAN opens the *Full page name* dialog, whereby every identifier has its own input field with a corresponding selection button ⊟, which can be used to branch to the selection dialog for the selected identifier.

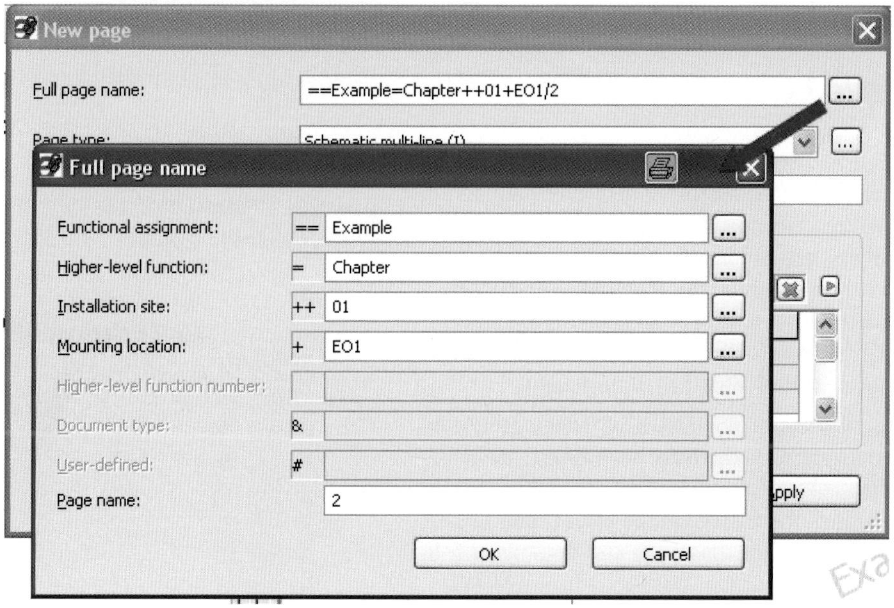

Fig. 4.12 Full page name dialog

Depending on the selected page structure, a *higher-level function*, *mounting location*, etc., can now be selected per mouse click and the desired identifiers applied via the OK button.

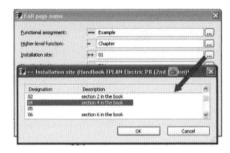

Fig. 4.13 Identifiers dialog

 NOTE: If an existing identifier cannot be applied, then the new identifier can also be directly entered into the [Identifier type] input field in the **Full page name** dialog.

With identifiers that do not yet exist, EPLAN only asks for the sorting sequence after saving the page. The query depends on the OPTIONS / SETTINGS / USER / DISPLAY / IDENTIFIER setting.

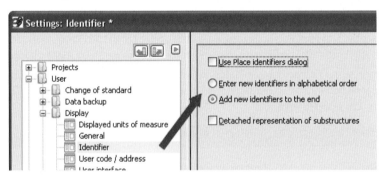

Fig. 4.14 Activate Place identifiers dialog

If (e.g.) the **Place identifiers** dialog should be used, then the corresponding check box in the settings must be set. In the example, new identifiers are automatically placed at the end. With this setting, the **Place identifiers** query dialog is not used. The sorting can then be performed manually or semi-automatically later in the structure identifier management.

If all settings and entries have been completed and entered in the **New page** dialog, and if any missing structure identifiers have been newly created, the dialog can be closed, and EPLAN will create the new page and then show it in the page navigator in the structure selected.

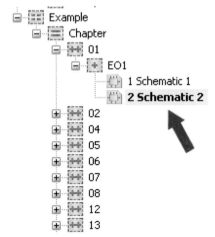

Fig. 4.15 Newly created page

4.1.2.2 Open pages

As the name of the function indicates, the PAGE / OPEN menu item opens a page. This is not unusual. The OPEN IN NEW WINDOW menu item in the page navigator popup menu is more interesting and useful. It is this menu item that allows you to open several pages or - and this is a special feature - the same page *several times* .

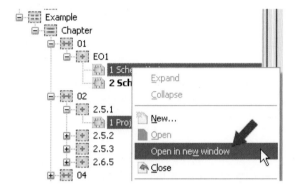

Fig. 4.16 Open in new window

The procedure is simple. The page or pages (it is possible to open several pages at once) is/are selected in the page navigator, and the OPEN IN NEW WINDOW menu item in the popup menu is called. The CTRL + ENTER keyboard shortcut can also be used. EPLAN then opens all selected pages.

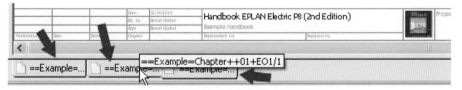

Fig. 4.17 Several pages opened

If the WORKBOOK menu item in the VIEW menu is activated, then all opened pages are displayed in Excel style (tabbed page list) at the lower edge of the workspace. The mouse can then be used to switch back and forth between these pages - or the CTRL + TAB keyboard shortcut can be used to switch between the pages, whereby this always cycles through all the pages. With three pages, the sequence starts at page 1 followed by page 2, page 3, back to page 1, and so on.

If several pages have been opened, you can select the desired page directly via the WINDOW menu.

To close the open pages again, simply select the pages in the page navigator and select in the PAGE menu the CLOSE menu item. EPLAN then closes all selected pages.

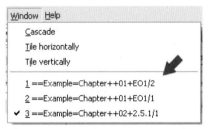

Fig. 4.18 Window menu

4.1.2.3 Copy pages

Pages can be created anew, but they are created then without any page content (devices, graphics, etc.). To apply the content of a page to a new page, the page or pages must be copied in EPLAN.

This occurs in different ways in EPLAN. You can use the standard Windows functions like CTRL + C (copy page), CTRL + V (insert page), or the menu items of **Page** (or in the popup menu of the page navigator) can be used.

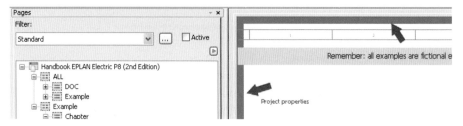

Fig. 4.19 Select a page via CTRL + M

To copy a page in EPLAN, the page is selected via CTRL + M in the graphical editor. EPLAN then draws a *thick gray border* around the page (a type of selection).

When the CTRL + C keyboard shortcut is pressed (standard Windows function for copying), EPLAN copies the page into the clipboard. The CTRL + V (paste) keyboard shortcut is then pressed to paste copied pages from the clipboard. EPLAN then opens the **Adapt structure** dialog.

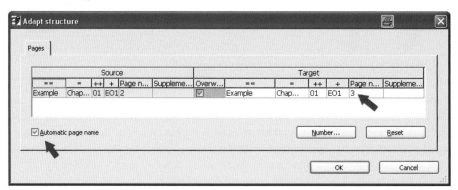

Fig. 4.20 Adapt structure dialog

Appropriate entries such as the assignment of identifiers or a new page name can now be edited in this dialog. However, EPLAN can also determine the new page name automatically (*Automatic page name* setting). It is always the highest free page name available in this structure that is suggested.

Once all entries are correct, the page can be adopted via the OK button.

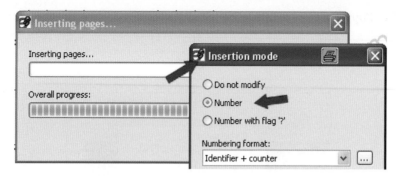

Fig. 4.21 Select insertion mode

This is followed by the **Insertion mode** dialog (depending on the settings) to allow the existing devices on the copied page to be immediately numbered.

If the dialog is exited with the corresponding settings by clicking the OK button, EPLAN copies the (source) page, generates a new (target) page, and sorts this into the page structure.

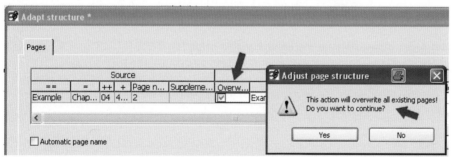

Fig. 4.22 Overwrite option

At this point in the **Adapt structure** dialog, it is also possible to overwrite existing pages (in the *Target -Overwrite* column). EPLAN displays a query message before actually overwriting pages, which must be confirmed with the YES (copy action continues) or the NO button (copy action is canceled and EPLAN opens the page navigator once more).

Again the following applies when copying pages: with new identifiers, the **Place identifiers** dialog may be displayed, depending on the user settings.

More than one page can, of course, be copied. In the page navigator, multiple pages can be selected and also copied to the clipboard via the CTRL + C keyboard shortcut.

With CTRL + V (paste), EPLAN once more opens the already-familiar **Adapt structure** dialog. After making all desired entries and confirming any subsequent dialogs, these pages are inserted into the EPLAN project in the selected structure.

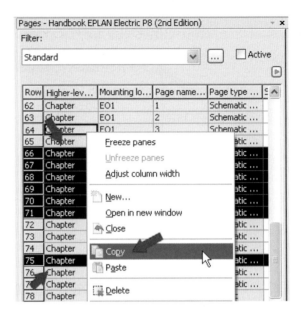

Fig. 4.23 Copy in list view

But pages cannot only be copied and pasted using the CTRL + C, CTRL + V keyboard shortcuts. These commands also exist in the popup menu of the page navigator. They are called COPY and PASTE and always relate to the currently selected pages. It makes no difference whether the pages are selected and copied / pasted from the tree view or the list view.

 TIP: You should take note of the CTRL + M (select page) keyboard shortcut. Everything relating to a page, i.e. copying a page, deleting a page, or editing the page properties, is done using this CTRL + M keyboard shortcut and the following commands.

Aside from copying pages within a project, it is, of course, also possible to copy the pages in a project into the clipboard, to switch to another project (which, naturally, should be open in the page navigator) and then to insert the pages there from the clipboard.

4.1.2.4 Copy pages from/to

The counterpart to simple page copying (within a project but also in other projects) is the COPY FROM / TO... function. Since it is possible to open and edit several projects at once in the EPLAN page navigator, a function allowing the copying of pages from one project to another *conveniently* should also be present.

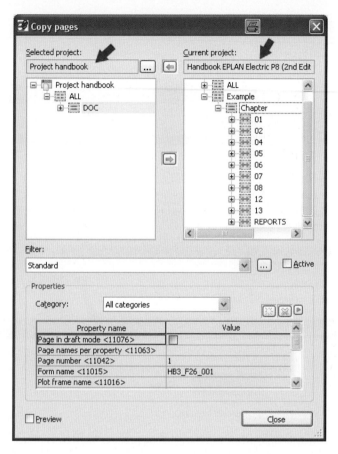

Fig. 4.24 Copy pages dialog

The COPY PAGE / PAGES FROM / TO menu also provides a separate function for copying pages from different projects (also from projects that are not open in the page navigator) into the current project. Upon selection of the menu item COPY PAGES FROM/TO, EPLAN opens the **Copy pages** dialog. In order to copy pages to another project, the project does not need to be open in the page navigator, as already mentioned.

The lower area of the **Copy pages** dialog shows additional information on the respectively selected page. The preview for selected pages can also be activated. The right field in the **Copy pages** dialog is fixed and always relates to the *current project*. It cannot be changed. The project selection in the left field is variable and the ▭ button can be used to select and open any other project.

By clicking the ▭ button, EPLAN opens the **Project selection** dialog. Initially, only projects currently open in the page navigator are listed.

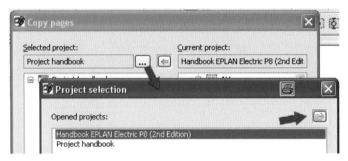

Fig. 4.25 Project selection dialog

When the (Open) button is clicked, EPLAN opens the **Open project** dialog. In this dialog, one or several projects can be selected and opened for adoption in the **Project selection** dialog. To do this, the project(s) is(are) selected, and the OPEN button is used to add this to the **Project selection** dialog.

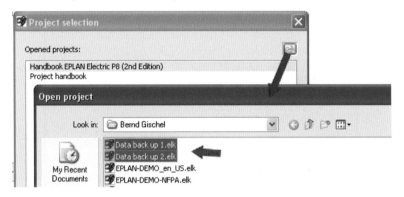

Fig. 4.26 Open project dialog

Once the project selection is finished, the selected projects are listed in the **Project selection** dialog and are ready for page copying and are opened simultaneously in the page navigator.

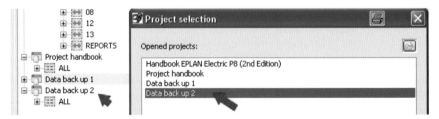

Fig. 4.27 Opened projects

The desired project can now be selected in the **Project selection** dialog and adopted in the **Copy pages** dialog via the OK button.

 TIP: You should take note of the CTRL + M (select page) keyboard shortcut. Everything relating to a page, i.e. copying a page, deleting a page, or editing the page properties, is done using this CTRL + M keyboard shortcut and the following commands.

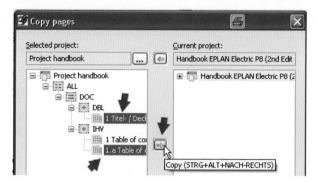

Fig. 4.28 Copy pages from a project

In the **Copy pages** dialog, the pages to be copied from the external project are *selected* and copied (moved) into the current project via the center button. EPLAN then opens again the **Adapt structure** dialog.

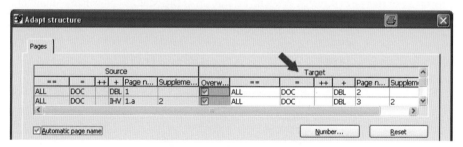

Fig. 4.29 Insert pages into the page structure

The page structure is adjusted as desired in the dialog.

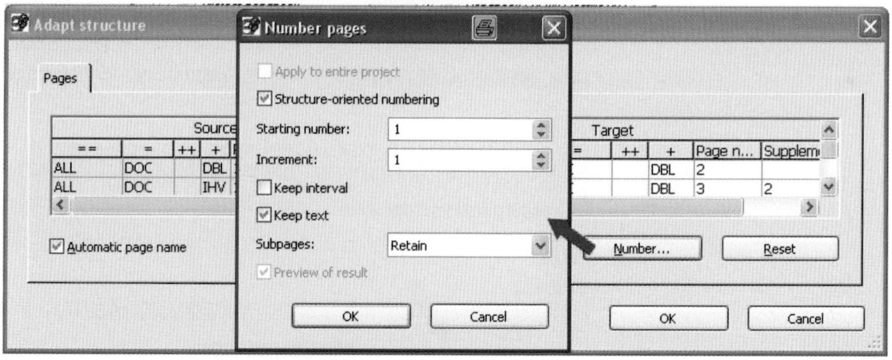

Fig. 4.30 Number pages

It is also possible to re-number the new pages using the NUMBER button according to specific settings. This is an option, for example, when many subpages need to be copied and converted to main pages.

After the OK button in the **Adapt structure** dialog is clicked, EPLAN copies the pages into the current project. This may be followed by other dialogs, such as the **Insertion mode** or **Place identifiers** dialog, which must be confirmed as required.

This completes the copying of pages from external projects. EPLAN closes all dialogs and sets the focus back to the **Copy pages** dialog. Other pages from other projects can now be copied. If this is not required, then the **Copy pages** dialog can be exited via the CLOSE button. EPLAN then returns to the graphical editing.

4.1.2.5 Rename pages

Pages in EPLAN receive a unique page name (page number). This page name *may* also include alphanumeric characters. Thus, a page can be called (e.g.) =ANL2+ORT2/*14_Overview*, whereby the word "Overview" belongs to the page name and not the page description.

The page name (number) is renamed using the function in the RENAME PAGE menu. In the **page navigator**, it is also possible to select the page and press the F2 key.

Then, EPLAN allows for the page name (number) to be modified, without opening the page properties of this page directly. Pressing ENTER saves the changed page name (number).

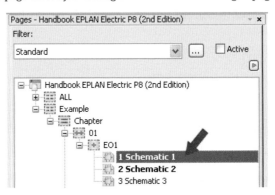

Fig. 4.31 Select page

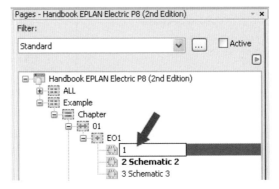

Fig. 4.32 Direct editing via F2

4.1.2.6 Delete pages

Pages are not only newly created or copied. It must also be possible to delete pages from time to time. In the popup menu of the page navigator, EPLAN offers the DELETE function for this.

The procedure is already familiar. The pages to be deleted are selected in the page navigator and then deleted using the Delete function in the page navigator popup menu or by pressing the DEL key.

Before the actual deletion, EPLAN displays a **Delete pages** dialog with a warning message asking if the page(s) should actually be deleted.

 NOTE: I recommend not to confirm *blindly* messages sent with the YES or YES TO ALL option, but instead to read the messages carefully before clicking!

The following is *very important* and should be noted when deleting pages. There is a difference in how pages are selected in the page navigator.

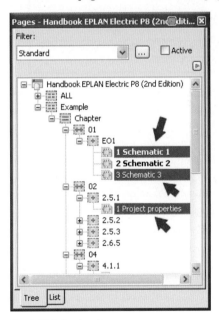 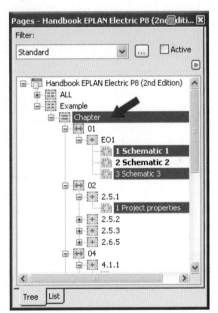

Fig. 4.33 Selection 1   Fig. 4.34 Selection 2

In Fig. 4.33, four pages are selected for deletion. Here only these three pages are deleted. The warning message in the **Delete pages** dialog is also very clear: "Do you want to delete page ... of 3 selected pages?"

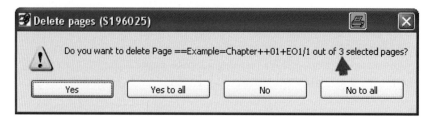

Fig. 4.35 Delete 3 pages?

In Fig. 4.34, however, the selection has been extended by mistake to the entire higher-level function =*Chapter*, because the higher-level function was selected alongside in the **page navigator**.

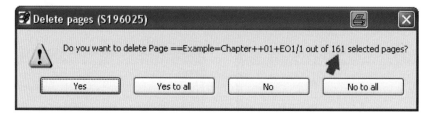

Fig. 4.36 Instead of 3 pages, 161 pages are deleted.

If these selected pages are then sent for deletion, EPLAN displays the familiar **Delete pages** dialog with an appropriate warning message, but with the minor difference that the *number of pages to be deleted* is no longer correct.

If this dialog is confirmed with YES without thinking, then EPLAN deletes the entire higher-level function =*Chapter* from the project. Unintentional deletion of pages is usually noticed immediately, since the page navigator is updated after a deletion and the missing pages, higher-level functions, or locations are (or should be) immediately obvious.

The usual UNDO command CTRL + Z, or the UNDO or UNDO LIST functions in the EDIT menu can then be used to restore the pages *immediately after deletion*.

NOTE: If the project was closed in the meantime, the deleted pages are irretrievably lost!

TIP: If the project name is selected, then the DELETE function is disabled by EPLAN. This way, the accidental deletion of entire projects from the popup menu of the page navigator is prevented.

4.1.2.7 Close pages

This function is self-explanatory. It closes open pages. The CLOSE function is accessible in the PAGE menu and also via the page navigator popup menu.

It is possible to select several open pages in the page navigator and close them all at once. The CLOSE function does not close the project, even if it is selected!

4.1.2.8 Edit page properties

To edit the page entries, the so-called *Page properties*, the appropriate page is selected in the page navigator and the PROPERTIES function is called via the popup menu.

Of course, this can also be done without the page navigator directly on the opened page using the CTRL + M and CTRL + D shortcut keys. CTRL + M selects the page and then CTRL + D calls the EDIT PROPERTIES function. EPLAN then opens the **Page properties** dialog.

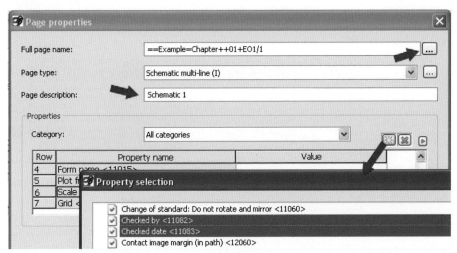

Fig. 4.37 Edit page properties dialog

All page properties can now be edited in the **Page properties** dialog. The *Full page name* can be changed (for example, you could change the page number from 2 to 3, which will move the page) or the *Page description* can be adjusted.

The *Properties* area displays additional properties, such as page scale (*Scale <11048>*) or the currently set grid (Grid <11051>). These properties can all be edited when they are not grayed out.

New properties are added via the [icon] button. After the button is clicked, EPLAN opens the **Property selection** dialog. Here you can select the desired property or properties (multiple selections are possible) and apply it or them in the **Page properties** display by clicking the OK button.

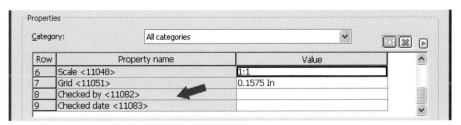

Fig. 4.38 Newly added properties

In contrast to the CTRL + M and subsequent CTRL + D keyboard shortcuts directly on a page, in the **page navigator** more than one page can be edited at once.

If several pages are selected in the page navigator and the popup menu is then used to call the PROPERTIES function, then the properties of all selected pages can be edited in a single step.

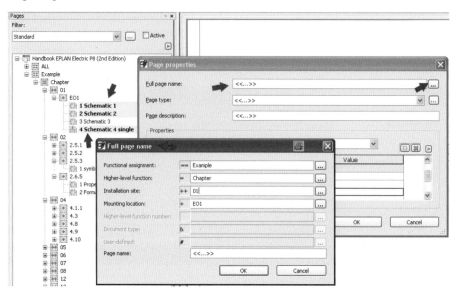

Fig. 4.39 Edit the properties of multiple pages

All properties that are the same in the selected pages are displayed in *"plain text"*. An example would be the scale or grid of the pages. If the page properties are different from each other, then they are displayed visually using the <<...>> string in the corresponding input fields.

 NOTE: Care should be taken when changing these fields (containing the <<...>> string) because, when new entries are made, all other entries that are not the same on other pages are overwritten.

When all entries are complete, the **Page properties** dialog can be exited by clicking the OK button. EPLAN saves all entries and then closes the dialog.

4.1.2.9 Edit page properties directly

Apart from the method described for adjusting or modifying *Page properties* such as page description, etc. by calling the dialog of the same name (via the page navigator popup menu or via the CTRL + M and CTRL + D shortcut keys), it is now possible to simply double-left-click an existing object (page description, customer name) in the plot frame to let EPLAN open the page properties.

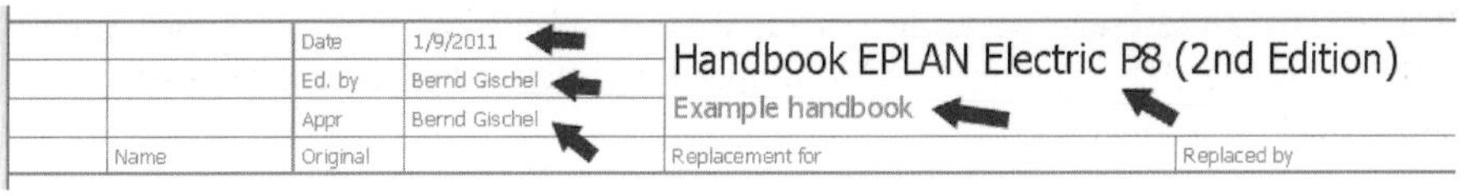

Fig. 4.40 Double-click any object in the plot frame to open page properties

But it has to be an area that contains something. Double-clicking an empty area will **not** open the page properties!

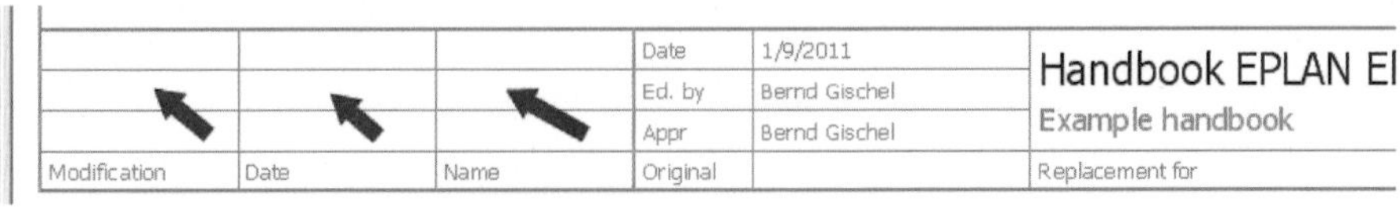

Fig. 4.41 Empty areas do not work

4.1.3 Page navigator filter

In addition to the functions previously described, it is now possible with EPLAN Electric P8 2.0 to directly activate a filter for previously created schemes in the page navigator, as well as other navigators.

This way, one can quickly and easily filter, for example, the selection of the page display for a specific page type. For this purpose, as usual, one can create and easily activate a filter scheme.

The filter function known from previous versions is now located in the popup menu of the **page navigator** and is called DETAILED SELECTION. This can be used to search and filter, as before, filters for search results or specific structure identifiers.

■ 4.2 General functions

The graphical editor *can* be divided into several areas by default. The main area is the working area for editing the schematic (the graphical editor). A number of dialogs can also be associated with this. These range from the page navigator to user-defined toolbars.

The graphical editor is not restricted to a single representation. The familiar idea of *workspaces* can be used to rapidly switch the entire EPLAN user interface from one representation to another, including all the associated dialogs, toolbars, etc.

4.2.1 The title bar

Aside from the actual working area, there is the *Title bar*, which displays only little, but crucial, information. The most important function of the title bar is the display of the currently open project name.

This display can be customized by the user. To do so, go to Settings (OPTIONS / SETTINGS / USER / DISPLAY / USER INTERFACE menu) and modify the *Display in title bar* item.

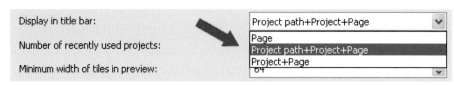

Fig. 4.42 Adjustable display of the title bar

The following entries are possible here: *Project+Page*; *Project path+Project+Page* and only *Page*. In Fig. 4.43, the *Project path+Project+Page* setting has been selected.

Fig. 4.43 Sample setting

The selection of the title bar display cannot be changed or expanded.

4.2.2 The status bar

The status bar in the graphical editor constantly shows various current information.

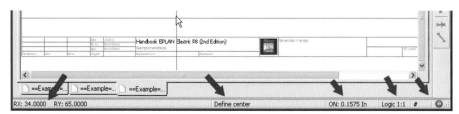

Fig. 4.44 Status bar

For example, the coordinates RX and RY of the current cursor position, information on the current grid (here 4.00 mm), and whether or not the grid is switched on (here ON), are all displayed here. The information logic 1:1 shows the respective page type and the scale of this page. The example shows an open logical page at a scale of 1:1.

In addition, EPLAN displays in the status bar supporting information for the function currently in use (in this case, a circle is to be drawn). This does not apply to all functions, but looking at the status bar can be of great help in some EPLAN actions.

The last characters in the status bar are defined as follows.

#	The hash means that the project contains connections that have not been updated. [1]
*	The asterisk means that the current page contains connections that have not been updated. [1]
	If a new error occurs since the last opening of the system messages, this icon will be displayed again. [2]
+	If the „advanced" support-log mode has been enabled (via the „ELogFileConfigToolu.exe" tool supplied by EPLAN), this will be indicated by this character.

[1] Since it is possible to assign menu commands to keyboard shortcuts in EPLAN, I would recommend directly assigning the PROJECT DATA / CONNECTIONS / UPDATE menu command to the F11 keyboard shortcut. This is easy to remember and the command then only requires a single keystroke.

[2] The icon can be double-clicked with the mouse to open the **System messages** dialog. The current system messages can then be read here. When the **System messages** dialog is closed, the icon disappears until the next system message occurs.

The UTILITIES / SYSTEM MESSAGES menu entry can be used to call up the **System messages** dialog at any time. All system messages are visible until EPLAN is exited. The current system messages are only deleted from the list when EPLAN is exited.

■ 4.3 Coordinate systems

The EPLAN coordinate system has the following structure. There are different points of view: the graphical system and the logical systems for the areas of electrical engineering, fluid power and process engineering.

4.3.1 Graphical coordinate system

The graphical system has its origin (X,Y = 0,0) at the lower left of the page.

This means that when the page scale changes, the page is enlarged upwards and to the right.

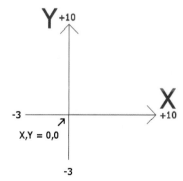

Fig. 4.45 Graphic

Fig. 4.46 Display in the status bar

4.3.2 Logical coordinate system

The logical system has its origin (X,Y = 0,0) at the upper left.

These positions are thus always measured in the different directions from the original element and its origin (usually the so-called insertion point).

In this case, these are negative values, since the property is moved downwards and to the left with the coordinate input X/Y in the property dialog.

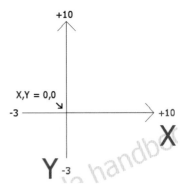

Fig. 4.47 Logic

Fig. 4.48 Display in the status bar

4.4 Grid

EPLAN can assist the user when placing (e.g., symbols) or drawing (e.g., panel structures) via a fixed *grid*. The grid is visually displayed with small dots. The display of the grid can be enabled or disabled via the VIEW / GRID menu or the CTRL + SHIFT + F6 shortcut key.

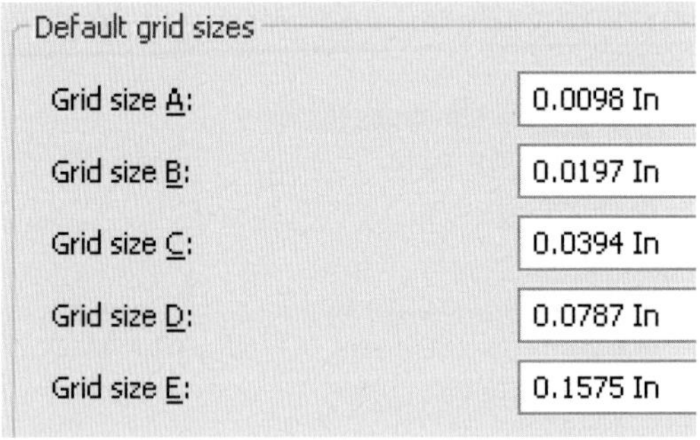

Default grid sizes	
Grid size A:	0.0098 In
Grid size B:	0.0197 In
Grid size C:	0.0394 In
Grid size D:	0.0787 In
Grid size E:	0.1575 In

Fig. 4.49 Grid settings

4.4 Grid

Five different grids can be entered in the OPTIONS / SETTINGS / USER / GRAPHICAL EDITING / 2D settings. My recommendations for the grids are the following values: Grid A = 0.25 mm; grid B = 0.5 mm; grid C = 1 mm; grid D = 2 mm and grid E = 4 mm.

These grid settings can then be selected in the EDIT / OTHER / GRID [A to E] menu or via the VIEW toolbar.

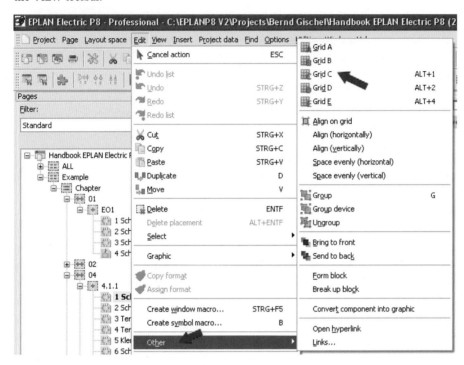

Fig. 4.50 Other with grid settings menu item

This VIEW toolbar has other buttons that will be briefly explained here. They can all be assigned to a keyboard shortcut as well.

Fig. 4.51 Toolbar view

 Using this button, the grid can be made visible or disabled.

This function enables or disables Snap to grid. This means that when Snap to grid is enabled, elements can be placed only on the activated grid. Intermediate positions or free placements are not possible. When the grid snap is switched off, then all elements can be freely placed (this also applies to logic elements such as symbols).

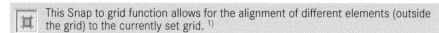 This Snap to grid function allows for the alignment of different elements (outside the grid) to the currently set grid. [1]

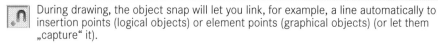 During drawing, the object snap will let you link, for example, a line automatically to insertion points (logical objects) or element points (graphical objects) (or let them „capture" it).

The design mode allows you to align graphical elements to specific points and thus to place them on specific coordinates. If the design mode is enabled, first the action - e.g., Move - is selected. Then, you select the object and define its starting and end points.

[1] Proceed as follows: Click on the button in the **View** toolbar. The mouse is then used to pull a window around the misaligned elements. The window can be pulled in any direction. When finished (all elements are contained in the window), the mouse button is released and EPLAN aligns all elements inside the window to the grid.

■ 4.5 Increments, coordinate input

In addition to the grid settings, EPLAN also allows for the input of increments and coordinates. These two features relate to the cursor position.

4.5.1 Increment

The increment is set via the S key or in the OPTIONS / INCREMENT menu. EPLAN opens the **Select increment** dialog.

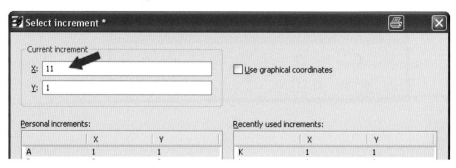

Fig. 4.52 Integer increment setting

The increments for the cursor can be entered into the *Current increment X and Y* fields of this dialog. If the *Use graphical coordinates* option is deactivated, then it is only possible to enter integer values, such as 1, 3 or 17.

If the *Use graphical coordinates* option is activated, then it is also possible to enter real numbers such as 1.75 or 3.88 because the graphical coordinates are then evaluated.

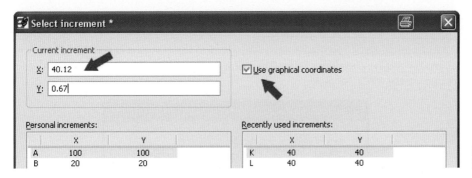

Fig. 4.53 Real-number increment setting

EPLAN supplies the information also visually in the status bar, regardless of whether in the **Select increment** dialog the *Use graphical coordinates* option has been activated or deactivated.

4.5.2 Coordinate input

As well as the increment, coordinates can be directly entered into EPLAN to allow jumping to a particular point by clicking OK.

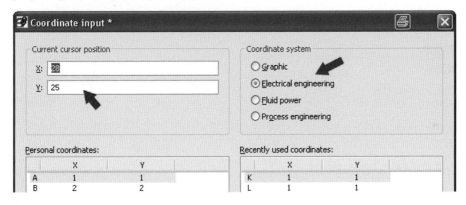

Fig. 4.54 Coordinate input

The **Coordinate input** dialog is started with the P key or via the OPTIONS / COORDINATES menu. The X and Y positions can be directly entered in this dialog. As soon as the dialog is closed using the OK button, EPLAN jumps to the specified X/Y coordinates.

 TIP: The mouse should not be used here, otherwise it is rather likely that the position to jump to may shift.

During the coordinate input, too, it is possible to switch between the different coordinate systems (graphic, electrical engineering, fluid power and process engineering).

4.5.3 Relative coordinate input

Apart from the *static* coordinate input (described in the previous chapters), EPLAN also allows for *relative* coordinate input. The RELATIVE COORDINATE INPUT function is accessed via the OPTIONS menu.

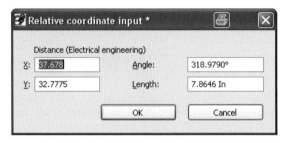

Fig. 4.55 Relative coordinate input

Using this function, the cursor, which is dragging a line for placing, for example, can be placed in a new position. This means that you could start drawing a new line and then call the relative coordinate input.

In the simplest case, you could enter the length of the line, and EPLAN would adopt this length, place the cursor in the position and terminate the drawing of the line.

In contrast to the static coordinate input, the input here is always relative to the current cursor position.

 TIP: If a grid is enabled and the jump to the cursor end point is not on the grid, EPLAN will still jump to the next grid point! You should note this when drawing.

4.5.4 Move base point

The MOVE BASE POINT function, accessible via the OPTIONS menu or the O key, sets a new base point (X,Y = 0), which can be manually set to be different from the *standard base point* (depending on the coordinate system graphic, electrical engineering).

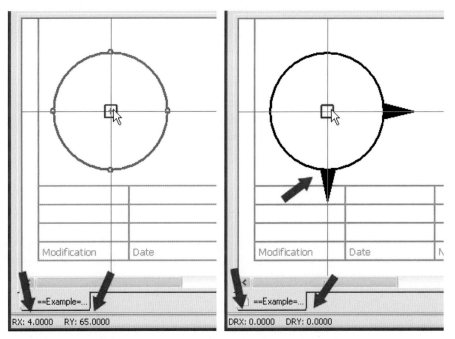

Fig. 4.56 Standard base point   Fig. 4.57 New base point

In the example (center of the circle), the base point is X = 256.00 mm and Y = 244.00 mm. In order to redefine the base point now, for example, to calculate the horizontal values from the center of the circle, the cursor is placed on the *center*, and the O key is pressed, followed by ENTER.

EPLAN now generates a new current base point for this page of X,Y = 0. All positions are immediately calculated using this new base point and displayed in the status bar. To differentiate from the usual values displayed (X = n; Y = n), EPLAN displays the values of the base point shift as marked with a D. In the status bar then, the X and Y values are displayed visually as follows - DX = n and/or DY = n.

The base point shift can be reset using the O key. Or you can simply change the page.

■ 4.6 Graphical editing functions

In addition to the creation of schematics, EPLAN allows for the creation of a type of CAD drawing (CAD = Computer Aided Design). This is rounded off by exact scaling, many dimensioning possibilities, and many additional functions such as stretching, trimming (modifying length) and grouping of elements that are common in the CAD world.

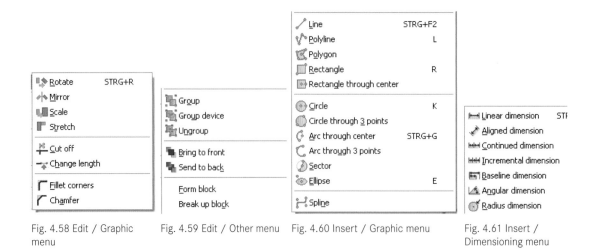

Fig. 4.58 Edit / Graphic menu

Fig. 4.59 Edit / Other menu

Fig. 4.60 Insert / Graphic menu

Fig. 4.61 Insert / Dimensioning menu

These graphical functions are accessible in the INSERT / GRAPHIC [function type] menu. Special functions such as stretching, grouping, or converting a component to a graphic are located in the EDIT / GRAPHIC menu or the EDIT / OTHER [function type] menu, such as Group, Bring to front, etc.

The dimensioning functions are located in the INSERT / DIMENSIONING menu.

EPLAN remains a pure CAE system (CAE = Computer Aided Engineering), but the ability to work with CAD functions makes certain pure drawing tasks much easier.

4.6.1 Graphical objects - lines, circles, rectangles

With the *Circle*, *Rectangle*, *Line* or *Polyline* functions and many other options, EPLAN enables you to create graphical drawings. This will not be explained in great detail here, because the calls and the functions are very simple and easy to understand.

TIP: When using the graphical functions, it is a good idea to switch on the *Grid* or to enable the *Snap to grid* function.

Also, you should „play" around with the set grid of the page. That is, if you require 1-millimeter increments, then set the grid for this page to 1 (1 millimeter); you can do this either by using the buttons of the grid settings or directly in the page properties (i.e., via the *Grid <11051>* page property).

For example, a rectangle can be called up via the INSERT / GRAPHIC / RECTANGLE menu or using the R key, with the starting point then being placed on the drawing. Now there are three options.

Option 1. Simply drag the rectangle open using the mouse or the keyboard. The width and height of the rectangle can be adjusted either by grid increments counted or, subsequently, via the properties of the rectangle.

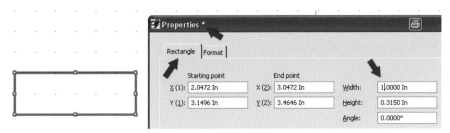

Fig. 4.62 Draw a simple rectangle with manual width and height

Option 2. Begin drawing the rectangle; its size is defined on the basis of the relative coordinate input.

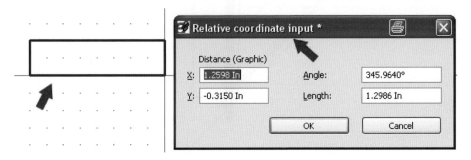

Fig. 4.63 Draw a rectangle on the basis of relative coordinate input

Option 3. Via the OPTIONS menu, switch on the INPUT BOX (subsequently always attached to the cursor). Here, by entering values directly, such as width and height, you can let EPLAN draw the rectangle.

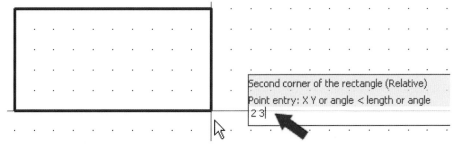

Fig. 4.64 Draw a rectangle by direct input in the input box

4.6.2 Trim, chamfer, stretch and more...

In addition to the pure graphical functions, such as drawing a rectangle or circle, EPLAN also has options for editing these graphical objects subsequently.

Such editing could include the following: Changing the length of graphical elements, cutting out parts of graphical elements, rounding corners and much more. A few examples of the edit function in the EDIT / GRAPHIC menu will briefly illustrate the options.

 NOTE: All these examples were carried out with the input box (in the Options menu) switched on.

4.6.2.1 Rotate

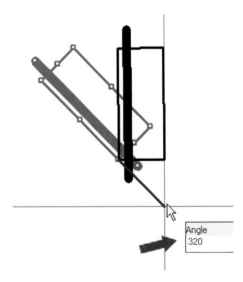

Fig. 4.65 Rotate

The *Rotate* function allows you to rotate a selected object by any angle. The function is called via the EDIT / GRAPHIC / ROTATE menu. Now, EPLAN waits for the selection of the object to be rotated. Then, the center of rotation (around which the object is to be rotated) must be defined. Now enter the rotation angle directly in the input box and confirm with ENTER. EPLAN has rotated the object.

4.6.2.2 Mirror

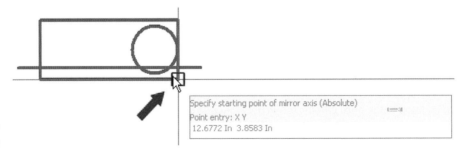

Fig. 4.66 Define start of the mirror axis

Another function is the *Mirror* function. The function is called via the EDIT / GRAPHIC / MIRROR menu. Then, the object to be mirrored is selected. Now EPLAN waits for your input and/or definition of the *Starting point of the mirror axis*.

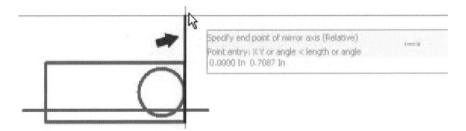

Fig. 4.67 Define end point of the mirror axis

Once this is done, you must define the *End point of the mirror axis*. When you click on it, EPLAN will mirror the object along the mirror axis defined.

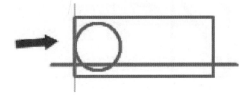

Fig. 4.68 Mirrored object

4.6.2.3 Scale

Aside from rotating and mirroring, objects can also be enlarged or reduced. In EPLAN, these actions are summarized under the term „Scale". The Scale function is called via the EDIT / GRAPHIC / SCALE menu.

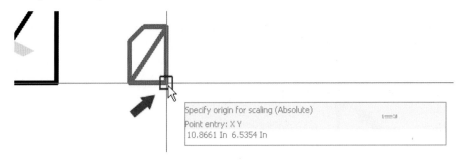

Fig. 4.69 Define origin for the scaling

Once the *Origin* has been selected, you can click on it. EPLAN opens the **Set scale** dialog.

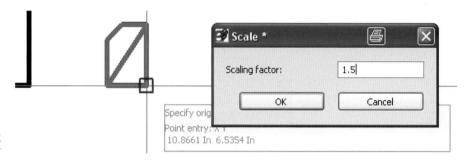

Fig. 4.70 Set scaling factor

If this dialog is confirmed with OK, EPLAN scales (here: reduces) the selected object.

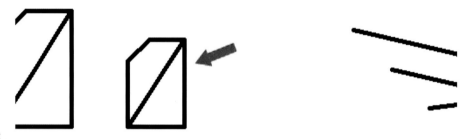

Fig. 4.71 Scaled object

Regarding the *Scaling factor*, the following bears mentioning. For a reduction, EPLAN expects a factor smaller than 1; for an enlargement, a factor greater than 1. All factors must be entered with a comma to take effect.

4.6.2.4 Stretch

The Stretch function allows you to lengthen and shorten objects. The function is called via the EDIT / GRAPHIC / STRETCH menu. Then, (in this example) the right-hand section of the rectangle is selected with a window. EPLAN places round markers in the corners and waits for the starting point of the stretch to be defined.

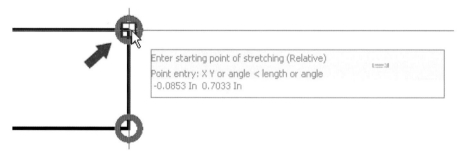

Fig. 4.72 Define starting point of the stretch

Once the *Starting point* is defined, use the mouse or keyboard to set the end point.

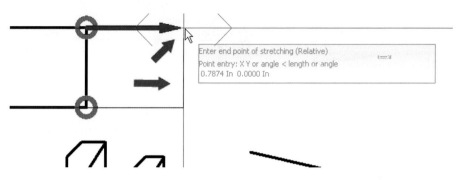

Fig. 4.73 Definition of the end point

Once the *End point* has been defined, confirm it with OK, and EPLAN will execute the stretch.

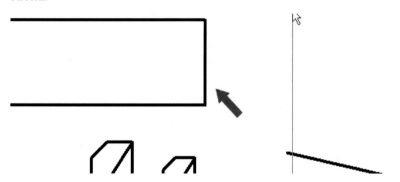

Fig. 4.74 Stretched object

Of course, you can change the length of more than just simple rectangles.

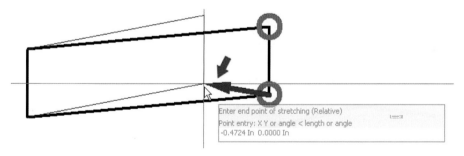

Fig. 4.75 Lengthen or shorten a different object

4.6.2.5 Cut off

The Cut off function, also called Trim in the world of CAD, is used to „cut" a so-called section out of elements. The function is called via the EDIT / GRAPHIC / CUT OFF menu. Once the function has been selected, EPLAN waits for the selection of the section to be removed.

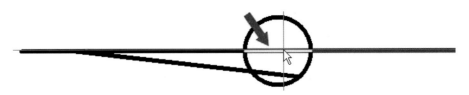

Fig. 4.76 Select element to be removed

If, as shown in Fig. 4.76, you click on the small section within the circle, EPLAN will cut it out.

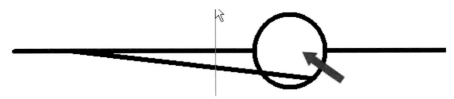

Fig. 4.77 Cut-off section

A typical example of this function would be the cutting off of sections of an equipped horizontal rail or a mounting rail in a mounting panel layout.

 NOTE: The cut-off does not work just like that in connection with grouped objects. If you want to cut sections off grouped objects, you will have to keep the SHIFT key pressed and then cut the relevant section!

4.6.2.6 Change length

The **Change length** menu item does exactly what it is supposed to do: It changes the length of objects. The function is called via the EDIT / GRAPHIC / CHANGE LENGTH menu. Once this function has been called, you click on the relevant object to change its length via the input box or by using the keyboard and/or mouse.

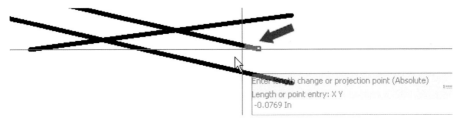

Fig. 4.78 Change length of a line

4.6.2.7 Fillet corners

The Fillet corners function does exactly what its name implies. It is called via the EDIT / GRAPHIC / FILLET CORNERS menu. After launching the function, you can click on the corner of your choice and define the *fillet radius* in the input box.

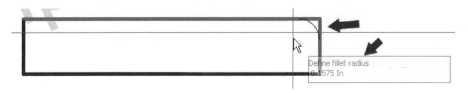

Fig. 4.79 Fillet corners

If you press the ENTER key, EPLAN will apply the radius and fillet the corner.

Fig. 4.80 Fillet the next corner

If you then click on the next corner, it will be filleted automatically using the radius previously defined.

4.6.2.8 Chamfer

The last function is Chamfer. Corners here are not filleted, but given a sloping edge. Like all other functions, it is called up via the EDIT / GRAPHIC / CHAMFER menu.

Fig. 4.81 Chamfer a corner

After starting the function, selecting the object and clicking on the first corner, you enter your chamfer parameters directly in the input box. EPLAN adopts these after you have hit ENTER or left-clicked the object. The same chamfer setting can be applied immediately to the next corner.

Fig. 4.82 Several chamfered corners

4.6.3 Group and ungroup

With several graphical elements, moving or (e.g.) scaling (enlarging, reducing) all these individual graphical elements can be somewhat difficult.

Every element must be manually adjusted to its new size via the **Properties** dialog, or with multiple selections, the elements that are not to be scaled must be manually removed from the selection.

With the GROUP and UNGROUP functions in the EDIT / OTHER menu, EPLAN allows particular elements that belong together to be grouped and then be subsequently regarded as a single element. In the process, all elements that belong together are selected with the mouse, followed by the Group command from the menu (or alternatively the G key).

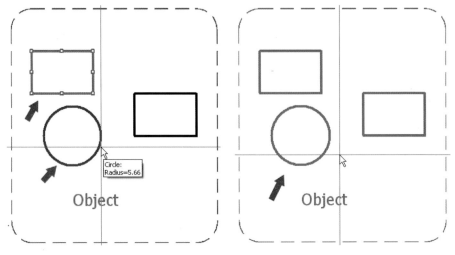

Fig. 4.83 Individual objects   Fig. 4.84 Grouped objects

The grouping can, of course, be removed again. To do this, the grouped elements are selected (EPLAN automatically selects the remaining elements), and the UNGROUP function in the EDIT / OTHER menu is selected. EPLAN removes the grouping and all elements are once more individually accessible.

Individual elements within a grouping can also be edited independently. To do this, the SHIFT key is pressed (and held), and the desired object within the grouping is double-clicked with the mouse.

The corresponding property dialog of this object is opened, and the properties can be edited – without removing the entire grouping. When all entries are complete, EPLAN adopts the changes for this object. The remaining elements in the grouping are not affected.

4.6.4 Copy, move, delete, ...

In addition to pasting and drawing new elements, the functions for copying or deleting the same elements are still missing.

✂ Cut	STRG+X
📋 Copy	STRG+C
📋 Paste	STRG+V
Duplicate	D
Move	V
Delete	ENTF

Fig. 4.85 Editing functions

These functions can be accessed via the EDIT menu, which contains the COPY, MOVE functions, etc. These functions can also be accessed via the keyboard: CTRL + C, for example, for *Copy*, or CTRL + X for the *Cut* function.

To use the functions on elements, these are first selected (clicked with the left mouse button), and then the function is selected.

4.6.5 Dimensioning

In addition to drawing, the drawn elements can also be dimensioned. EPLAN offers accurately scaled dimensioning, depending on the page scale that is set.

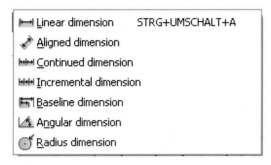

Fig. 4.86 Dimensioning functions

EPLAN comes with a number of dimensioning functions. Dimensioning types include, for example, continued dimension or radius dimension. The procedure for dimensioning elements is just as simple as drawing the elements themselves. The dimensioning types can be accessed in the INSERT / DIMENSIONING [DIMENSIONING TYPE] menu.

The various dimensioning methods will be illustrated with an example.

 TIP: It is recommended that you enable the object snap function (accessible via the OPTIONS / OBJECT SNAP menu) when dimensioning. This allows EPLAN to start precisely at the ends (or the midpoints) of elements, without you having to tediously try to reach the ends of a line, for example. As a rule, this cannot be done precisely.

A rectangle is to be dimensioned. To do this, the INSERT / DIMENSIONING menu is used to call up the LINEAR DIMENSION function.

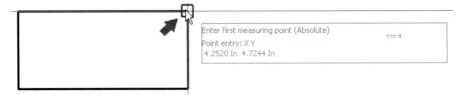

Fig. 4.87 Start of the dimensioning

The first *Dimension point* (corner of the rectangle) is selected and clicked with the left mouse button. EPLAN sets the start of the dimensioning.

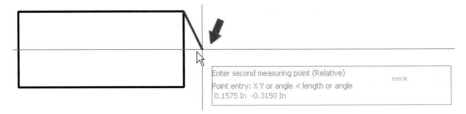

Fig. 4.88 Define second dimensioning point

Then the second point is sought; the first dimension hangs on the cursor and is placed by clicking the left mouse button.

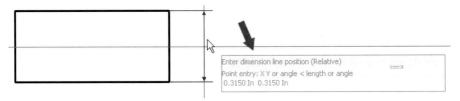

Fig. 4.89 Define position of the dimension line

Once the position of the dimension line has been set, EPLAN will dimension the object. For the dimensioning itself, there are a number of properties that can be called by double-clicking on the dimension itself.

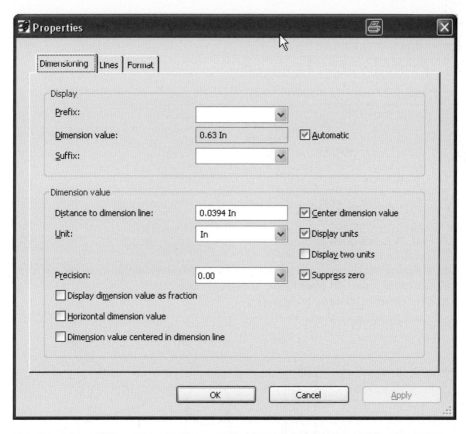

Fig. 4.90 Settings for current dimension

There are also additional tabs for *lines* and the *format* of the labeling of the dimension.

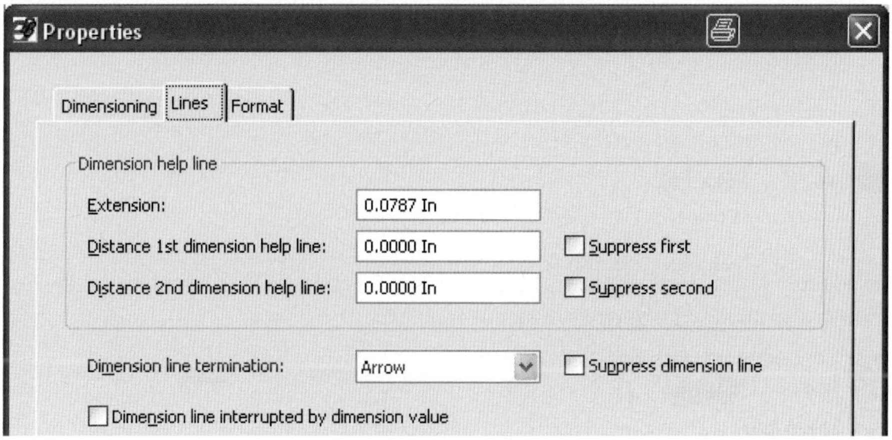

Fig. 4.91 Settings for dimension lines

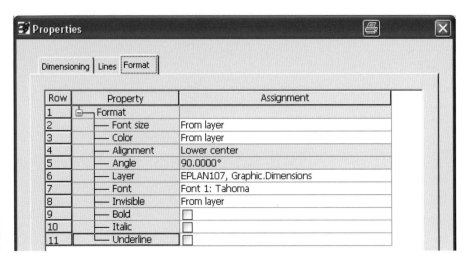

Fig. 4.92 Format (dimension values, etc.) settings

On the *Format* tab, the settings for dimension values, etc. are defined. This does not affect the dimension lines. To control the dimension lines directly, you must modify the *EPLAN107* layer in the layer management.

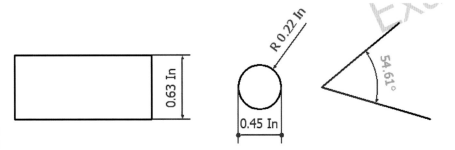

Fig. 4.93 Sample dimensions

4.7 Texts

In addition to graphical or electrical engineering functions such as symbols or devices, EPLAN also allows for the entry of simple free texts.

Fig. 4.94 Free graphical text

Fig. 4.95 Function text

This is not really a new feature but rather a necessity to allow various schematic functions to be clarified with text entries, and to provide a better explanation of the schematic and its functions, or to fill reports with additional information (or to have EPLAN fill them).

Plain text is always used in EPLAN. No numbering or coding system is used, so that the user can always see what the text is and does not need to look up a number in a text file to know which text entry this is supposed to be.

4.7.1 Normal (free) texts

EPLAN allows the entry of normal (free) texts. Normal texts are texts that have no further functionality (except their visual display).

Translation of this type of text is possible without problems. Normal text should usually only be used to (e.g.) place text in a black box (to provide extra information), which does not need to be used in any other way. Normal texts are inserted via the INSERT / GRAPHIC / TEXT menu. Or the T key is used for normal text.

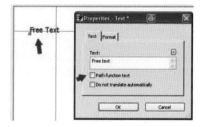

Fig. 4.96 Free text

4.7.2 Path function texts

In addition to normal texts, EPLAN also has path function texts. These are equipped with „intelligence". Path function texts should generally be used to provide corresponding devices with automatically filled function texts, for example derived from the page path.

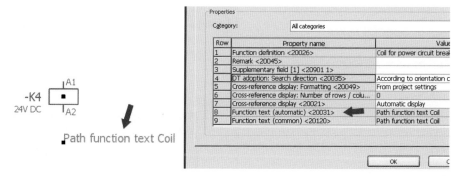

Fig. 4.97 Path function text

EPLAN uses the path function text for embedding in different reports (depending on the structure of the form).

It is therefore advantageous to always work with path function texts, since subsequent conversion from normal texts to path function texts is possible, but requires unnecessary extra work.

Path function text is inserted via the INSERT / PATH FUNCTION TEXT menu. If path function texts are generally to be used, the keyboard shortcut of free text should be redefined for the path function text. In this case, the keyboard shortcut for normal (free) text should be set to a different keyboard shortcut.

Fig. 4.98 Path function text

TIP: Normal (free) texts can be converted to path function texts by simply setting the *Path function text* setting in the **Properties - Text** dialog. After leaving the Properties - Text dialog, the layer is also set automatically to the standard layer for path function texts LAYER110.Graphic.Path function texts.

To generate a report for a path function text for a device, it is normally necessary that it be directly (in the path) below or above the devices (the insertion points of the path function text and device are then located „in a sequence").

But it may also be that this is not desired, for reasons of space or because the path function text is to apply equally to several devices. To avoid having to align the path function text several times in the path of the individual devices, it is possible to set the GRAPHICAL EDITING / GENERAL project setting and, in this case, *Extend path function text on the schematic path*. This way, it will not matter whether the insertion point is directly below the device or not. It is sufficient if both - i.e. the path function text and the device - are located in the same path (of the column)!

4.7.3 Special texts

Special texts are another type of text known to EPLAN. Special texts can be *page property texts* or *project property texts*. They are generally used in forms (project properties) or in the plot frame (page properties) and less so as text on project pages.

Page property or project property texts are inserted via the INSERT / SPECIAL TEXT / PROJECT PROPERTY menu or

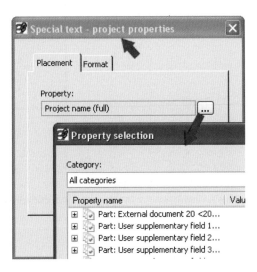

Fig. 4.99 Special texts

PAGE PROPERTY. This way you can place text on pages taken from the project properties or the page properties.

4.7.4 The Properties – Text dialog

To not only insert texts but also format them according to the requirements, the **Properties - Text** dialog for the text must be called up. To do this, the text is selected and the dialog is opened with the ENTER key or a left mouse click.

The dialogs for normal text and path function text are the same. Only the window title bar of the text input field has a different text type name.

4.7.4.1 Text tab

This *Text* tab contains the actual *Input field* for the text. The text is entered into the *Text* field and it is translated at this point, or the translation can be removed.

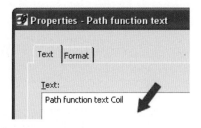

Fig. 4.100 Text input field

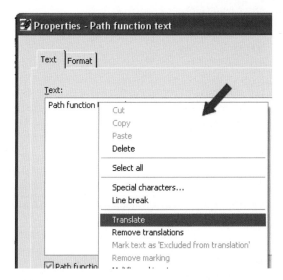

Fig. 4.101 Popup menu

To do this, the popup menu is called up using the right mouse button or via the ▪ button. In addition to the usual Windows text functions such as *Copy*, *Paste*, *Delete*, or *Cut*, the popup menu also contains other functions such as *Special characters* or *Translate*, *Remove translation*, *Remove marking* or *Multilingual input*.

4.7.4.2 Format tab

The *Format* tab contains all functions needed for formatting the text (*Font size* or *Font*) or (e.g.) assigning it a different layer. The formatting options and settings can be divided roughly into the three nodes Format, Border and Value/unit.

Properties - Path function text

Text | Format

Row	Property	Assignment
1	Format	
2	Font size	From layer
3	Color	From layer
4	Alignment	Basis left
5	Angle	0.0000°
6	Layer	EPLAN110, Graphic.Path function texts
7	Font	Font 1: Tahoma
8	Invisible	From layer
9	Row spacing	From layer
10	Language	All displayed languages (arrange vertically)
11	Bold	☐
12	Italic	☐
13	Underline	☐
14	Border	
15	Draw text box	From layer
16	Size from project setting	☐
17	Activate leader line	☐
18	Activate alignment box	☐
19	Draw alignment box	No
20	Width	0.0000 In
21	Height	0.0000 In
22	Text width fixed	☐
23	Text height fixed	☐
24	Remove breaks	☐
25	Never hyphenate words	☑
26	Fit graphic	Do not fit
27	Value / unit	
28	Displayed measuring unit	Unchanged
29	Display	All
30	Decimal places	0
31	Decimal places, variable	☐

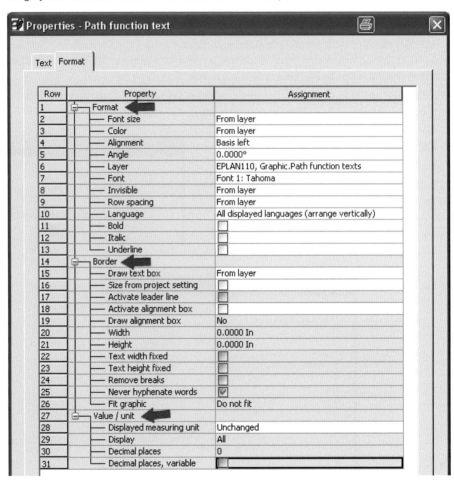

Fig. 4.102 Formatting options and more for texts

Most fields are selection fields, meaning that when (e.g.) the Alignment field is opened, then all selection options are displayed. A number of fields are input fields, such as the angle property. Here, you can select default values or enter your own.

4	Alignment	Basis left
5	Angle	12.1200°
6	Layer	EPLAN110, Graphic.Path function texts

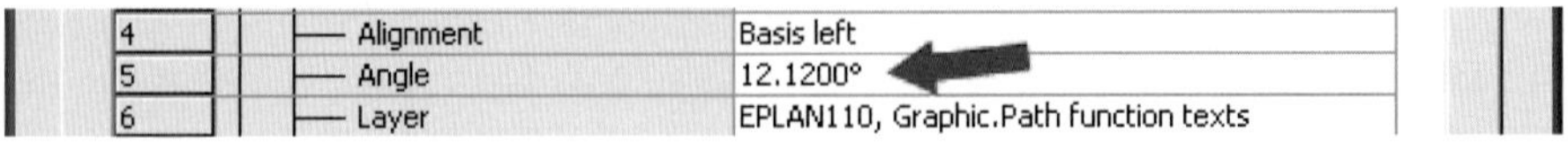

Fig. 4.103 Free input fields

The next section contains a few explanations regarding selected properties of the **Format** tab.

4.7.4.2.1 Format tab - Format node - Language

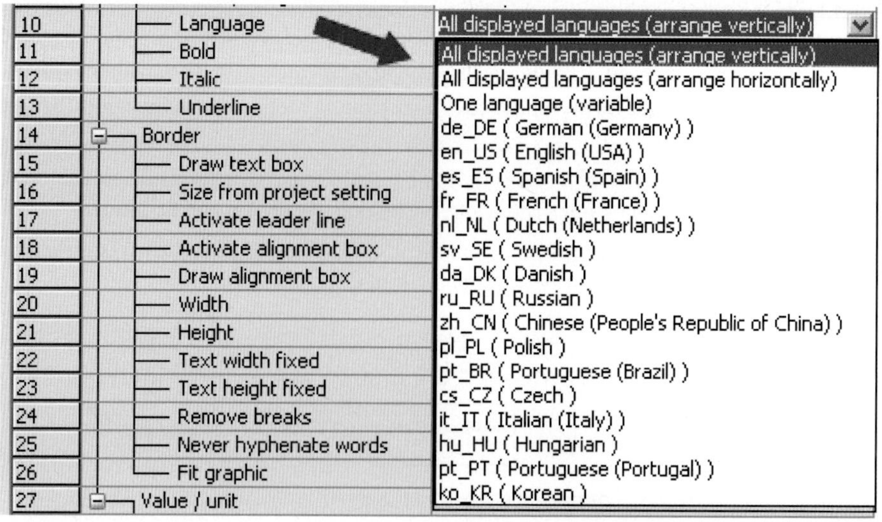

Fig. 4.104 Language selection list

The *Language* node contains the functions for displaying translated texts in the project. The ONE LANGUAGE (VARIABLE) selection entry depends on the UTILITIES / TRANSLATION / SETTINGS settings. In these settings (**Settings / Translation [project name]** dialog), the general project languages and the languages for display in the project are also defined.

Fig. 4.105 A few translation settings

From the existing *project languages*, you can select the *displayed languages*. Depending on the existing project languages, you can then select in the *One language (variable)* selection field the desired *Displayed language*.

4.7.4.2.2 Format tab - Border node - Activate alignment box

Apart from the previous Format and Language nodes, there is an additional *Border* node on the **Format** tab. This Border node contains properties, for example, to always limit the extension of text to a specific *Width* or *Height*. Other options include removing breaks from texts or defining that words (in the case of automatic breaks due to the width/height limit) must not be split.

These functions are very useful, especially for translations. This prevents translations from overlapping each other since the width of the text expansion can be limited.

Fig. 4.106 shows a typical example. In this case, it does not matter whether it is pure text or text of a device. The procedure is always the same.

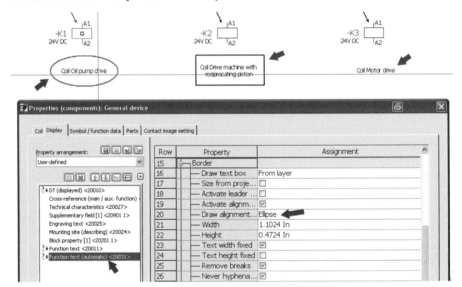

Fig. 4.106 Examples of an activated alignment box

 TIP: In order to read the oftentimes fairly large number of properties and their values in dialogs better, it is possible to change the height of lines. To do so, click on the table and, while keeping the CTRL key pressed, turn the SCROLL WHEEL of the mouse. Depending on the direction, the lines will increase or decrease in height, thus improving legibility.

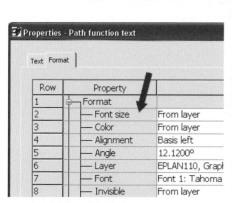

Fig. 4.107 Large row height   Fig. 4.108 Small row height

EPLAN remembers the settings. But EPLAN handles all dialogs differently. This means that, if so desired, you could set different line heights everywhere.

■ 4.8 Components (symbols)

Components, also called symbols, are among the most important elements in EPLAN and, naturally, of any schematic. Components and their entered properties or assigned function definitions allow for the generation of logical connections or creation of reports.

4.8.1 Insert components (symbols)

Symbols can be inserted in EPLAN in different ways. The usual way is via the INSERT / SYMBOL menu or the INS key.

EPLAN then opens the **Symbol selection** dialog. In this dialog, the appropriate symbol can now be selected. Either direct entry (*List view*) is used, in which case you must have a certain amount of knowledge of identifiers, and the symbol names are directly entered into the *Direct entry* field, or the symbol is selected from the list using the mouse.

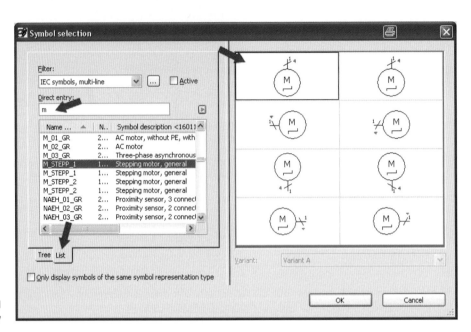

Fig. 4.109 Symbol selection via list view

Next to the symbol list is the *Symbol preview*. EPLAN (generally) provides symbols with eight variants. The normal four variants: Variant A = 0°, variant B = 90°, variant C = 180°, and variant D = 270°. The remaining four variants E to H are 180° mirrored variants of the first four variants.

As with many other dialogs, it is possible to set a filter in this dialog via the [...] selection button. Filters can thus be used to display only *Multi-line symbols* or only a special *Symbol library*.

For example, the *Only symbols of the same symbol representation type* parameter allows for the display of only multi-line symbols from all displayed symbol libraries.

In addition to the *List view*, the **Symbol selection** dialog also allows the selection of symbols from a *tree view*. In contrast to the list view, symbols here are listed according to a structure. Depending on the application, a symbol can be found more quickly when you search for the *motor overload switch* using plain text.

4.8 Components (symbols)

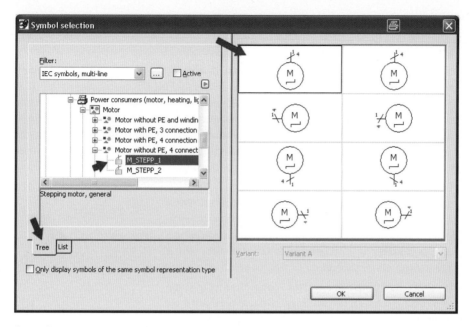

Fig. 4.110 Symbol selection via tree view

Once the component (symbol) has been selected, it can be applied to the schematic with ENTER or the OK button.

4.8.2 Properties (components) dialog – [device] tab

Once a symbol has been adopted from the symbol selection, it hangs on the cursor and can be placed in the schematic.

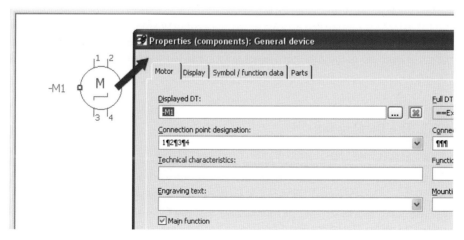

Fig. 4.111 Properties (components) dialog

After placement, EPLAN opens the **Properties (components) [device type]** dialog. The dialog consists of the different tabs *[Device type]*, *Display*, *Symbol / Function data* and *Parts* (in case of a main function).

Here we find the fields for the *Displayed DT* (can be edited), the *Full DT* field (can be edited indirectly via the [...] selection button and the following **Full DT** dialog, where the DT is split into its elements, which can be edited individually), the fields *Connection point designation* (selectable from the selection list), *Connection point description* (selectable from the selection list), *Technical characteristics*, *Function text* (of the symbol!), the *Engraving text* field (can be edited and selected from a selection list if there are already other entries in the project for this) and the *Mounting site (describing)* field, which can also be selected from a selection list.

Aside from these items, there is also an important check box: *Main function*. This way, you decide whether this symbol is to be a main function (thus enabling it to carry a part), or whether it is to be an auxiliary function, which does not contain parts data.

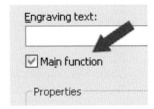

Fig. 4.112 Main function

The lower half of the **Properties (components) [device type]** dialog consists of the *Property name* and *Value* columns (of the properties). The [...] button can be used to add new properties.

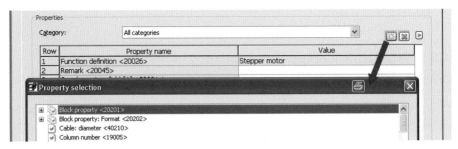

Fig. 4.113 Add new properties

After the [...] button is clicked, EPLAN opens the **Property selection** dialog. Here, a new property can be selected and then adopted in the **Properties (components)** dialog via the OK button. All properties that are not gray in the *Value* column of this dialog can be changed.

4.8.3 The Display tab

The *Display* tab allows new symbol properties to be added to the existing default symbol properties at the symbol, or existing properties to be deleted, *for display*.

The *Display* tab also contains all elements for changing the display of the properties. Here the format, font size or font can be individually changed for every property, including the connection points of devices. For this purpose, for example, you should select the *Connection points* tab in the property arrangement.

The toolbar can be used to insert new properties, move the sequence of properties, or dock and undock them.

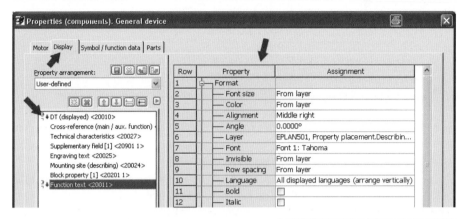

Fig. 4.114 Display tab and its options

This section explains *docking and undocking of properties* in more detail. *Docked* properties are those that do not have a symbol in front of the property name. These properties are assigned to the next upper property that has a symbol.

In Fig. 4.115, all properties framed in red would make up a "block". For example, if you move one of the properties out of this "block", all other properties in this "block" will be moved as well.

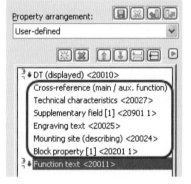

Fig. 4.115 Related properties

If you undock a property, it will be removed from this "block" and is now "free". Now it is independent of the other properties and can be placed freely, for example.

You undock a property by selecting it and then undocking it using the 🗁 button. Then, using the direction buttons, move it to the end of the property arrangement. If you do not do this, the undocked property and all other properties that follow it in the display will form another „block".

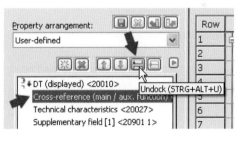

Fig. 4.116 Undocking a property

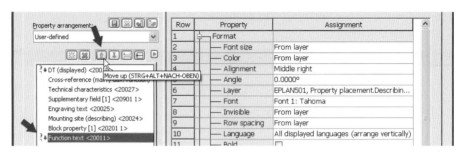

Fig. 4.117 Undocked and moved property

In addition, the *Display* tab also contains all the functions you need to format or position text. Be it for the purposes of placing a border around text, setting a different font size for a specific property or defining/adjusting a position (X/Y).

The procedure is the same as with texts and will not be discussed further here, and the control of the language display will also not be discussed again.

4.8.3.1.1 Activate leader line property

There are some differences from formatting text when it comes to certain properties. One example is the *Activate leader line* setting.

The *Activate leader line* setting allows you to activate a leader line for each property displayed on a device. This way, for example, you can undock or move properties, while the leader line ensures a type of "connection" with the actual device.

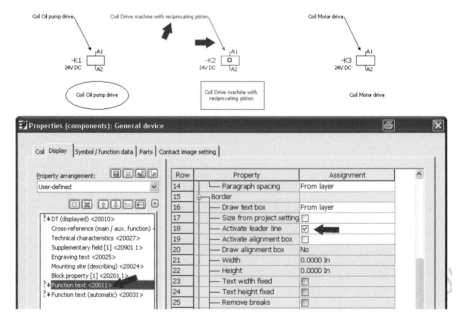

Fig. 4.118 Example of activated leader line

4.8.4 The Symbol/function data tab

The *Symbol / Function data* tab contains important information for assigning **logic** to the graphical symbol.

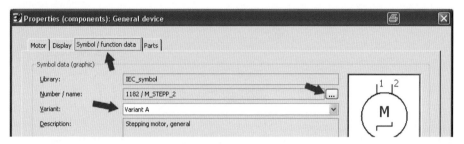

Fig. 4.119 Graphical information for the symbol

The upper area contains the graphical symbol data. In the *Number / Name* field, the ⃞ button can be used to exchange the existing symbol for a different one. A different angle variant of the symbol can be selected in the *Variant* field. To do so, simply open the selection list and select the desired variant. All other fields (*Library*, etc.) are derived from the graphical symbol data and cannot be edited.

The lower area of this tab contains the logical symbol information, such as which *Function definition* is attached to the symbol, the *Representation type*, or whether it is a MAIN FUNCTION, etc..

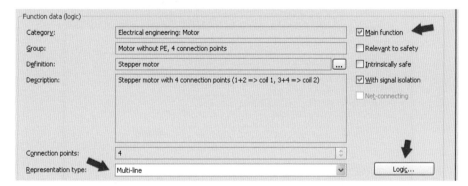

Fig. 4.120 Logical symbol functions

In the *Definition* field, the ⃞ button can be used to assign a different function definition to the symbol if necessary. All other fields are then derived automatically from the adopted function definition.

The *Representation type* here can be modified from a selection list. For example, this may be necessary when you wish to create or represent an overview on a multi-line schematic page.

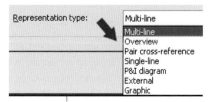

Fig. 4.121 Representation type selection

Aside from the previous graphical and functional information, there are additional settings options under the LOGIC button. In a targeted manner, you can change or adjust logical information such as *Connection point type* or the *Number of targets*, *Potential type* and much more for the individual connection point of a device.

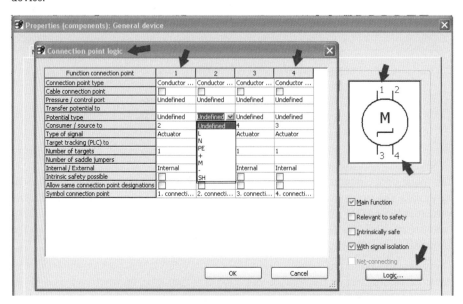

Fig. 4.122 Logical functions of individual connection points

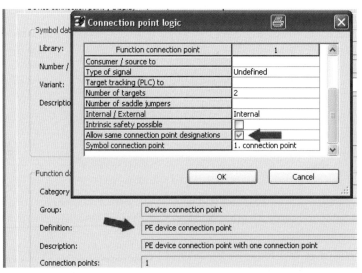

Fig. 4.123 Allow same designation

A typical example of changing the *Connection point logic* is the enabling of the *Allow same connection point designation* setting when using device connection points (here PE). This way it is possible, like with terminals, to use the same designation several times (PE rail), and the check run will not find any duplicate connection point designations.

4.8.5 The Parts tab

The *Parts* tab allows the assignment of one or more parts to the symbol. To do this, the *Part number* field is clicked. The ⋯ button that then appears can be used to switch to the parts management.

From the *Parts management* that then opens, you can select the necessary part and apply it to the symbol via the OK button. Parts management automatically closes after adoption of the part.

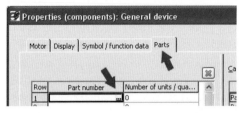

Fig. 4.124 Simple part selection

Apart from the simple part selection, this dialog also has a DEVICE SELECTION button. In contrast to simple part selection, this will render only devices for selection that fit at least the existing functions of the symbol in the project. This ensures that EPLAN selects from Function a coil for a coil, rather than a motor overload switch. The *Device selection* should be given preference.

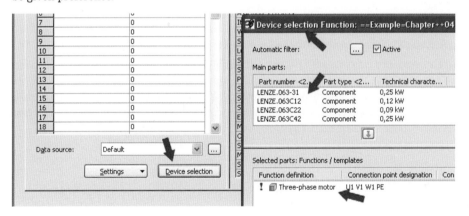

Fig. 4.125 Device selection

4.9 Cross-references

Cross-references are one of the most important elements in EPLAN. When automatically generated, they create visual and logical connections, e.g. between *main and auxiliary functions*.

 NOTE: The insertion and filling out of the corresponding symbols for the cross-reference display examples will not be explained at this point. Please read the corresponding chapters of this book for more information.

4.9.1 Contact image on component

For example, a motor overload switch or a pushbutton switch can have auxiliary contacts. The auxiliary contacts physically belong to the motor overload switch or pushbutton switch symbol and, for the sake of clarity, should also be displayed at the switch.

A motor overload switch is inserted into the schematic. The motor overload switch initially has no auxiliary contact at the symbol. As such, this is not a problem, because EPLAN creates a report on the auxiliary contacts **used** in the schematic on the basis of the settings of the *Display* tab in the *Contact image* selection field (the selection here is set to ON COMPONENT). This way, the **contacts used** and **their cross-references** are displayed automatically.

Fig. 4.126 Automatic cross-referencing

This does not matter provided that both or all auxiliary contacts that have such a motor overload switch are also used in the schematic.

Not all auxiliary contacts are always used, and EPLAN would display the motor overload switch only with the auxiliary contacts used (as graphical representation at the motor overload switch). The other auxiliary contacts physically exist but are „suppressed" with this setting. To display those contacts anyway, a device with the appropriate *Function definition* must be assigned to the *Motor overload switch* symbol.

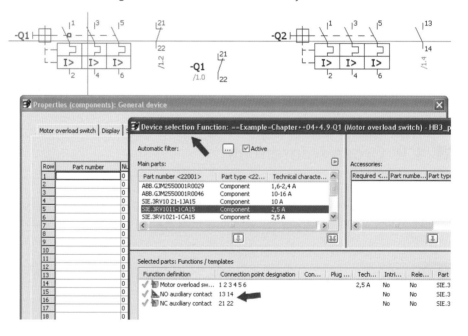

Fig. 4.127 New device selection

This is best done using the Device selection. For this purpose, open the properties of the symbol, switch to the *Parts* tab and click on the DEVICE SELECTION button.

> **TIP**: The SETTINGS button provides access to the settings defining how EPLAN should behave during device selection: Should existing function data be used or not; should any other criteria of devices, functions be taken into account, etc. The same applies to device selection.

This is important to know because in certain situations this can result in no display and no selection of devices in the device selection. This may then be related, for example, to "false" or inappropriate settings of the device selection, or there may not be a part with the corresponding functions in parts management.

Back to the device selection. EPLAN recognizes that the motor overload switch consists of the motor overload switch itself and a placed auxiliary contact (considered a function). Now all parts are offered for device selection that fit the selected motor overload switch (including its functions used) and the parts with their function definitions.

In the upper area, you can now select the corresponding fitting part and adopt it via the button. EPLAN „moves" the part across the existing functions of the motor overload switch and adds to the function template the functions not yet placed.

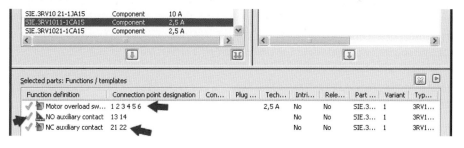

Fig. 4.128 Selected part with its function template

The **Device selection function** dialog can now be closed with the OK button. EPLAN transfers the part into the *Parts* tab. If you leave the *Symbol properties* dialog now by clicking on the OK button, EPLAN will use the stored function definitions to recreate graphically the entire contact image for the auxiliary contacts. It now makes no difference whether the auxiliary contacts have been used in the schematic or not.

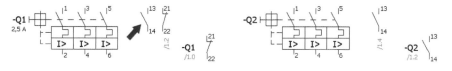

Fig. 4.129 Complete contact image

But this representation can be achieved only if the display of the contact image has been set accordingly in the symbol properties of the Display tab.

Fig. 4.130 Contact image setting

This display of the placed contacts is set by EPLAN automatically at first. But if you wish to, or must, deviate from this automatic setting, you can modify the item manually.

To do so, press the MORE BUTTON to edit manually the position for the Y and X values in the **Contact image position** dialog.

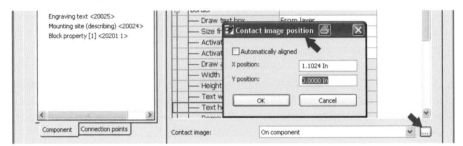

Fig. 4.131 Manual editing of the contact image position

4.9.2 Contact image in path

In contrast to a contact image at the component, which is usually displayed to the right of the symbol, for (e.g.) contactors the contact image is generated by EPLAN below the contactor coil in the column (path).

Similar to the previous example, at first only the **contacts used** are **displayed** automatically in the contact image.

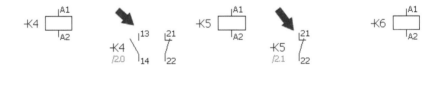

Fig. 4.132 Automatically generated contact image of the functions in use

To obtain a complete *contact image*, a *device selection* is performed in the same way as was done in the example of the motor overload switch.

The procedure is identical to that already described in the motor overload switch example. After the device selection the full contact image is displayed. If this does not occur, then the *Contact image* entry in the **Display** tab of the *Symbol properties* dialog must be checked. This must be set to the *In path* entry.

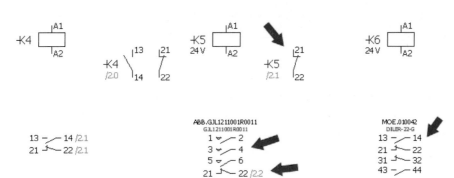

Fig. 4.133 Complete contact image following device selection

 TIP: EPLAN also allows for the moving of contact images. Every contact image can be individually moved (it makes no difference whether this is a contact image of the *On component* or *In path* type).

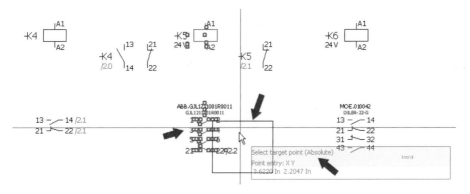

Fig. 4.134 Move contact image manually

To do this, the symbol is first selected, and then the CTRL + B keyboard shortcut or the EDIT / TEXTS menu is used to call the MOVE PROPERTY TEXTS function. EPLAN then displays a small blue rectangle for "grabbing" with the mouse. The contact image is then clicked with the mouse, hangs as a blue rectangle on the cursor, and can be placed as desired (while keeping left mouse button pressed). This is then placed at the desired position by clicking the left mouse button.

At this point, it should be mentioned that the positions of the contact images for the motor overload switch and/or contactor coils in the respective plot frame (or only for this page, in which case, in the Page properties) can be defined, for example, as *Contact image margin (on component) <12059>* and *Contact image margin (in path) <12060>* plot frame properties.

3	Contact image margin (in path) <12060>	0.0000 In
4	Contact image margin (on component) <12059>	0.0000 In
5	Contact image offset <12061>	0.0000 In

Fig. 4.135 Properties of contact image offsets

Regardless of these default values, all contact images can, of course, be individually and manually set with the desired values and placed at the respective devices in the schematic.

4.9.3 Special feature: pair cross-reference

In addition to the previous examples, the functionalities of EPLAN have been used on the basis of the device selection and its functions. But situations can arise where there are no parts with correct function templates. It is, after all, only because of the function templates that EPLAN knows, for example, that an illuminated pushbutton consists of several functions.

Using an illuminated pushbutton as an example, I will show how contact images can be referenced to the inter-connected functions even without parts. In the process, this will involve the pair cross-reference representation type (in the **Properties (components)** dialog on the *Symbol / Function data* tab in the *Representation type* selection list).

Fig. 4.136 Insert and designate symbols

A pushbutton is inserted. It is assigned a DT and is designated as a main function. Apart from the pushbutton, a lamp is inserted as well. It is **not assigned a DT**, nor is it **designated as a main function**. Additionally, the representation type of the lamp is set to *Pair cross-reference*.

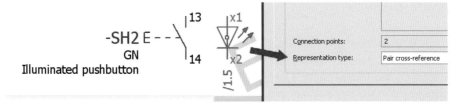

Fig. 4.137 Set pair cross-reference representation type

Now, at the auxiliary function (of the lamp), the *Cross-reference (main / aux. function)* <20300> property is set to the optimal position of the display of the cross-reference. For this purpose, the following settings are made: the *angle to 90°*, the *alignment to center right*, the *position* of the X coordinate to *0.00 mm* and of the Y coordinate to *-6.00 mm*.

4.9 Cross-references

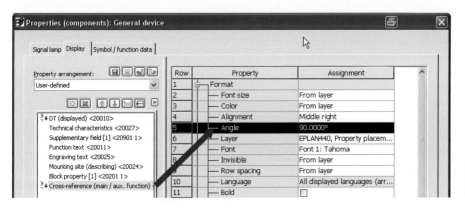

Fig. 4.138 Set appropriate values for the cross-reference

If all values have been set accordingly, the cross-reference will be rotated subsequently by 90° directly below the lamp.

Now you can place the lamp, either from the **device navigator** via the PLACE / MULTI-LINE popup menu or via the symbol selection.

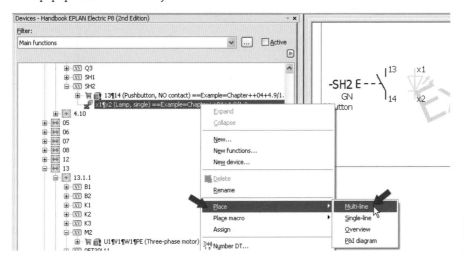

Fig. 4.139 Place multi-line function from the device navigator

After this function has been placed, EPLAN immediately displays the cross-reference on both pages: on the complete device illuminated pushbutton and on the device of the lamp itself, which is displayed as distributed.

Fig. 4.140 Cross-reference with pair cross-reference representation type

TIP: The placing of functions from the device navigator is always preferable to manual placing (insert symbol, fill out DT, etc.), because EPLAN implements all key settings automatically, which means that everything fits all the time!

4.9.4 Device selection settings

There are a number of settings for the device selection. Depending on how they are set or adjusted, they can influence the *device selection* and thus the selection of parts.

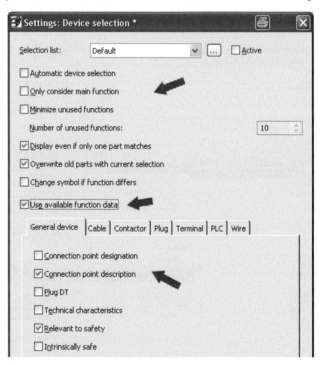

Fig. 4.141 Device selection settings

It helps if the *Use available function data* setting and the desired properties have been enabled. This way, only parts will be offered for selection that match the identifying function data. Identifying function data are those that are located in the *General devices*, *Cable*, *Contactor*, etc. tabs and that are also active.

If the *Use available function data* setting is disabled, it is possible to select devices freely. This means that the function data available in the project (at the function) do not have to match the function data in the parts selection.

4.10 Macros

In addition to construction using devices (symbols), EPLAN also allows working with partial circuits, the so-called macros.

4.10.1 Types of macros

EPLAN has different types of macros. These can be window, page, and symbol macros. Special macros are macros with value sets. These can be window and page macros. When using or creating macros (exception: macros with value sets), it generally makes no difference whether this is a window macro on a multi-line page or a macro on a graphical page.

4.10.1.1 Window macros

Window macros are (can be) the smallest partial circuits in EPLAN. Window macros can include single or multiple devices and objects within an area, or several items within a page.

To create a window macro, the associated devices are first selected on the page using the mouse. These may be individual devices or unrelated parts circuits on a project page. These are „gathered" by selecting the first device, and then selecting the other devices that are to belong to the window macro while holding down the CTRL key.

The CTRL + F5 keyboard shortcut can then be pressed or the function can be started via the EDIT / CREATE WINDOW MACRO menu item. EPLAN opens the **Save as** dialog. A descriptive file name for the window macro should be entered here first of all. The macro directory is usually the company-specific macro directory. But this can be changed if desired.

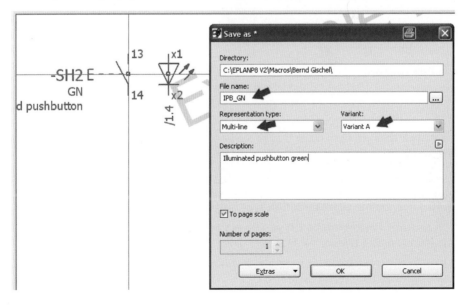

Fig. 4.142 Save macro

Window macros are inserted into a page with the M key or via the INSERT / WINDOW MACRO menu item. This is followed by the **Select macro** dialog. In the default directory (or a different macro directory – select a different directory in the *Search in:* selection field), the macro and the desired *variant* or *representation type* can be selected, and then the OPEN button clicked to hang the macro on the cursor and place it anywhere on the page.

To summarize, window macros are only possible on **one** (the same) page but with different variations such as the gathering of objects that do not all have to be within a single window. A window macro can also be the contents of the entire page (all objects on the page). Window macros have a file extension of *.ema.

4.10.1.1.1 Define handle setting

This **Save as** dialog contains further settings relating to the macro. The EXTRAS button allows for additional settings.

The menu item MOVE HANDLE allows for a (separately created) handle (base point) to be saved with the macro. A handle means that when the macro is inserted, it hangs on the cursor crosshairs using this handle as the base point. To do this, the MOVE HANDLE menu item is clicked, EPLAN temporarily closes the **Save as** dialog, and you can define the handle by clicking with the left mouse button.

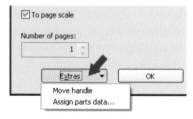

Fig. 4.143 Extras settings

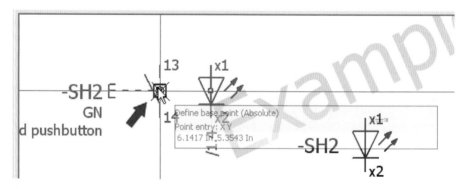

Fig. 4.144 Define a new handle

The cursor changes when doing this, as can be seen in the figure. It is a good idea to enable the OBJECT SNAP setting here. Once the handle has been defined, EPLAN returns to the **Save as** dialog, and the macro can be saved.

4.10.1.1.2 Define representation type setting

The selection **Representation type** allows you to define the representation type for the macro.

The individual representation types are usually defined via the devices at the symbol (SYMBOL / FUNCTION DATA

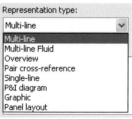

Fig. 4.145 Representation types

tab). However, EPLAN allows a different representation type to be defined for each individual macro. This way, you can save many representations under one macro name.

It is possible to have a macro for the multi-line representation type, one for the part placement and one for the single-line display. When selecting a macro, EPLAN chooses the right representation type based on the page type defined. Of course, this must be present in the macro.

4.10.1.1.3 Define variant setting

The **Variant** selection box provides the next possibility of working more effectively.

This allows different variants (so far, sixteen variants A to P) to be stored within a macro.

This just means that the same macro name (file name of the macro) can have different content. Depending on the variant used and in connection with the representation types, this means that you have **up to 128 possible variants** (macro variants) within **one macro**. For example, you could save different macro variants of a PLC card in one macro.

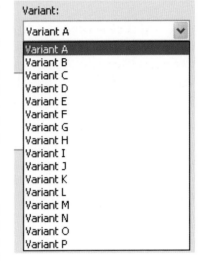

Fig. 4.146 Variants

4.10.1.1.4 To page scale setting

In addition to all the other settings for a window macro, there is also the *To page scale* option. If this option is activated, then EPLAN reduces or enlarges the macro to suit the page scale when it is inserted into a page.

4.10.1.2 Page macros

Page macros are the counterpart to window macros. Page macros can comprise one or several pages within an EPLAN project.

Page macros are created using the CTRL + F10 keyboard shortcut (at least one page must be open and the user must be on this page), or the CREATE PAGE MACRO popup menu item is used on selected pages in the **page navigator**.

The familiar **Save as** dialog, as described in the window macro chapter, is displayed. The *file name* and *description* can be entered here. *No* handles, *no* representation types, and *no* variants are possible with page macros. However, page macros store all the information relating to a page or pages (page properties). These are (e.g.) the structure identifiers or the form pages stored in the page properties.

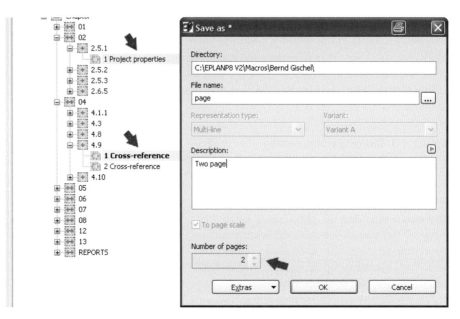

Fig. 4.147 Create page macros

In addition to the individual page macro (directly on the page), in the page navigator it is possible to write multiple pages (also pages that are not related) into a page macro. To do this, the page navigator is opened. Now the desired pages are selected and the right-click popup menu is used to call the CREATE PAGE MACRO command.

Custom keyboard shortcuts can be used for inserting page macros, e.g. CTRL + ALT +F10 or the popup menu in the **page navigator** can be called and the INSERT PAGE MACRO command selected. EPLAN then opens the **Select macro** dialog. Here the desired macro can be selected and then adopted with the OPEN button.

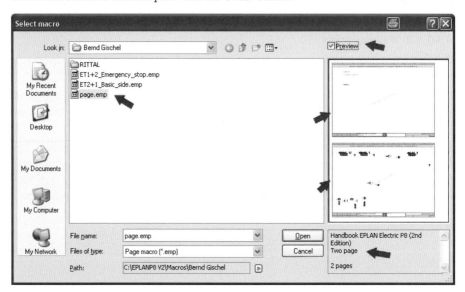

Fig. 4.148 Insert page macro

If the preview is switched on in this dialog, then all pages are displayed in a small preview. In addition to the preview display (which has a fixed size), the GRAPHICAL PREVIEW window can also be permanently displayed (VIEW menu). The advantage of this window is that the size can be changed, and you can see the details better.

After adoption of the macro, EPLAN opens the **Adapt structure** dialog. Here, the page(s) can be stored in the project and sorted into the existing page structure. At this point, it is, of course, possible to adjust the structure of the pages as desired.

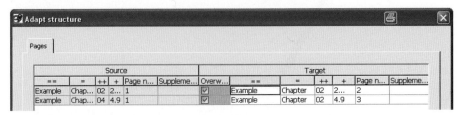

Fig. 4.149 Adapt structure

To summarize, page macros are macros containing one or more pages including their page information (page properties). Page macros have the file extension *.emp.

4.10.1.3 Symbol macros

Symbol macros are an additional type of macro. Generally, symbol macros are completely identical to window macros in terms of handling and creation. Further explanation will therefore not be provided.

4.10.2 Macros with value sets

Macros with value sets are a special function type in EPLAN Electric P8. These are usually window macros equipped with additional functionality that can make project planning much easier.

In addition to the basic macro (partial circuit) and its properties, such as technical characteristics, part numbers, etc. of the various objects, this type of macro also contains additional sets of properties such as technical characteristics, part numbers, etc. These sets of properties are called **value sets**. Value sets are activated via a particular symbol, the placeholder object. This symbol switches the value set.

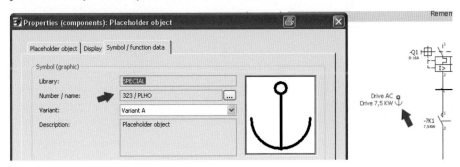

Fig. 4.150 Placeholder object symbol

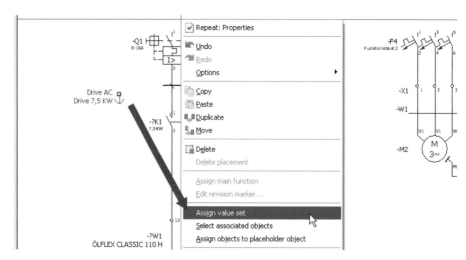

Fig. 4.151 Assign value set

For example, all devices of the window macro could be equipped with Siemens parts. A second value set could then contain Moeller parts data for all devices. The value set can then be used to switch between Siemens and Moeller parts data for the same macro with a mouse click.

Once these macros and the properties they contain have been developed and tested, they provide a source of error-free data.

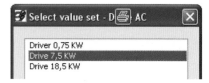

Fig. 4.152 Select value set

4.10.2.1 Placeholder object

The most important object in a macro with value sets (subsequently called a value set macro here) is the *placeholder object*. A placeholder object can be inserted via the INSERT / PLACEHOLDER OBJECT menu and allows for subsequent switching between value sets. An anchor is used as the visible symbol for a placeholder object.

The placeholder object is a symbol and, accordingly, has settings and options similar to other symbols.

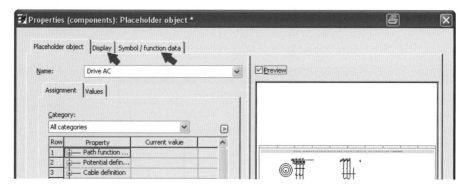

Fig. 4.153 Placeholder object symbol properties

4.10.2.2 Value set

A *value set* is a collection of variables of selected objects stored in a window macro. Value sets are managed in a type of table "behind the placeholder object" (*Placeholder object* tab) and, in addition to all the device properties (filled or empty), they contain additional information such as the actual values or the variables for the values.

4.10.2.3 Variables of a value set

Variables bring a value set to life. Without these variables, there would be no table for the actual values used for switching the value set. Every property of a device can be provided with any desired variable name. Variable names are always surrounded by "less than" (<) and "greater than" (>) characters on the keyboard. For example: <Variable name>.

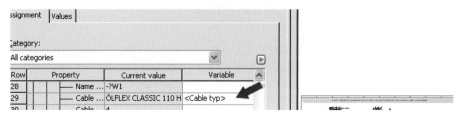

Fig. 4.154 Example of variables

Apart from one exception, all characters can be used. Square brackets are the exception. These have a special purpose within a variable. They are used to define a line break.

For example, a variable <Variable name [12]> is split at the twelfth position (line break) and the words are simply split at this twelfth position. If the additional option 1 is inserted into the square brackets: <Variable name [12,1]>, then a split also occurs at the twelfth position (i.e., a maximum of twelve characters) but an attempt is made to recognize the ends of words. In other words, the words are not „torn apart" but remain intact, and line breaks are only inserted between the words.

 NOTE: The square brackets **must** be entered directly after the variable name. Spaces are allowed, but then the defined line break option does not take effect!

4.10.2.4 Create a macro with a value set

Only a few steps are needed to create a value set macro. Firstly, the partial circuit is created with all the required or desired devices and their associated functions such as part numbers, technical characteristics, or function texts. After this, the placeholder object is inserted via the INSERT / PLACEHOLDER OBJECT menu. There are several ways of assigning objects to a placeholder object in EPLAN.

are only used for the purpose of these exercises. They are generally not usable in a real world application

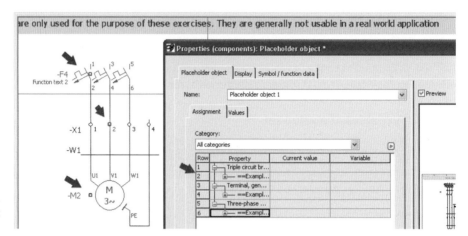

Fig. 4.155 Assign individual objects to the placeholder object (option 1)

Option 1: All objects are first selected, and then the INSERT / PLACEHOLDER OBJECT menu is used to insert the placeholder object. After the placement of the placeholder object, EPLAN opens the **Properties (components) placeholder object** dialog, and the data can be edited. This approach is recommended, because it allows you to select objects accurately that are to be applied to the placeholder object.

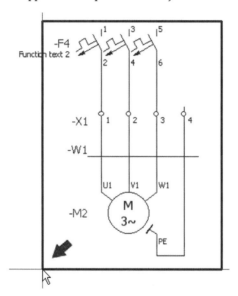

Fig. 4.156 Select objects with a window (option 2)

Option 2: The placeholder object can first be inserted in the page via the INSERT / PLACEHOLDER OBJECT menu. But before EPLAN places it, all objects must be selected with a window (pulled open with the mouse).

 Disadvantage of option 2: remote objects lying outside a window cannot be included in the value set.

Once one of the methods has been chosen, the placeholder object can be placed. It is recommended that you place the placeholder object close to the macro. To do this, the placeholder object is moved to the desired position using the left mouse button and placed with a mouse click.

EPLAN then immediately opens the **Properties (components) placeholder object** dialog. A descriptive name for the placeholder object should be entered into the *Name* field here.

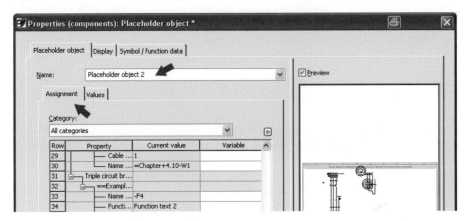

Fig. 4.157 Edit placeholder object

After this, you can open on the *Placeholder object* tab the *Values* tab and define the variables that are to be later switched when the value set is selected. To do this, the right mouse button is clicked in the free field, and the *Name new variable* function is selected.

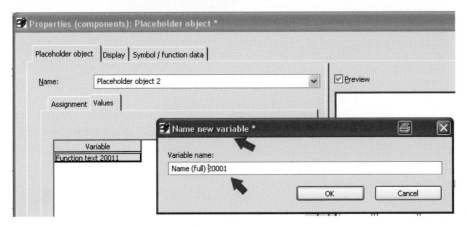

Fig. 4.158 Name a new variable

EPLAN opens the **Name new variable** dialog. The *Variable name* is now defined. After clicking the OK button, the variable is adopted in the *Values* tab.

If the variable name contains characters that are not permitted, EPLAN will display a message to this effect and will not adopt the variable.

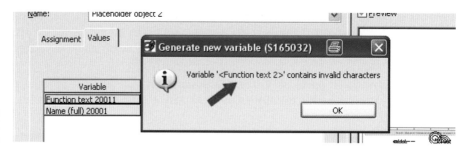

Fig. 4.159 Message alerting to invalid characters

The value sets, which will be available for selection later on, must now be defined so that data is entered into the variables for the different value sets.

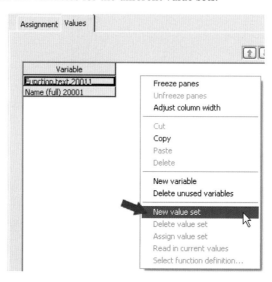

Fig. 4.160 Create a new value set

To do this, click the right mouse button again, and select the NEW VALUE SET function. EPLAN inserts a new column next to the *Variable* column. Enter the desired designation of the value set into the header of the column. Proceed in the same way for the next new value set. Thus, you will have several value sets at your disposal for the subsequent selection of "switchable" values.

Data (the actual values) for the variables can now be entered below the value set name.

To now achieve an assignment of the variables (and their values) to the device properties, a switch is made to the *Assignment* tab, and the property that is later to be switched is selected.

Fig. 4.161 Fill value sets with data

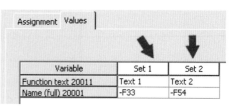

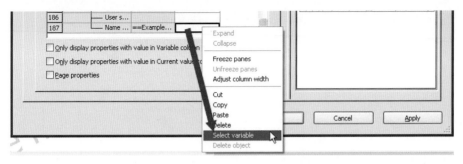

Fig. 4.162 Select variable

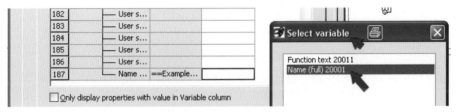

Fig. 4.163 Apply variable

Now the variable can be selected from the **Select variable** dialog and be applied. EPLAN applies the variable and enters it correctly in the *Variable* column.

 TIP: This approach is to be recommended, since variables can be simply adopted from a dialog, and EPLAN enters these with the correct syntax. This excludes the possibility of incorrect entries, which is not always the case with manual entry.

Once the variable has been applied, EPLAN establishes the assignment between the variable value and the value set (name).

The dialog can now be saved and closed with OK. To now switch between the individual value sets, the placeholder object is selected, and the right mouse button is used to select the ASSIGN VALUE SET command from the popup menu.

EPLAN opens the **Select value set** dialog. Select the desired entry here and apply it by clicking on OK. EPLAN now switches over to the variables / values inserted into this value set. In the example, that could be the function text.

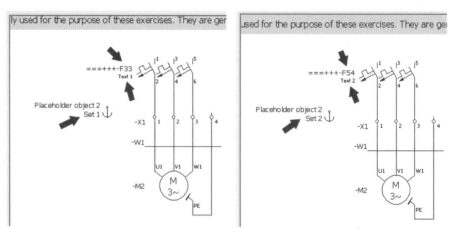

Fig. 4.164 Example value set 1   Fig. 4.165 Example value set 2

In conclusion, it can be said that this type of macro can be used in many different ways in projects. The example shown here was deliberately simple, merely pointing out the option of macros with value sets. The number of potential ideas is unlimited and possible sources of errors, e.g. incorrect part data or wrong cable cross-sections when changing motor power, are eliminated once and for all.

5 Navigators

The graphical editor, with its many aids such as the navigators, is the central focus of project editing. Whatever approach you use, whether graphically-oriented, object-oriented, or a mixture of both, you always remain in the graphical editor. Only the way in which you use it can differ.

What is a navigator? A navigator provides a particular 'viewpoint' for looking at project data. As the name indicates, the cable navigator only 'looks' at devices with the 'cable' function. The PLC navigator only looks at devices with the 'PLC' function.

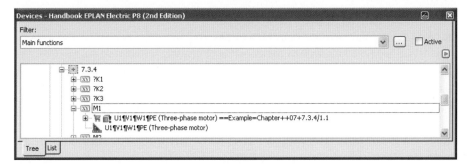

Fig. 5.1 A navigator and its view of devices

This chapter will deal with navigators, their basic functions, and the additional functions in the PROJECT DATA menu. Numerous explanatory examples are used to demonstrate the correct use of devices, their settings, and their properties.

To work quickly and efficiently with EPLAN, you must make use of the various navigators, because they allow certain project actions to be performed much faster and more effectively – the key phrase here is *bulk editing*.

This chapter will not deal with all navigator functions, nor will it deal with all of the navigators in EPLAN Electric P8. The essential purpose of this chapter is rather to explain functions that are often used in daily work.

5.1 Overview of possible navigators

When possible, it is a good idea to keep some navigators open all the time, such as the **device navigator** (this is the most comprehensive because it displays all of the project data).

But the device navigator is not the only navigator. EPLAN has a separate navigator for each of the different *device types* in the project data. Each of these navigators has specific tasks and features relating to the selected project data.

In addition to this, every PROJECT DATA / [DEVICE TYPE] menu item contains extra functions with a functional scope similar to the navigator menus. EPLAN has the following navigators.

The **Device navigator** – shows all *project data* and can (e.g.) directly edit the properties of a device. However, it cannot (e.g.) number terminals; this can only be performed by the special terminal strip navigator.

Fig. 5.2 Various navigators

The **Terminal strip navigator** – shows only project data having the *terminal* function and has terminal editing functions such as renumbering or creating terminal strips.

The **Plug navigator** – shows only project data having the *Plug/Socket* function and provides functions for editing plugs and sockets.

The **PLC navigator** – contains only project data relating to *PLC* functions such as (e.g.) creating new PLC cards, and provides an overview of other PLC box functions (these can also be frequency converters with an integrated EA layer). The PLC navigator provides different views (displays) of the project data.

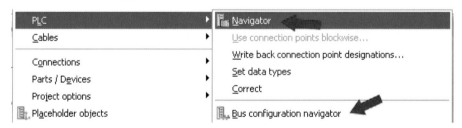

Fig. 5.3 Additional navigator: Bus configuration navigator

The PLC navigator also contains the **bus configuration navigator** (an additional navigator for PLC objects). *Master / slave bus configurations* and *rack / module* assignments are managed here.

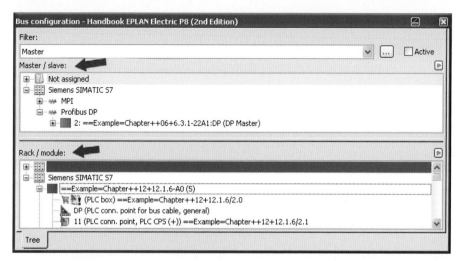

Fig. 5.4 The bus configuration navigator

The **Cable navigator** – contains all functions for cables and their properties. You can use this to (e.g.) place or collect conductors.

The **Connection navigator** – contains all project data on the connections in a project. You can use this to (e.g.) use existing connections.

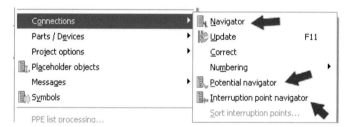

Fig. 5.5 Additional navigators for project data on connections

The project data on connections also includes the **Potential navigator** and the **Interruption point navigator**! These are used to manage potentials and interruption points respectively.

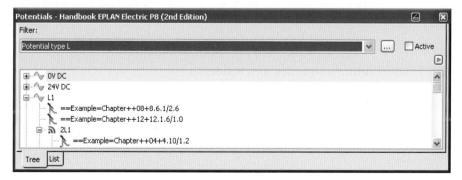

Fig. 5.6 Potential navigator

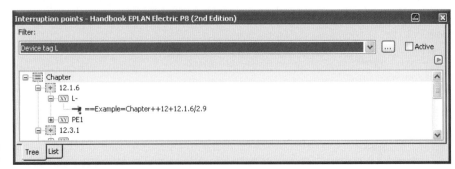

Fig. 5.7 Interruption point navigator

There are additional navigators in the PROJECT DATA / PARTS / DEVICES menu. The **Bill of materials navigator** contains all functions for performing parts editing from a central place, such as swapping parts or selecting a contactor. In contrast to other navigators, the bill of materials navigator also provides special views of the project data in a tree view.

Furthermore, there are two more navigators in the PARTS / DEVICES menu.

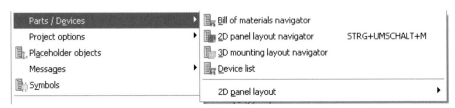

Fig. 5.8 Navigators in the parts / devices area

The **2D panel layout navigator** – among other functions, it allows the creation of a 2D structure diagram of the devices used in the project, based on technical data (at least width and height) from the parts master data or manually entered data. The **3D mounting layout navigator** – enables a 3D diagram of the devices used in the project based on technical data (at least width, height and depth) from the parts master data.

The **Project options navigator** – allows creation of schematics using various options. The options can be easily switched on and off, by page, across pages, or in page sections.

The **Placeholder objects navigator** – conveniently manages macros with value sets from a single location, whether global value sets need to be changed or only those in selected objects need to be edited.

5.1.1 Additional navigators and modules

The following entries in the PROJECT DATA / SYMBOLS menu are not navigators in the sense of those described above. This entry allows symbol selection to stay open all the time like a navigator.

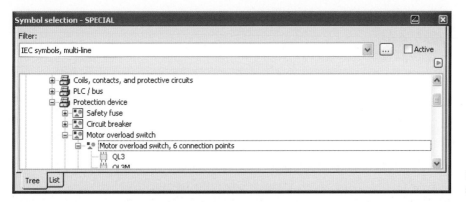

Fig. 5.9 Constantly open symbol selection

An additional tool can also be found in the PROJECT DATA / DEVICES / PROPERTIES menu. This dialog enables direct editing of currently selected components without having to open the properties dialog for the component.

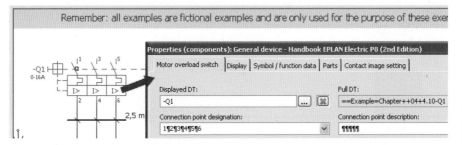

Fig. 5.10 Constantly open Devices / Properties dialog

The following navigators are found at other locations: the **Part master data navigator** and the EPLAN **Data portal navigator**. The part master data navigator is started in the UTILITIES / PARTS / PART MASTER DATA NAVIGATOR menu, and the EPLAN data portal navigator is started in the UTILITIES / DATA PORTAL NAVIGATOR menu.

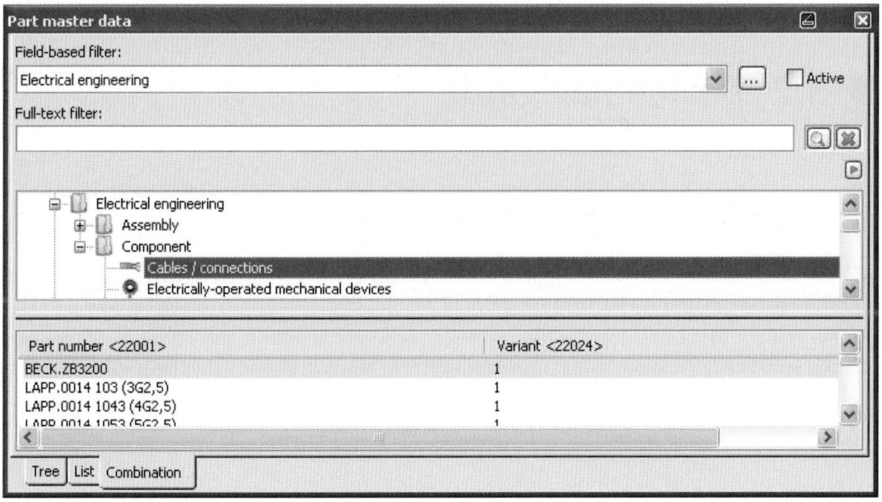

Fig. 5.11 The part master data navigator

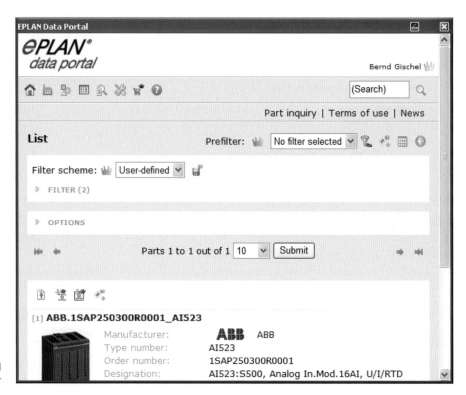

Fig. 5.12 The EPLAN data portal navigator

These navigators are independent dialogs that can remain open in the same way as 'true' navigators. For example, the **Parts master data navigator** always allows access to the parts management without having to open the parts management via UTILITIES / PARTS / ADMINISTRATION.

■ 5.2 Device navigator

All information and project data come together in the **device navigator**. The device navigator can be accessed via the PROJECT DATA / DEVICES / NAVIGATOR menu item, and can then be positioned anywhere on the desktop.

The device navigator can do everything that can be done manually to each device in the project while device editing remains possible, meaning changing device tags, the display format, a device's various properties, and much more. However, the major advantage is that everything is done from a central point with a view of the complete project.

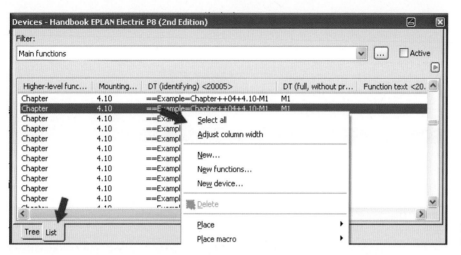

Fig. 5.13 List view of the device navigator

The *filter* and *sorting features* of the device navigator can be used to swap device navigator function definitions or connection point designations simply and conveniently. The device navigator can do more that just display and edit the properties of a device. You can also use it to edit or number several devices at once via the block functions (bulk editing).

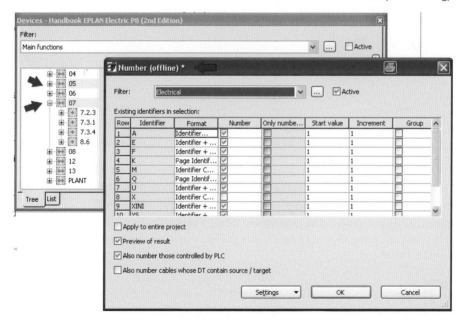

Fig. 5.14 Bulk editing with the device navigator

This chapter not only deals with the device navigator and its functional scope, but also with a series of other functions in the PROJECT DATA / DEVICES menu. For reasons of space, this chapter does not provide an overview of all of the menu items and their functions.

For the device navigator, it is a good idea to create your own keyboard shortcut for opening and closing the navigator, since it is used most often and must be closed and opened constantly (unless it remains open all the time anyway). My recommendation would be CTRL+SHIFT+B.

Fig. 5.15 Keyboard shortcut for the device navigator

5.2.1 Devices / Assign main function

EPLAN distinguishes between a **main function**, for example a coil, and the **auxiliary function**, such as the contacts belonging to a coil.

The *main function* setting in the **Properties (components)** dialog on the first tab, or the *Symbol / function data* tab, turns a device into a main function.

Fig. 5.16 Main function setting in the Properties (components) dialog

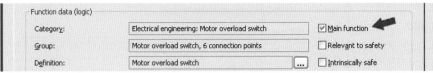

Fig. 5.17 Main function setting on the Symbol / function data tab

This is very important because, for example, parts can only be stored and items can only be selected from devices that have a main function. If devices do not have the main function characteristic, then (e.g.) they cannot be numbered.

Therefore, every independent item should initially be a main function. However, if other items with the same device tag exist in the project, then these are forced to become auxiliary functions.

 NOTE: Only one main function per device is allowed at all times. Double main functions are not allowed and are (usually) detected with a check run and reported.

In the **Properties** (components) dialog, both ways of activating a main function are always set when one of them is set! Regardless of where you set the check box for the main function, the other input check box is also automatically activated or deactivated.

The advantage of the ASSIGN MAIN FUNCTION menu function here is that the main function need not be manually set for every device where this is necessary. It is entirely possible that copy operations or other similar actions can cause a device to have no main function.

 NOTE: Usually, EPLAN automatically checks the main function checkbox after inserting a new symbol. ∎

In the example, all of the devices on the schematic page were selected in order to renumber them. As an alternative, of course, these devices can also be selected in the device navigator.

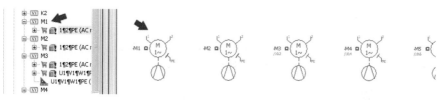

Fig. 5.18 Selected devices

After this, the NUMBER function in the PROJECT DATA / DEVICES / NUMBER menu was executed. EPLAN, however, would not want to number all of the devices. Some of the devices are not numbered because they do not have the *main function* characteristic.

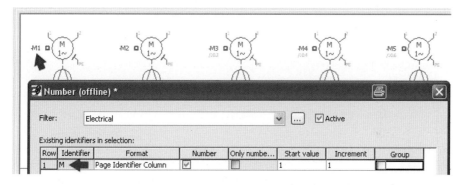

Fig. 5.19 Definitions before numbering

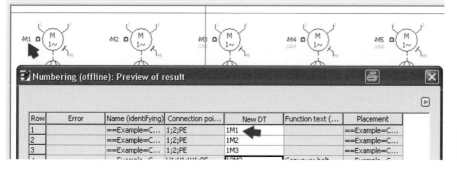

Fig. 5.20 Preview of result before final numbering

TIP: I recommend wherever possible to display the preview of result in the subsequent dialogs. Even if this may take a moment longer, this preview is advantageous because you can cancel immediately if necessary without something happening to the project.

The results of the numbering would then appear as follows. The devices, excepting a few of them, were numbered. All devices without the main function were not numbered.

Fig. 5.21 Result of numbering

You can now use ASSIGN MAIN FUNCTION to transfer the characteristic to devices without the main function. To do this, you first select the devices in the schematic, one after the other, and then select the ASSIGN MAIN FUNCTION item in the PROJECT DATA / DEVICES menu.

After performing this action (i.e. by clicking the menu item), all devices now have the main function. If you now start the **Number** function again for comparison purposes, the preview dialog will show all devices waiting to receive the new device tag.

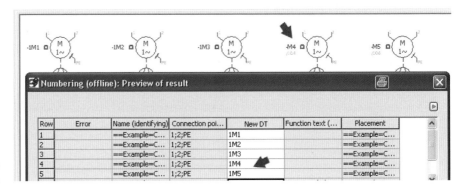

Fig. 5.22 All devices are now numbered.

ASSIGN MAIN FUNCTION can, of course, be assigned to a keyboard shortcut.

5.2.2 Devices / Cross-reference DTs

This function cross-references the device tags of the selected devices. To put it more precisely, the device tag is set to the same value for all the selected devices.

The important phrase here is *'selected devices'*. This does not include the main or auxiliary functions belonging to the devices, rather only the *really* selected devices. You should note this when you use this function.

All devices in the image should receive the device tag –K2. All devices are selected (at least two must always be selected, otherwise the CROSS-REFERENCE DTS function cannot be executed) and the CROSS-REFERENCE DTS function is selected from the PROJECT DATA / DEVICES menu.

Fig. 5.23 Desired devices are selected.

The **Cross-reference DTs** dialog is then displayed. Here you click **only** the device tag that is to be transferred to all the selected devices. In the example, the row –K2 is clicked. This transfers the device tag –K2 to the other device tags.

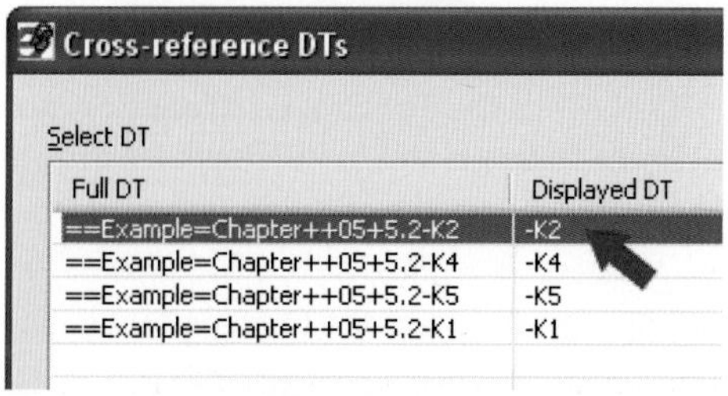

Fig. 5.24 Cross-reference DTs dialog

Fig. 5.25 Cross-referenced devices

All other devices retain their original device tags because they were not selected.

5.2.3 Devices / Synchronize function texts

This function allows the user to synchronize the function texts of the selected devices, including all auxiliary functions and the main function. It makes no difference whether the main function or one of the auxiliary functions is selected.

This must always be the same device, though. Different devices cannot be edited with this function.

All devices should receive the same function text. To do this, an element is selected. This can be the main function or any auxiliary function in the schematic.

The SYNCHRONIZE FUNCTION TEXTS function is then executed via the PROJECT DATA / DEVICES menu item.

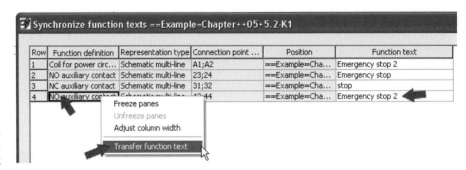

Fig. 5.26 A selected element and the Synchronize function texts dialog

EPLAN opens the **Synchronize function texts** dialog. In this dialog, only the row containing the function text to be transferred to all other functions is selected. In the example, the function text of row 2 will be transferred to all other elements. To do this, the right-click popup menu is opened and the TRANSFER FUNCTION TEXT function is selected.

EPLAN now transfers the function text to all other entries in the dialog.

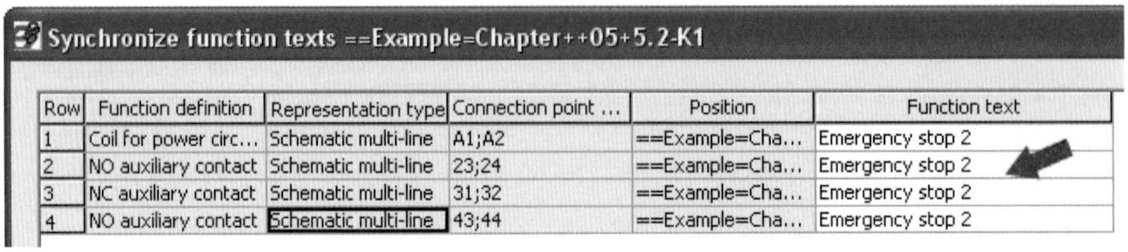

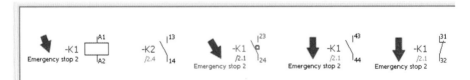

Fig. 5.27 Transferred function text in the dialog

After the transfer, the OK button is clicked. EPLAN exits the dialog and transfers the function text to all elements and writes this text to the *function text <20011>*, *function text (automatic) <20031>*, and *function text (common) <20120>* symbol properties.

Fig. 5.28 Transferred function text in the schematic

5.2.4 Devices / Interconnect devices

The EPLAN INTERCONNECT DEVICES function allows interconnection of devices. In this case, interconnection means the generation of connections between the selected interconnected devices.

These devices can already be placed, i.e. exist graphically in the schematic, but these can also be unplaced devices, i.e. they only exist in the device navigator. A mixture of both types of placement is also possible, of course, and they can be any devices.

In the example, two terminal strips (with several unplaced terminals and isolated cable ends) are interconnected with an existing cable and its reserve conductors without being placed graphically on a page.

| 🛒📄 | Icon for a placed terminal in the device navigator |

| 🛒📄 | Icon for an unplaced terminal in the device navigator |

The INTERCONNECT DEVICES function is started via the menu item PROJECT DATA / DEVICES. EPLAN then opens the **Interconnect devices** dialog.

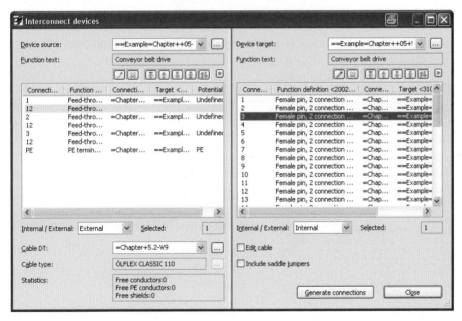

Fig. 5.29 Interconnect devices dialog

The **Interconnect devices** dialog can be filled with various data after it is called up. If these entries are not required, then you can simply delete them. You delete entries by selecting them and then clicking the delete button.

If this is not possible, then you can also delete the entries in the fields *device source*, *device target*, and *cable DT*. To do this, you click the field, select the desired entry, and then press the DEL key. All entries are deleted after this, and new devices can be selected and interconnected.

New devices are selected in the *device source* and *device target* fields using the button. EPLAN then opens the **Select device** dialog. In this dialog, the desired devices for the source and target (one of each) are selected, and the selection is applied by clicking the OK button.

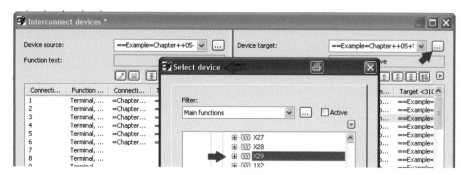

Fig. 5.30 Select device source and target

EPLAN then loads the selected device into the *device source* and *device target* fields.

An additional cable is to serve as a connection between the two terminal strips to be interconnected. An existing cable is selected in the *cable DT* field for this.

 TIP: At this point, however, a new, non-existent cable can be created easily. Just click the field and enter the cable DT.

If an existing *cable DT* should be used, it is selected with the ... button in the **Select cable** dialog and applied with the OK button.

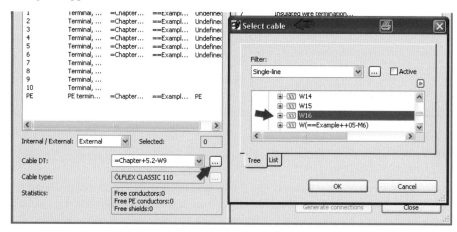

Fig. 5.31 Selecting cables for interconnection

Once the preparatory measures have been completed, both terminal strips, including the intended functions (unplaced terminals and unplaced isolated wire terminations), can now be interconnected.

To do this, *terminals* and *isolated wire terminations* are selected in the fields (click and then hold down the CTRL button and the mouse to select the desired terminals and isolated wire terminations) and then push the GENERATE CONNECTIONS button.

5.2 Device navigator

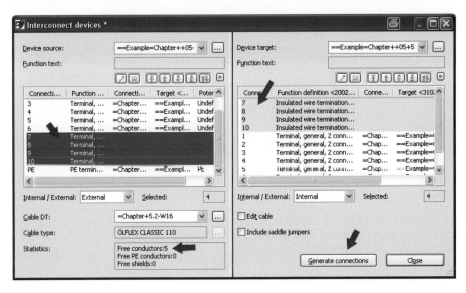

Fig. 5.32 Generate connections

EPLAN now generates the unplaced connections between the terminals and the isolated wire terminations and assigns the cable and cable conductor.

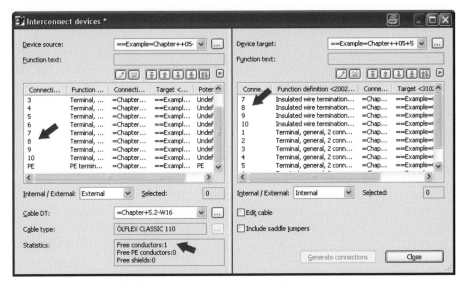

Fig. 5.33 Connections were generated.

You can also still edit these connections directly in the **Interconnect devices** dialog. To do this, just double-click the connection and EPLAN opens the **Properties (components)** dialog.

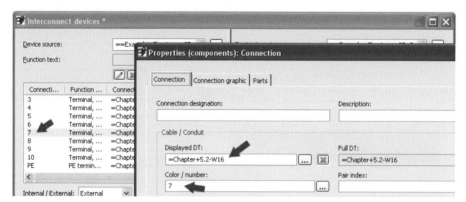

Fig. 5.34 Direct editing of individual connections

This basically completes the creation of connections. If you had created a report before interconnecting the devices (in this case, a cable diagram for the affected cable), EPLAN would have only listed the devices used, including the cable conductors used, in the report.

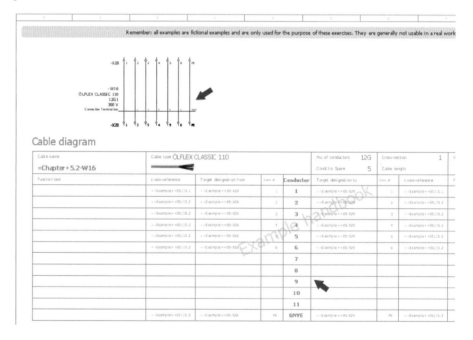

Fig. 5.35 Report before interconnecting devices

After updating the cable diagram report, EPLAN then adds the interconnected devices to the report. The devices don't receive a cross-reference (page / path) because they are not placed anywhere.

Cable diagram

Cable name =Chapter +5.2-W16	Cable type ÖLFLEX CLASSIC 110				No. of conductors 12G / Condctrs Spare 1	Cross-s / Cable l
Function text	cross-reference	Target designation from	Conn. pt	Conductor	Target designation to	Conn. pt
	==Example+ +05/15.1	==Example+ +05-X28	1	1	==Example+ +05-X29	1
	==Example+ +05/15.2	==Example+ +05-X28	2	2	==Example+ +05-X29	2
	==Example+ +05/15.2	==Example+ +05-X28	3	3	==Example+ +05-X29	3
	==Example+ +05/15.2	==Example+ +05-X28	4	4	==Example+ +05-X29	4
	==Example+ +05/15.2	==Example+ +05-X28	5	5	==Example+ +05-X29	5
	==Example+ +05/15.2	==Example+ +05-X28	6	6	==Example+ +05-X29	6
Reserveklemme		==Example+ +05-X28	7	7	==Example+ +05-X29	7
=		==Example+ +05-X28	8	8	==Example+ +05-X29	8
=		==Example+ +05-X28	9	9	==Example+ +05-X29	9
=		==Example+ +05-X28	10	10	==Example+ +05-X29	10
				11		
	==Example+ +05/15.3	==Example+ +05-X28	PE	GNYE	==Example+ +05-X29	PE

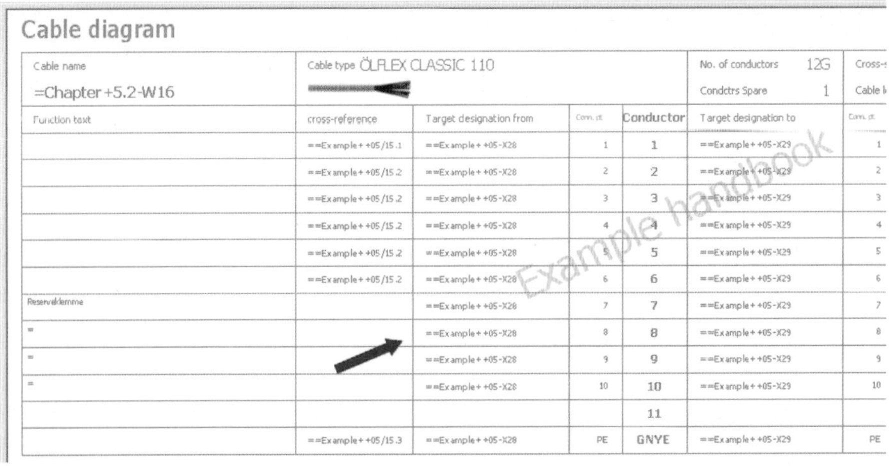

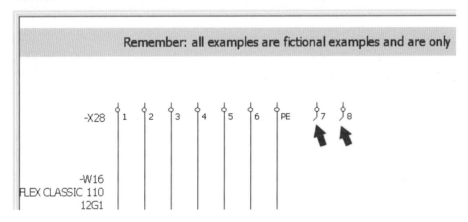

Fig. 5.36 Updated report

The devices themselves are displayed with a red check on their symbol to visually identify them. This lets you know that this is an unplaced connection.

Fig. 5.37 Visual identification of unplaced connections

5.2.4.1 Place interconnected unplaced devices

Of course, it is also possible to place precisely these interconnected devices, including connection information (cable DT and their conductors), without having to redraw everything.

To do this, either the *device source* or the *device target* from the device navigator are placed in multi-line mode on a schematic page.

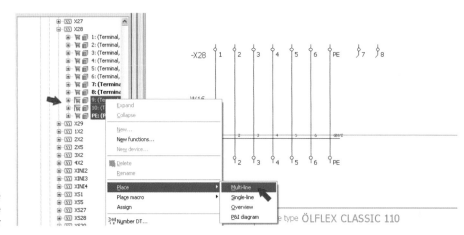

Fig. 5.38 Place multi-line functions from the device navigator

After the devices are placed, the CONNECTED FUNCTIONS menu item can be called up via the INSERT menu.

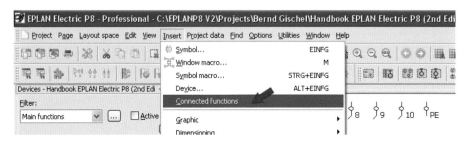

Fig. 5.39 Insert connected functions

Then EPLAN will ask for which devices the connected functions should be displayed. These should be selected with a window.

Fig. 5.40 Select the affected functions

If the window's second corner is defined (mouse-click), EPLAN immediately displays the corresponding function on the schematic page.

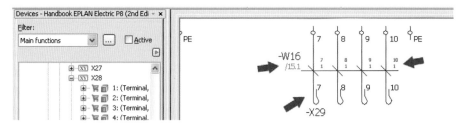

Fig. 5.41 Placed functions

If you again update the corresponding report, then the data of the previously unplaced functions will be updated in the cable diagram with a cross-reference specification.

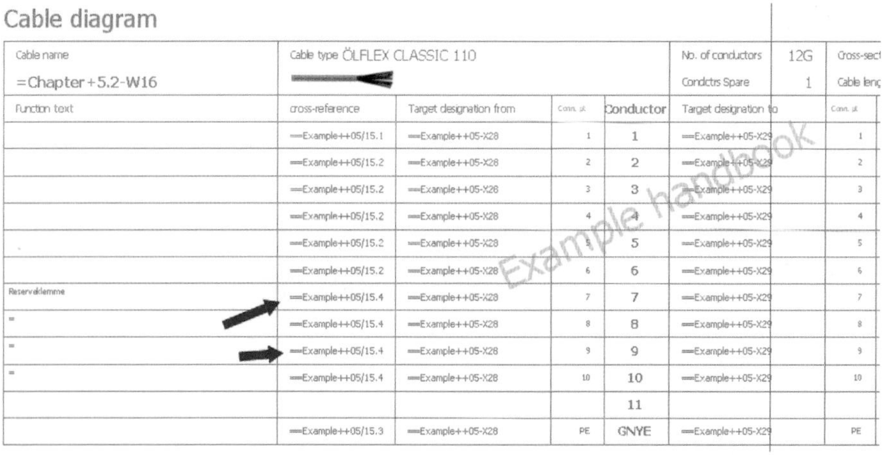

Cable diagram

Cable name		Cable type ÖLFLEX CLASSIC 110			No. of conductors	12G	Cross-sect
=Chapter+5.2-W16					Condctrs Spare	1	Cable leng
Function text	cross-reference	Target designation from	Conn. at	Conductor	Target designation to		Conn. at
	==Example++05/15.1	==Example++05-X28	1	1	==Example++05-X29		1
	==Example++05/15.2	==Example++05-X28	2	2	==Example++05-X29		2
	==Example++05/15.2	==Example++05-X28	3	3	==Example++05-X29		3
	==Example++05/15.2	==Example++05-X28	4	4	==Example++05-X29		4
	==Example++05/15.2	==Example++05-X28	5	5	==Example++05-X29		5
	==Example++05/15.2	==Example++05-X28	6	6	==Example++05-X29		6
Reserveklemme	==Example++05/15.4	==Example++05-X28	7	7	==Example++05-X29		7
"	==Example++05/15.4	==Example++05-X28	8	8	==Example++05-X29		8
"	==Example++05/15.4	==Example++05-X28	9	9	==Example++05-X29		9
"	==Example++05/15.4	==Example++05-X28	10	10	==Example++05-X29		10
				11			
	==Example++05/15.3	==Example++05-X28	PE	GNYE	==Example++05-X29		PE

Fig. 5.42 Updated report

5.2.5 Devices / Number (offline)

EPLAN can number devices according to a definable numbering pattern as soon as they are inserted (according to the personal setting). If the schematic and its devices need to be renumbered in a different format later, a fast and effective un-numbering method is important.

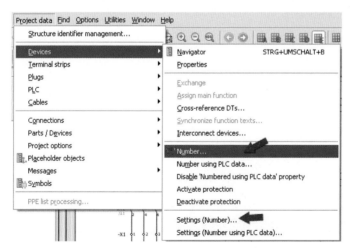

Fig. 5.43 Number devices

EPLAN offers the NUMBER DEVICES function for this. In contrast to numbering when symbols are inserted (*online numbering*), subsequent renumbering is so-called *offline numbering*. Certain filters and schemes can be used for numbering, and predefined schemes can also be applied, exported or imported.

One important thing should be noted about offline numbering. Auxiliary functions cannot be numbered. If these items are selected and numbering is then attempted, EPLAN generates an appropriate message.

How does the numbering function work? This is really simple. The devices to be numbered are selected and the NUMBER function is started from the PROJECT DATA / DEVICES menu.

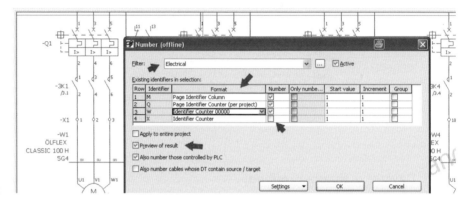

Fig. 5.44 Settings in the Number (offline) dialog

EPLAN opens the **Number (offline)** dialog and lists the *identifiers* in the selection that are found in the table. In this dialog, you can set your own *numbering format* for every identifier entry, as well as a start value and the increment. Entire identifiers can be excluded here from numbering so that you are not forced to select only very specific devices when selecting devices.

NOTE: Important: Always use the *Preview of result* setting. This enables one more check to make sure that everything is numbered as desired before EPLAN writes the new numbering back to the project.

A different numbering format can simply be selected in the *format* selection list, or the EXTRAS button in the lower area of the dialog can be used to create a new numbering scheme.

Fig. 5.45 Settings: Numbering (offline) dialog

After selecting the numbering format, you can click the OK button. EPLAN numbers all devices according to the specified format and displays this numbering in a *preview of result*.

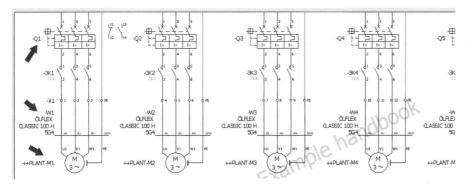

Fig. 5.46 Preview of result dialog

If the *preview of result* (if used) is confirmed, then EPLAN writes the modified device tags back to the schematic.

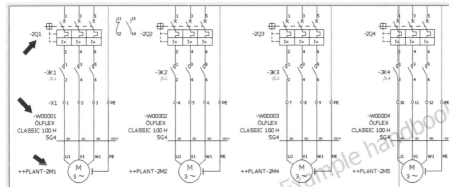

Fig. 5.47 Before numbering

Fig. 5.48 After numbering

In the example, it is obvious that EPLAN only numbers devices that are also a main function. In the example, that would be everything except the coil contacts. These were already not included in the numbering dialog.

This is why it is important to know that although you can number or re-number devices very quickly by selecting them on a page, these devices must be main functions. If you want to avoid this, you can perform numbering via the device navigator.

To do this, the devices are selected and then the NUMBER DT menu item is chosen via the popup menu.

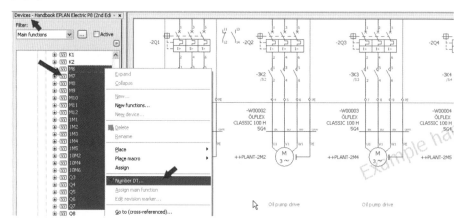

Fig. 5.49 Number DT in the popup menu

Then EPLAN restarts the **Number (offline)** dialog, and numbering can proceed from here as described above. This time, however, all functions are included because the contacts **and** the coil have been marked (in the device navigator).

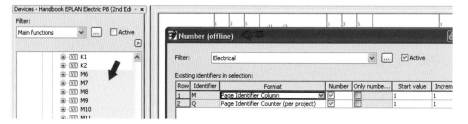

Fig. 5.50 Number (offline) dialog from the device navigator

5.2.6 Device navigator / New...

Fig. 5.51 New... menu item

The NEW... function inserts a new function definition, for example an additional auxiliary NC contact for a motor overload switch.

To do this, the motor overload switch is selected in the device navigator. The NEW... function is then called up from the popup menu in the device navigator. EPLAN opens the **Function definition** dialog.

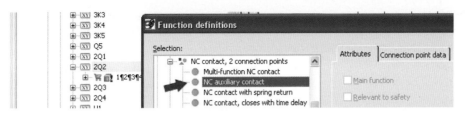

Fig. 5.52 Function definition dialog

In this dialog, the function definition **NC auxiliary contact** is selected and applied by clicking the OK button. In the subsequent **Properties (components)** dialog, EPLAN asks for the *device tag*, among other information.

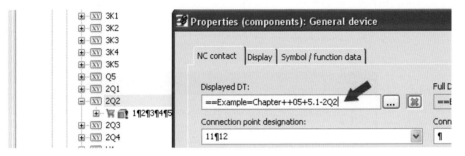

Fig. 5.53 Properties (components) dialog

After exiting the dialog via the OK button, the *NC auxiliary contact* is assigned to the selected motor overload switch in the device navigator.

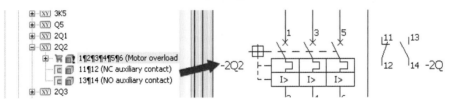

Fig. 5.54 New function on motor overload switch

5.2.7 Device navigator / New device...

To insert a new device in the device navigator you select the NEW DEVICE... function in the popup menu of the device navigator.

When the NEW DEVICE... function is selected, EPLAN starts the **parts management**. The corresponding device is selected here and applied by clicking the OK button.

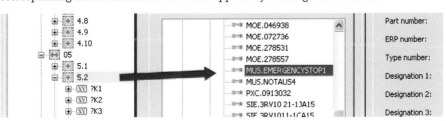

Fig. 5.55 Select device

EPLAN sorts the device according to the *identifier* into the device navigator. It can now be edited subsequently. To do this, you select the PROPERTIES function in the popup menu, which then opens the **Properties (components)** dialog.

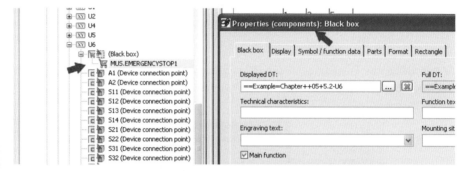

Fig. 5.56 Subsequent device editing

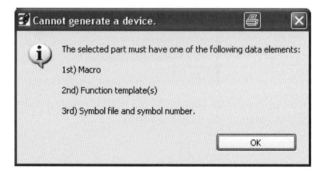

Fig. 5.57 Query message if a device cannot be generated

Only new devices with at least one of the following properties (macro, function template, symbol file or symbol number) can be assigned in the device navigator. If this is not the case, then EPLAN displays an appropriate message.

5.2.8 Device navigator / Place / [Representation type]

In the last example, a device was inserted into the navigator. It is inserted in the navigator as an unplaced device. To obtain this device in the schematic (i.e. to place it), you select the PLACE function and MULTI-LINE in the popup menu of the device navigator.

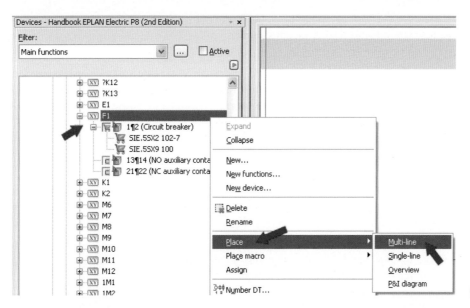

Fig. 5.58 Place device

Then the device hangs on the cursor and can be placed on the schematic page. It is not only possible to pull a single device onto the page; you can also select several devices in the device navigator and place them on the page. EPLAN then places an item every time the mouse is clicked.

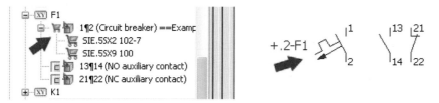

Fig. 5.59 Placed device

The major advantage of using the device navigator for placement is that EPLAN updates and enters all important information and data. This means that you won't have any incorrect DTs on the device because EPLAN enters the correct one from the navigator!

■ 5.3 Terminal strips

Fig. 5.60 Terminal strips menu

There is not much to say about terminal strips. The **terminal strip navigator** supports the user in performing many repetitive actions, for example numbering terminals.

In addition to the actual terminal strip navigator, the PROJECT DATA / TERMINAL STRIPS menu contains three additional functions. These can be executed directly on the selected elements on a page or on selected pages in the **page navigator**. The terminal strip navigator also has a popup menu, accessible via right-clicking the mouse or via the ⬛ button.

5.3.1 Terminal strips / Edit...

The EDIT function in the PROJECT DATA / TERMINAL STRIPS menu opens the **Edit terminal strips** dialog. This dialog contains different functions from the popup menu for editing terminals. Examples of these functions are *Number terminals*, *Move*, *Add strip accessories*, and delete or *Ungroup multi-level terminals*.

As with everywhere in EPLAN, the popup menu in the terminal editor can be called up via the right mouse button or the ⬛ button.

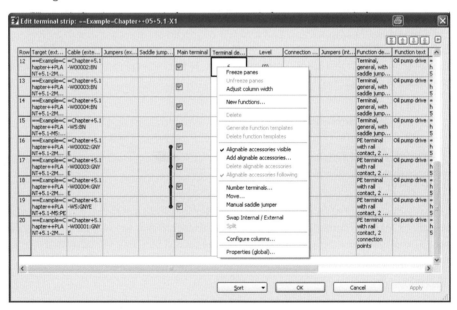

Fig. 5.61 Edit terminal strips dialog

Additional functions for editing terminals are located in the lower area of the dialog after clicking the SORT button.

Fig. 5.62 The Sort button and its functions

The SORT button opens another menu containing functions for sorting terminals. The DEFAULT function, for example, restores the existing sorting of the terminals back to the default. The NUMERIC function sorts the terminals by number, and the ALPHANUMERIC function sorts terminals by the alphanumeric components in the terminal designation.

The other buttons, OK and CANCEL, speak for themselves. The APPLY button applies the changes in the **Edit terminal strips** dialog and does not close the dialog yet.

To edit a terminal strip using the terminal editor, you first select the terminal strip in the schematic or in the terminal strip navigator (you can also just select one terminal in the terminal strip to be edited) and then select the EDIT function in the PROJECT DATA / TERMINAL STRIPS menu.

 TIP: To change the row height in the **Edit terminal strips** dialog, position the cursor on a row and then pull the row while holding down the left mouse button to make it smaller or larger.

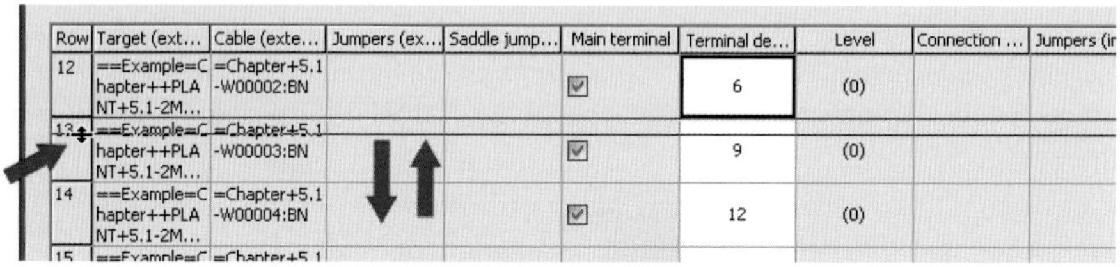

Fig. 5.63 Change row height

EPLAN opens the **Edit terminal strips** dialog. The terminals are sorted here according to the standard. To do this, click the SORT button and select the STANDARD item. EPLAN immediately sorts the terminal strip according to the standard.

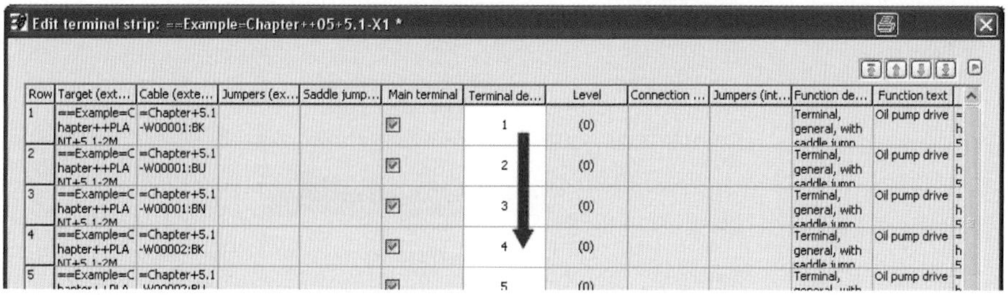

Fig. 5.64 Resort terminal strip

For example, to change the sequence of terminals 1, 2 and 3 (from 1-2-3 to 3-2-1), each of the terminals is selected and moved to the desired position by using the buttons or the associated keyboard shortcuts.

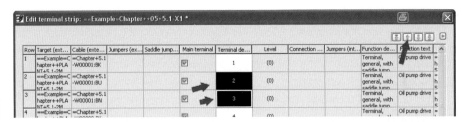

Fig. 5.65 Manually moving a terminal

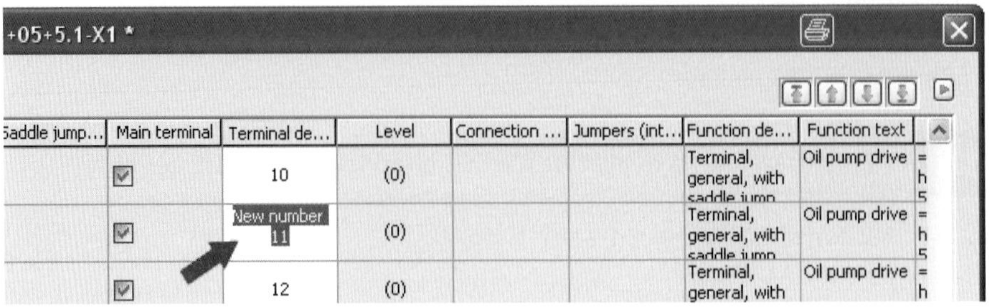

Fig. 5.66 Rename terminal designation

To give one or more terminals another terminal designation, you need only edit the *terminal designation* field directly. EPLAN would write the changed terminal designation to the corresponding terminals immediately upon leaving (and saving the changes) in the dialog.

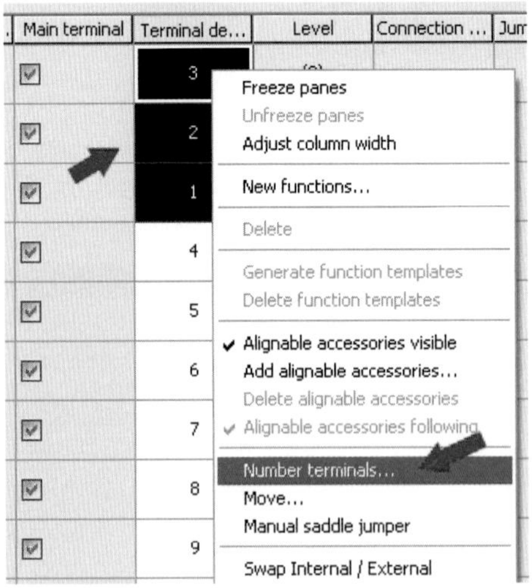

Fig. 5.67 Number terminals

Terminals can also be numbered in the dialog. To do this, simply select the desired terminals with the mouse and call up the NUMBER TERMINALS menu item in the popup menu. EPLAN then starts the **Number terminals** dialog, where the settings are made, and EPLAN can then number the terminals.

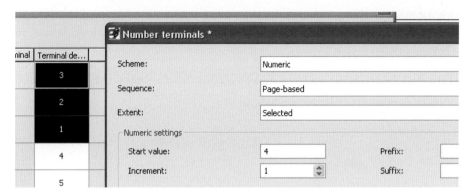

Fig. 5.68 Number terminals dialog

5.3.2 Terminal strips / Correct

Another important function in the PROJECT DATA / TERMINAL STRIPS menu is the CORRECT function. At least one terminal or terminal strip must be selected before this function can be called. When the CORRECT function is called, EPLAN opens the CORRECT TERMINAL STRIPS dialog.

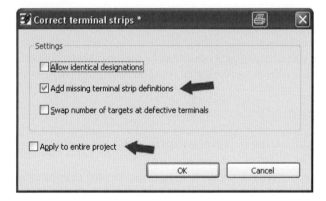

Fig. 5.69 Correct terminal strips dialog

The dialog offers four settings. The *Allow identical designations* setting means that terminals with the same terminal number are not logged as errors and also sets the *Allow multiple entries<20811>* property for all terminals with the same terminal number.

The *Swap number of targets at defective terminals* setting means that EPLAN corrects terminals with a different number of targets.

You can also use the *Add missing terminal strip definitions* setting so that EPLAN automatically generates all missing terminal strip definitions and adds them to the unplaced terminal strip functions in the terminal navigator.

What is a terminal strip definition, and why should you use them? Only through generated terminal strip definitions can terminal strips be (e.g.) excluded from various reports (*properties <20851, 20853 and 20857> Do not assign an output...*), or a descriptive terminal strip definition text (function text field) can be assigned to terminal strips, or the terminal strip shall be graphically output using a special form (terminal diagram and / or connection-point diagram form).

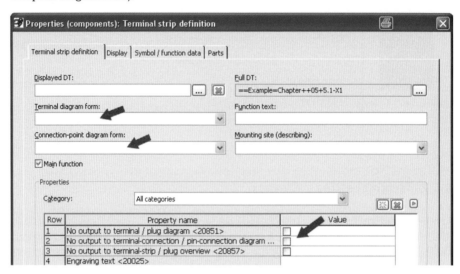

Fig. 5.70 Terminal strip definition and its properties

This information can only be entered when a terminal strip definition has been created. This information cannot be entered or stored in the properties of a terminal.

If the *Add missing terminal strip definitions* check box is set, and the *Apply to entire project* setting is set, and the dialog is confirmed by clicking the OK button, then EPLAN generates a terminal strip definition for all terminal strips, but only for those that do not have one.

5.3.3 Terminal strips / Number terminals

As a result of various project actions in EPLAN, e.g. through copying or copying of pages from other projects, terminals can end up with the wrong or completely incorrect designations.

To avoid having to manually renumber all these terminals, the PROJECT DATA / TERMINAL STRIPS menu offers the NUMBER TERMINALS function.

Fig. 5.71 Number terminals dialog

You can number terminals using more than a simple scheme such as numbering integer values from 1 to 10. EPLAN also allows you to define your own schemes, e.g. for initiator terminals.

The terminal strip to be renumbered is selected in the schematic (or in the terminal strip navigator). It is sufficient when any terminal belonging to this terminal strip is selected.

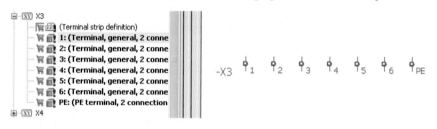

Fig. 5.72 Select terminals

The **Number terminals** dialog is then started via the NUMBER TERMINALS item in the PROJECT DATA / TERMINAL STRIPS menu.

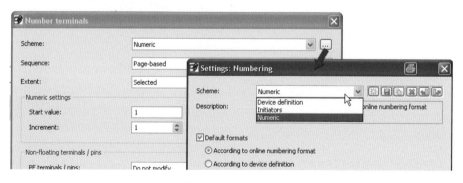

Fig. 5.73 Number terminals dialog with selection of numbering scheme

In the **Number terminals** dialog, the desired scheme is now selected in the Scheme selection field; in the example, this is the *Numeric* scheme. The scheme can only be changed before numbering by using the [...] button or by generating a new scheme in the **Settings: Numbering** dialog.

Before starting the actual numbering, you should check and adjust the other settings in the dialog, such as the numeric settings, the range, and the handling of PE terminals.

When the OK button in the **Number terminals** dialog is clicked, EPLAN numbers the terminal strips as specified.

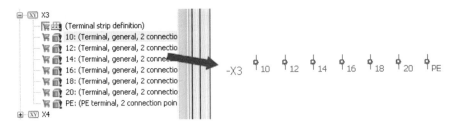

Fig. 5.74 Newly numbered terminals

This simple example is only meant to illustrate the process of using the Number terminal strips function. More complex numbering tasks can be accomplished by making use of the many possible settings in terminal schemes.

5.3.3.1 Number terminals settings

In the **Number terminals** dialog, there are, in addition to the basic numbering settings, the following options, which will be explained.

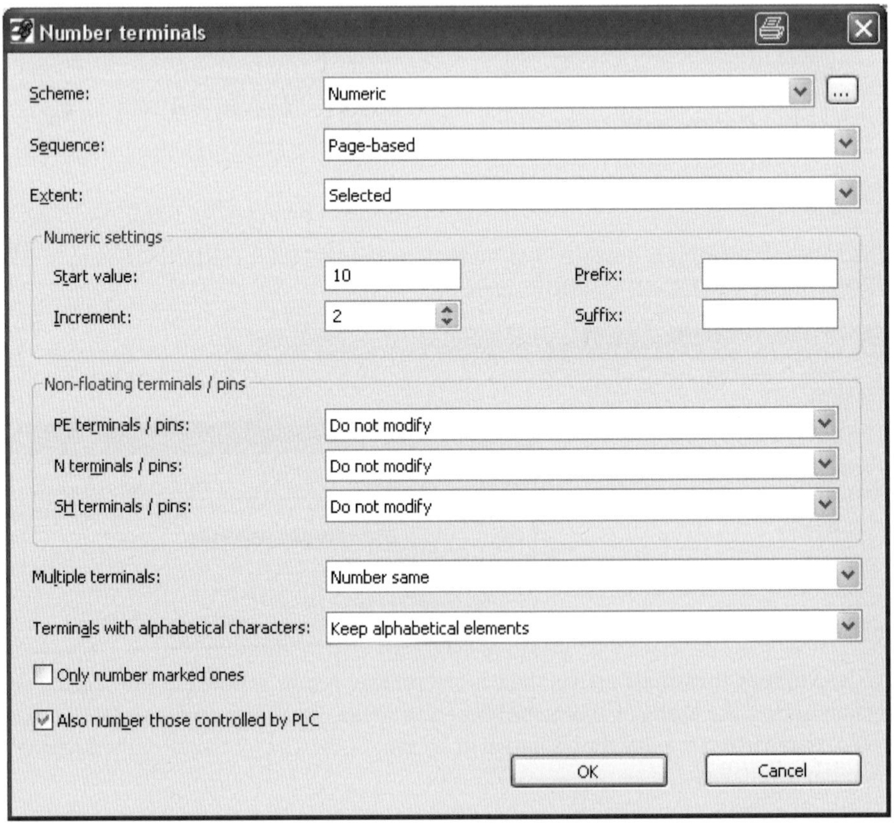

Fig. 5.75 Number terminals dialog

Scheme –The desired scheme can be set here from a selection list. The available schemes can be expanded.

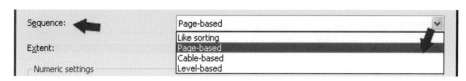

Fig. 5.76 Sequence setting

Sequence – The numbering sequence is set here in this selection list. Is the sequence supposed to be oriented to pages, cables or levels, or should the existing sorting method be applied?

Fig. 5.77 Extent setting

Extent – This setting determines the range of the terminals to be numbered. What should be numbered - only selected terminals, the entire terminal strip, or all terminals in the project?

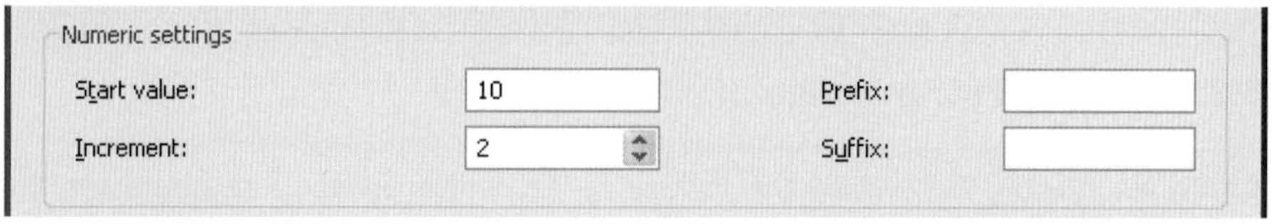

Numeric settings			
Start value:	10	Prefix:	
Increment:	2	Suffix:	

Fig. 5.78 Numeric settings settings

Numeric settings – This is where the start value, the increment, the prefix (a character preceding the terminal designation) and the suffix (a character after the terminal designation) are set.

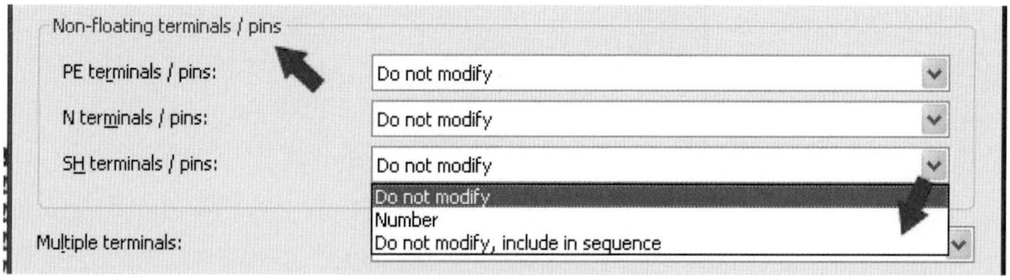

Non-floating terminals / pins	
PE terminals / pins:	Do not modify
N terminals / pins:	Do not modify
SH terminals / pins:	Do not modify
	Do not modify
	Number
Multiple terminals:	Do not modify, include in sequence

Fig. 5.79 Settings for certain terminal types

Non-floating terminals / pins – This is where settings for PE, N and SH terminals are set. Should they not be changed, numbered or modified, but rather maintain the set sequence? The potential type of the terminals is crucial here!

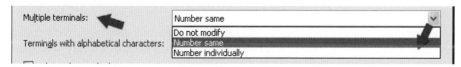

Fig. 5.80 Settings for multiple terminals

Multiple terminals – How should multiple terminals be handled during numbering operations? Should they be left unchanged, numbered all the same, or numbered individually?

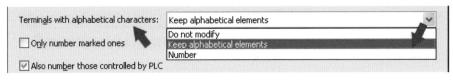

Fig. 5.81 Settings for alphabetically designated terminals

Alphabetically designated – This setting means that terminal designations that contain alphabetical components are taken into account. This setting determines whether these will be left unchanged and the components retained, or whether these will be completely renumbered. In the latter case, the alphabetical components would be lost.

In addition to these settings, there are two more options.

Fig. 5.82 Further options

Only number marked ones - only those terminals with the ? placeholder are numbered. The last setting, *Also number those controlled by PLC*, only numbers those terminals that are controlled by PLC.

5.3.4 Terminal strip navigator / New...

The NEW... function is basically the same as in the device navigator. In contrast to the device navigator, which can insert any function definition, in the terminal strip navigator only terminal or terminal strip function definitions can be inserted.

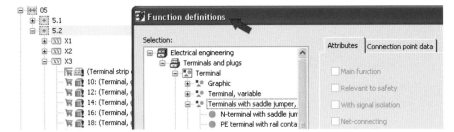

Fig. 5.83 Insert new terminals

5.3.5 Terminal strip navigator / New functions...

The NEW FUNCTIONS... function in the terminal navigator allows fast generation of complete terminal strips using particular (own) numberings.

The NEW FUNCTIONS... function is called up in the terminal strip navigator's popup menu. EPLAN opens the **Generate functions** dialog, which is initially empty, or a terminal strip designation is applied. The full DT is already displayed, depending on the current focus in the terminal strip navigator.

5.3 Terminal strips

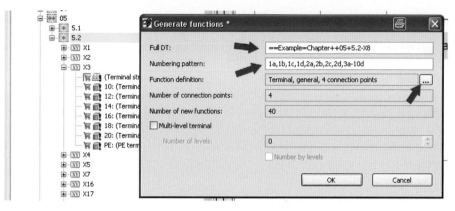

Fig. 5.84 Generate functions dialog

The full DT (the terminal strip designation), the numbering pattern, and the setting defining whether or not these are multi-level terminals should be selected / set in this dialog.

It is also possible to select a different function definition instead of the default terminal. To do this, just click the ... button. EPLAN switches to the **Function definitions** dialog, where you can select a different terminal and apply this by clicking the OK button.

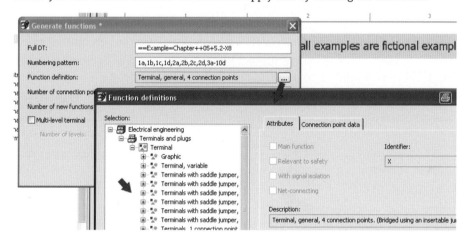

Fig. 5.85 Function definitions dialog

However, this is not a precondition. All function definitions can be easily changed later using block editing in the terminal strip navigator.

We return once more to generation of the functions. The functions can only be generated when all entered data is correct (the OK button is enabled by EPLAN).

If an error exists in the numbering pattern, e.g. a missing comma, then the OK button is not enabled. You must check for this. When the OK button is clicked, EPLAN generates a terminal strip according to the specified options.

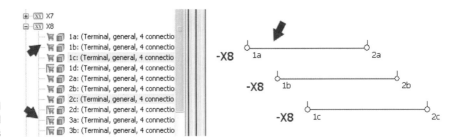

Fig. 5.86 New terminal strip with placed and unplaced terminals

NOTE: These newly created terminals do not have any parts data. They must be assigned later with a device selection.

5.3.6 Terminal strip navigator / New terminals (devices)...

In addition to the easy generation of new terminal strips and new terminals, it is also possible in the terminal strip navigator's popup menu to generate new terminals as devices on the basis of existing parts and their function templates.

To do this, the NEW TERMINALS (DEVICES)... menu entry is started from the popup menu.

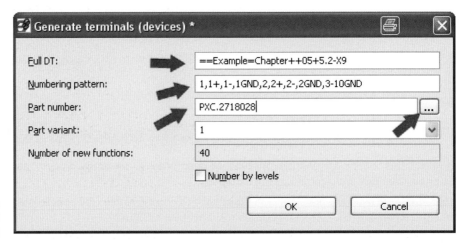

Fig. 5.87 Generate terminals (devices) dialog

EPLAN opens the **Generate terminals (Devices)** dialog. Similar entries are generated as in the previous section, meaning that a DT is set, a numbering pattern, and, crucially, a part (the device) is selected from parts management.

Once all entries are correct, the generation of devices can be started by clicking the OK button. EPLAN saves the functions in the navigator as unplaced functions.

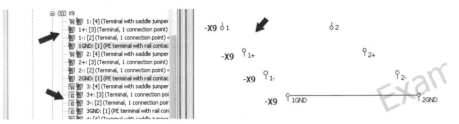

Fig. 5.88 Newly created terminals (devices)

In contrast to the last section, these new terminals are real devices, and they already have parts data.

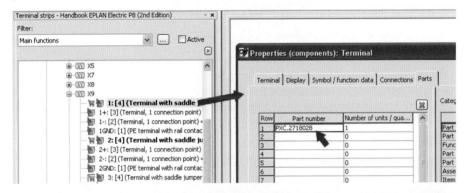

Fig. 5.89 Parts data for this kind of terminal

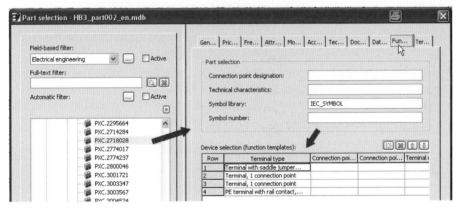

Fig. 5.90 Function template for a terminal part

5.3.6.1 Main terminals / Auxiliary terminals

It is important to know that, in addition to the generation of terminals or of terminals as devices, EPLAN has introduced new terminal functions and a new term in Version 2.0: the **main terminal**.

What is a main terminal? It is comparable to a main function; a terminal with multiple function definitions can only be identified once as a main terminal. All of this terminal's other functions must be **auxiliary terminals**.

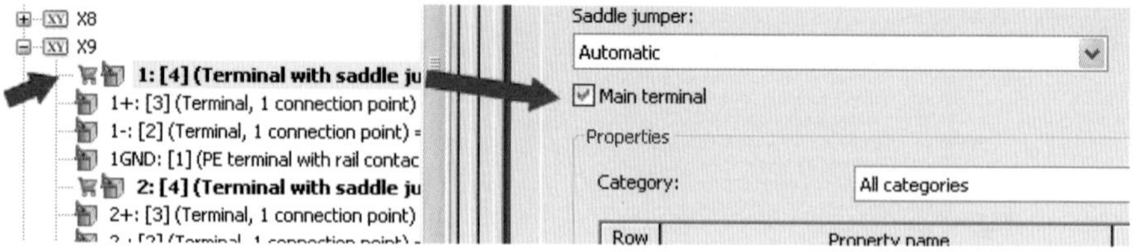

Fig. 5.91 Main terminal

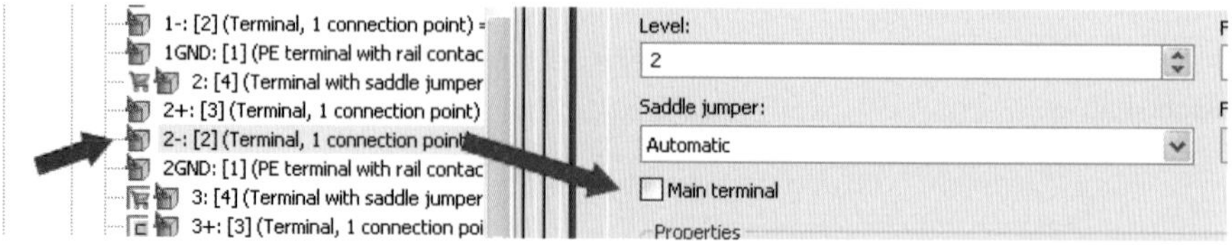

Fig. 5.92 Auxiliary terminal

Only main terminals can have *parts data*. These new main and auxiliary terminals are now handled like all other devices and can be used like other devices in EPLAN. You therefore have the option of using or placing some of this terminal's functions, e.g. as a grounding contact or a coil. Main and auxiliary terminals have the same relationship to each other. Device selection can be performed for these main terminals.

The terminal strip definition, however, remains a terminal strip's *main function*!

5.3.7 Terminal strip navigator / Place / [Representation type]

If, for example, unplaced terminals are needed in the schematic, you can use the PLACE function in the terminal strip navigator to put them there.

You use the mouse or the keyboard (the mouse is easier) to select the desired terminals in the terminal strip navigator. You then select the PLACE / MULTI-LINE function (for a multi-line representation) from the right-click popup menu in the terminal strip navigator.

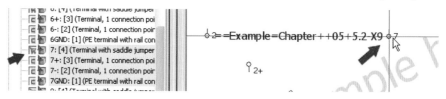

Fig. 5.93 Multi-line placement of terminals

EPLAN now 'hangs' all the selected terminals on the cursor, and you can place them one at a time by clicking the left mouse button.

You can also hold down the left mouse button and (e.g.) drag the mouse to the right over the page. EPLAN then places multiple terminals in a single operation until you release the mouse button. Counter targets must be present, though, so that EPLAN can generate an automatic connection here!

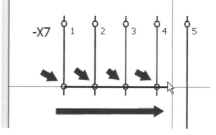

Fig. 5.94 Multiple placement

5.4 Plugs (Sockets)

In addition to terminals, there are plugs and sockets. In principle, the **plug navigator** offers the same functions as those used for editing terminals and terminal strips. This chapter will therefore only deal with the plug / socket functions that are different from the functions provided in the terminal strip navigator.

Plug is a generic term in EPLAN. Plugs are distinguished by male and female pins. Corresponding function definitions exist for these, and the various menu items are also labeled as such.

5.4.1 Plugs / Edit

In addition to the actual plug navigator, the plug project data also has a submenu with additional special functions. The EDIT PLUG function works in a way very similar to the EDIT TERMINAL function. In this case, you edit only male pins or female pins.

EPLAN then displays the **Edit plug** dialog. The functions available in the popup menu, such as *Number pins* or *Edit properties*, can be used here.

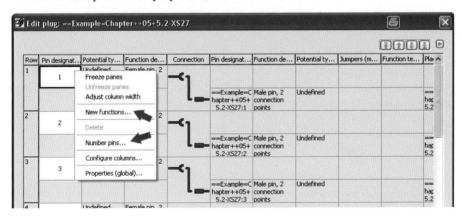

Fig. 5.95 Edit plug dialog

5.4.2 Plugs / Correct

Among other features, the CORRECT function in the PROJECT DATA / PLUGS menu creates correct plug definitions. Whereas terminals only have one definition, the terminal strip definition, plugs in EPLAN have three possible definitions.

There are *plug definitions* for female pins, male pins, and for a combination of male and female pins.

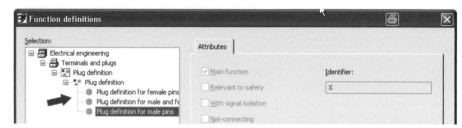

Fig. 5.96 Different types of plug definitions

It is very important that the correct *plug definitions* matching the corresponding plug, socket, or plug strip are generated. If for example, a socket strip with an entered report form (different from the default plug diagram form so that it can later be manually placed on a schematic page) is assigned the wrong plug definition, then EPLAN will not generate a manually placed report for this socket strip.

 NOTE: Since this is an important point, I would recommend to always automatically generate plug definitions. EPLAN then automatically assigns the different functions to the correct plug definition because EPLAN determines the required information, such as the function definitions, from the project.

5.4.3 Plugs / Number pins

This function is identical to the Number terminals function - with one exception. In the **Number pins** dialog, the **Multiple terminals** option is not possible for plugs.

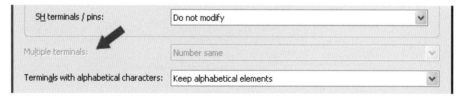

Fig. 5.97 Number plugs dialog

5.4.4 Plug navigator / New...

In the plug navigator popup menu (accessible via right-clicking or the ![] button) is the NEW... menu item.

The procedure here is identical to the NEW... function in the terminal navigator. A new function definition can be selected here as well. In contrast to terminals, this is only possible for plugs and sockets.

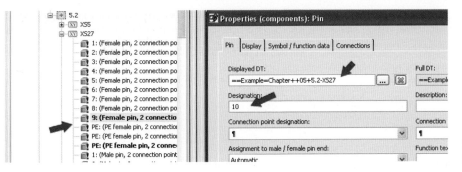

Fig. 5.98 Create new plug pin

5.4.5 Plug navigator / New functions...

New plug strips are created using the popup menu item in the plug navigator in exactly the same way as in the terminal strip navigator.

Whereas the terminal strip navigator creates new terminal strips, in the plug navigator you create new plugs, sockets, and plug / socket strips – depending on the function definition settings.

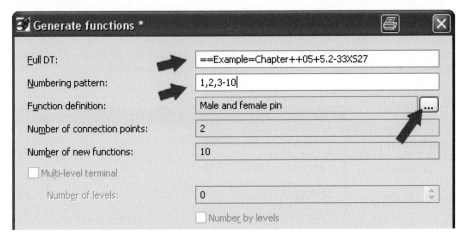

Fig. 5.99 Example: Generating new functions

In contrast to generating new terminal functions, here it is not possible to create multi-level terminals. The remaining procedures are the same as in the terminal strip navigator.

5.4.6 Plug navigator / Generate plug definition

As in the terminal strip navigator, you select new plugs in the plug navigator via the GENERATE PLUG DEFINITION item of the popup menu, and then select ONLY MALE PINS or ONLY FEMALE PINS in the MALE AND FEMALE PINS field.

If (e.g.) *Only male pins* is selected, EPLAN displays the **Properties (components) / Plug definition** dialog and enters the correct function definition. The remaining entries such as device tag or form can be entered and updated accordingly.

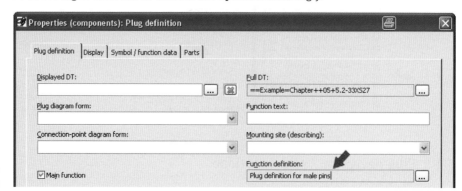

Fig. 5.100 New definition for male pins

5.4.7 Plug navigator / Generate pin

The GENERATE PIN function generates a new pin for the plug, socket, or plug strip selected in the plug navigator.

The GENERATE PIN function is selected from the popup menu in the plug navigator and (e.g.) the MALE PIN entry is selected. EPLAN then immediately opens the **Properties (components) / Pin** dialog. The device tag of the plug is automatically applied.

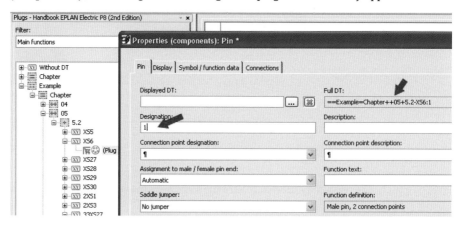

Fig. 5.101 Create individual male pin

Fig. 5.102 Unplaced male pin

You can use the ... button in the *Displayed DT* field to open the DT selection dialog. All pins currently belonging to the plug are listed here. The pin can be designated and is assigned to the plug as an unplaced pin when OK is clicked.

5.4.8 Plug navigator / Place / [Representation type]

In the same way as in the terminal strip navigator, to place unplaced plugs, sockets or combined plug/sockets in the schematic, you select the PLACE function in the popup menu and then select the representation type.

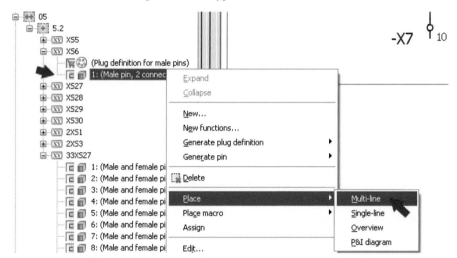

Fig. 5.103 Place an unplaced male pin

This function is again identical to how the terminal strip navigator works.

5.5 Cables

Cables and their conductors connect the individual devices. The cable navigator provides an overview of all used and unplaced cables in the project.

 NOTE: To edit cables with all the functions provided by the cable navigator, such as (e.g.) automatic cable selection, the logical connection data must always be complete and up to date!

In most cases, EPLAN automatically performs *Update connections* before such actions. But in the event that the connection data are not up to date, it is possible that subsequent actions will not work as desired or expected. This is why you should check the *status of connections* (up to date or not) in the status bar (symbol # or *), and update the connections yourself as required.

As I have mentioned before in this book, my personal recommendation is to assign the *Update connections* function to the F11 key. Then updating connections is only one keystroke away.

As with all the other navigators, the cable navigator also has a popup menu with additional functions for cable editing. As usual in EPLAN, the popup menu is called up via the right mouse button or the ▣ button.

5.5.1 Cables / Edit

The PROJECT DATA / CABLES menu has an EDIT menu item. This function allows manual editing of the assignments of the individual connections (which cable conductor – source / target) without requiring the individual conductors of the cable to be manually changed in the schematic.

After various copy and delete actions, the assignment of conductors / terminals for a cable no longer reflects the desired assignments. To conveniently change the assignment, the cable to be changed is first selected in the schematic (click the cable symbol) and then the EDIT function in the PROJECT DATA / CABLES menu is called.

EPLAN opens the **Edit cable** dialog. The left field lists the cable with its properties (stored *function templates* – conductor/wire). The cable's determined connections are shown in the right field.

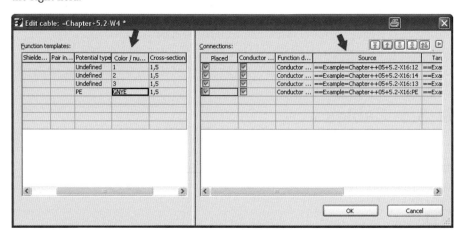

Fig. 5.104 Assignment of conductor to source and target DTs

The corresponding conductors can now be clicked in the *Connections* field and can then be moved using (e.g.) the ▣ ▣ buttons (move up or down) and assigned to the correct (desired) connections. You can also exchange connection places by using the ▣ button. To do this, you select both of the connections to be exchanged and push the button. EPLAN then exchanges the assignments in the table.

The other 'wrong' assignments can be corrected in the same way. After finishing the changes to the assignment, the **Edit cable** dialog can be exited. EPLAN then automatically updates the conductor – source / target assignments in the schematic, as can be seen in the following images.

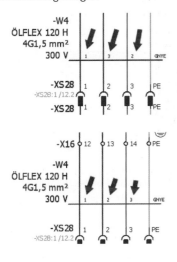

Fig. 5.105 Conductor assignment before editing

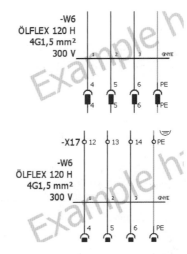

Fig. 5.106 Conductor assignment after leaving the Edit cable dialog

5.5.2 Cables / Number

Schematics can be composed of many different macros or other project pages. The *cable DTs* can be very different. The NUMBER function in the PROJECT DATA / CABLES menu allows these to be consistently numbered across the entire project.

The cables to be numbered are selected before numbering in the schematic or in the cable navigator.

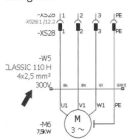 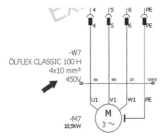 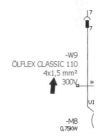

Fig. 5.107 Selected cables

The NUMBER function in the PROJECT DATA / CABLES menu is then called. EPLAN opens the **Number cables** dialog. The numbering options can be set in the *Settings* selection field.

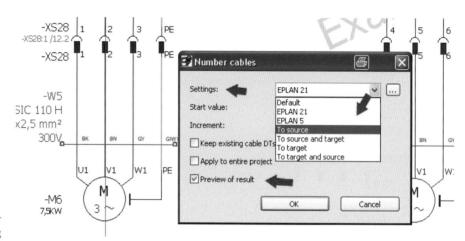

Fig. 5.108 Number cables dialog

You can use the ⊞ button to switch to the **Settings / Cable numbering** dialog and create your own scheme for cable numbering, or modify and save an existing scheme.

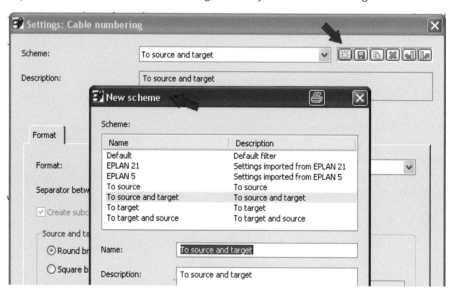

Fig. 5.109 Settings: Cable numbering dialog

In the example, the existing *Scheme: according to target* was selected in the **Edit cable** dialog for numbering the selected cables. Here too it is a good idea to activate the *preview of result* parameter. This allows the new cable DTs to be checked before they are written back to the schematic, and if necessary, the numbering can still be stopped via the CANCEL button.

After you click on OK, EPLAN renumbers the cable DTs according to the selected scheme and displays the result in the **Number cables: Preview of result** dialog.

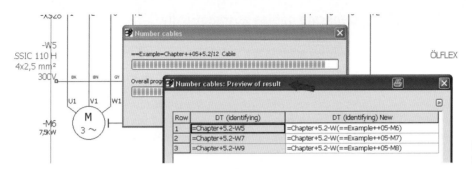

Fig. 5.110 Preview of result dialog

If the result is as desired, you can exit the **Number cables: Preview of result** dialog by clicking OK. EPLAN finishes the numbering process by writing back the new cable device tags.

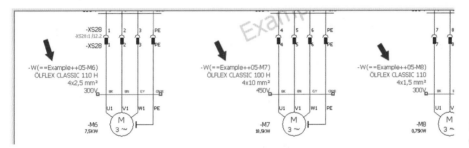

Fig. 5.111 Changed cable device tags

The default settings for the schemes for the different cable editing functions, such as (e.g.) the *Numbering function*, are defined in the project settings. The settings can be found under OPTIONS / SETTINGS / PROJECT [PROJECT NAME] / DEVICES / CABLE (AUTOMATIC).

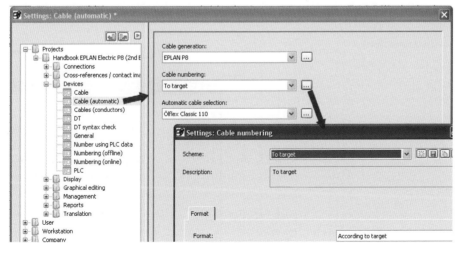

Fig. 5.112 Default settings for cable editing functions

5.5.3 Cables / Generate cables automatically

EPLAN can automatically generate cables and their associated functions such as cable type etc. and thus complete schematics for you.

Preconditions: The cable for which automatic generation is intended already has a cable definition line entered at the connection points in the schematic. The cable device tag is adopted exactly from the online numbering. The sequence of the cable device tags plays a subordinate role here. A subordinate role means that the format of the cable device tag can still be defined retroactively in the settings for automatic cable generation.

Before calling up the GENERATE CABLES AUTOMATICALLY menu item, the corresponding schematic pages or the individual cables in the schematic that are to be automatically generated must first be selected.

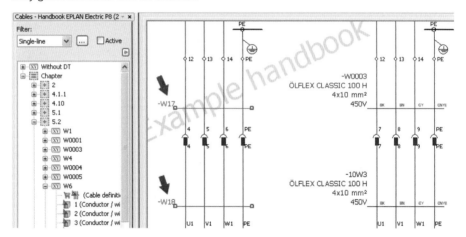

Fig. 5.113 Select existing cables

The GENERATE CABLES AUTOMATICALLY function in the PROJECT DATA / CABLES menu is then called up.

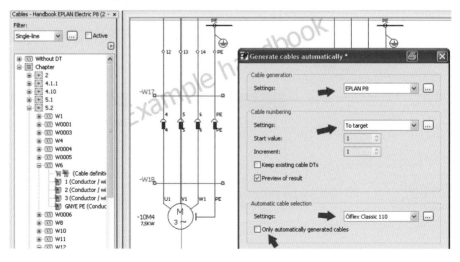

Fig. 5.114 Place cables automatically dialog, with its settings

EPLAN displays the dialog of the same name. It applies the project settings for automatic cable generation. All settings can still be changed for this particular automatic generation of cables. To do this, you must use the button to display a subsequent settings dialog. Here you can create new schemes or modify and save existing schemes.

It is also a good idea to activate the *Preview of result* setting when using this function. This allows you to cancel the operation before actually generating the cables.

> The *Automatic cable selection* and *Only automatically generated cables* settings must be deselected if they do not exist in the cable main function. Otherwise, EPLAN will search for the cable property *Automatically generated <20059>* and, if this isn't activated, will not automatically update the cable data, such as cable type, etc.!

Once all settings in the **Number cables automatically** dialog are correct, generation can be started by clicking OK. EPLAN then generates the cables and displays the future result in the **Preview of result** dialog.

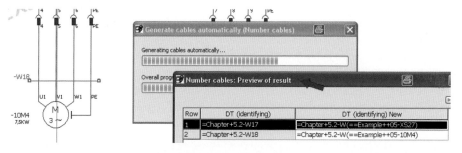

Fig. 5.115 Preview of result

As in other Preview of result dialogs, you can use the popup menu or the button to call the *Use original DT* function. To do this, you select the device tag and restore it back to the old DT. It is also possible to manually change the identifier directly in the *DT (identifying) New* field of the preview dialog.

Once all settings are correct, you can leave the **Preview of result** dialog by clicking OK. EPLAN writes the modified cable data back to the selected cable definition lines in the project.

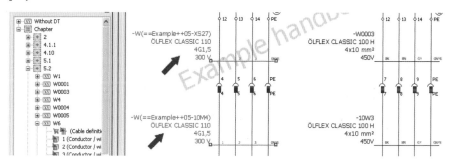

Fig. 5.116 Automatically generated cable data

5.5.4 Cables / Automatic cable selection

The AUTOMATIC CABLE SELECTION setting allows existing cables in the schematic to be updated with a specified cable type.

To automatically assign cables, the cables to be modified are again selected in the schematic and then called in the AUTOMATIC CABLE SELECTION function in the PROJECT DATA / CABLES menu. EPLAN opens the **Automatic cable selection** dialog.

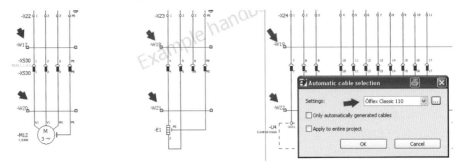

Fig. 5.117 Automatic cable selection dialog

The dialog offers three settings. The selection of which cable is to be used (you can use the existing scheme or use the ... button to define and then select a new scheme), the cable to which the automatic cable selection should refer, and whether the automatic cable selection is to apply to the entire project.

The *Scheme for cable types* (settings) can contain multiple cable types. EPLAN automatically determines the required number of conductors and applies this cable to the selected cable DT in the schematic.

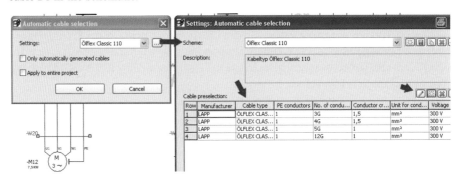

Fig. 5.118 Automatic cable (types) selection settings dialog

Once all settings in the **Automatic cable selection** dialog are correct, the automatic cable selection can be started by clicking OK. EPLAN compares the existing cable and connection data with the entered default cable types and assigns the new cable type(s) to the connections. After this, the cable data is also transferred to the cable definition lines and / or updated, depending on the settings.

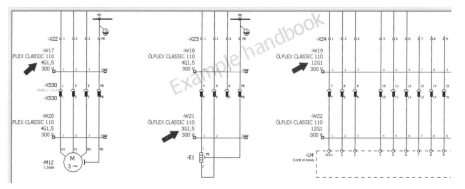

Fig. 5.119 Automatically generated cable data

5.5.5 Cables / Assign conductors

Fig. 5.120 Assign conductors menu item in the Cables menu

In the PROJECT DATA / CABLES / ASSIGN CONDUCTORS menu item, there are two more menu items, the importance and effects of which will be explained in the following sections.

5.5.5.1 Keep existing properties

The KEEP EXISTING PROPERTIES menu item enables the new assignment of conductors without affecting conductor assignments that already exist. Accordingly, they keep the cable and conductor assignments as previously assigned.

This would be necessary, for example, if cable conductors have already been generated for which assignments should not be changed later. Therefore, the new conductor assignments should use this menu item.

All conductors are normally designated during the *Device selection* of the cable. If components, like new cable connections, are to be added from the reserve conductors, the previous conductor assignments should not change. Nevertheless, it is preferable for you to be able to automatically assign the rest of the new conductors to the cable and its connections. This is done via the KEEP EXISTING PROPERTIES menu item.

The main function of the cable to be edited (to be updated) is selected.

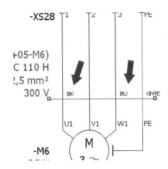
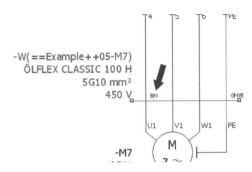

Fig. 5.121 Cables are selected by main function.

Now the KEEP EXISTING PROPERTIES function is called via the PROJECT DATA / CABLES / ASSIGN CONDUCTORS menu. Based on the function definitions for the conductor designations stored in the cable part, EPLAN writes to the previously undefined connection definition points (conductor designations), whereby existing conductor designations are not overwritten.

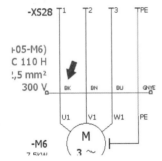
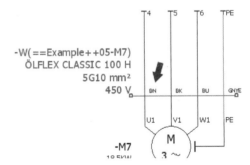

Fig. 5.122 Previously undesignated conductors are written.

5.5.5.2 Reassign all

The REASSIGN ALL menu item does the opposite of the function described in the previous section. In contrast to the previous function, this one (re)writes previously designated conductors with the (conductor) designations from the cable's function definition. This function completely replaces the old conductor designations.

The procedure is as follows. The cables are selected for which the conductor designations should be completely reassigned.

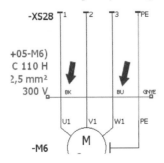
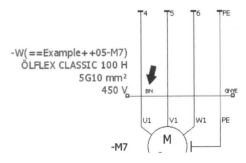

Fig. 5.123 Cables are selected.

Then you select the REASSIGN ALL menu item from the PROJECT DATA / CABLES / ASSIGN CONDUCTORS menu. EPLAN then reassigns the conductors for all of the selected cables without further questions.

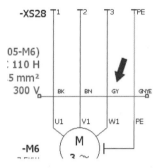 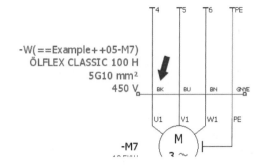

Fig. 5.124 All conductors were reassigned.

5.5.6 Cable navigator / New...

The **Cable navigator** offers similar functions as the other navigators. The NEW... function in the cable navigator popup menu creates a new *Cable definition* function definition.

EPLAN then opens the **Properties (components) / Cable** dialog. Here you make the usual entries, such as a device selection, and save the dialog. EPLAN then adds this cable definition to the cable navigator as an unplaced function.

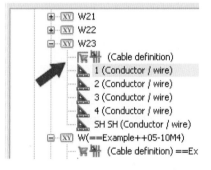

Fig. 5.125 New cable

5.5.7 Cable navigator / Place / [Representation type]

Devices, in this case cables, can of course also be inserted into schematics via the PLACE function in the cable navigator. The cable to be placed is first selected in the cable navigator. Now the PLACE function, here (e.g.) MULTI-LINE, is started via the popup menu.

EPLAN then 'hangs' the cable on the cursor and you can place it in the schematic. To do this, the starting point of the cable is defined by clicking the left mouse button at a position to the left of the first connection. The cable is then pulled to the right over the connections and then placed by clicking the left mouse button again to the right of the last connection.

After placement, EPLAN fills out the cable properties. The cable is now placed.

5.5.8 Cable navigator / Number cable DT

The function and procedure used by the NUMBER CABLE DT function in the **cable navigator** is identical to the function in the PROJECT DATA / NUMBER CABLES menu. Here too, the cables to be numbered are selected in the cable navigator and then the NUMBER CABLE DT function is called.

The familiar dialogs and settings are then displayed. This will not be explained further because they were described in the aforementioned section.

5.5.9 Cable navigator / Number DT

A cable in EPLAN is also a normal device, albeit with special properties. EPLAN not only allows *cable numbering* with (e.g.) source / target into cable device tags, but also allows normal *numbering formats* such as [Page Identifier Column].

The cables to be numbered are selected in the **cable navigator** or in the schematic, and the NUMBER DT function is started via the popup menu.

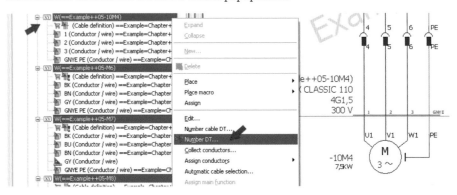

Fig. 5.126 Selected cables in the cable navigator

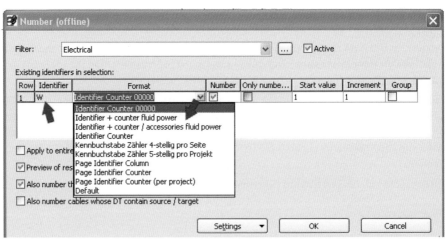

Fig. 5.127 Number (offline) dialog

EPLAN opens the familiar **Number (offline)** dialog. This is where the desired *numbering scheme* and the other possibly required settings are defined. After clicking on the OK button, EPLAN numbers the cables like normal devices.

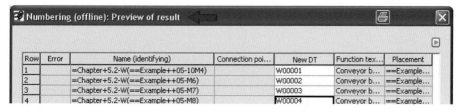

Fig. 5.128 Preview

Fig. 5.129 Final result, numbered cables

5.6 PLC

The **PLC navigator** allows the data of PLC cards and other PLC components used in the project to be managed conveniently from a central place. The PLC navigator shows the individual functions of the PLC cards, such as *addresses*, *power supply connections*, and *channels*.

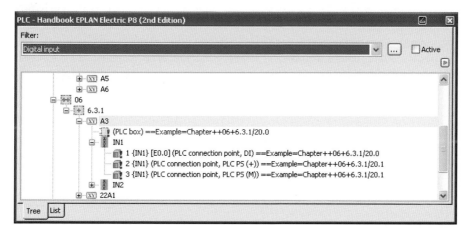

Fig. 5.130 The PLC navigator

In addition to the PLC navigator, the PROJECT DATA / PLC menu contains additional functions for editing PLC data, such as SET DATA TYPES or ADDRESS CARDS.

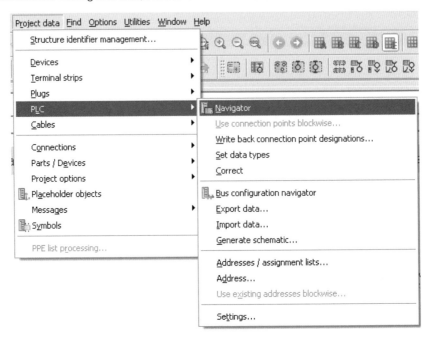

Fig. 5.131 Project data/ PLC menu

 NOTE: A very important fact is that EPLAN uses the **connection point designation** as the **identifying property** of a PLC function. This means that (e.g.) the designation E1.0 or A23.7 is not the identifying property, but rather the connection point designation (PIN number or terminal number) of the function of the PLC item.

The representation of PLC items can vary. One option is „dividing by three." This means that there is a *graphical PLC overview* (rack layout), a *logical PLC overview* (card with cross-references), and the *distributed representation* in the schematic (inputs and outputs used, power supply connections, etc.). This representation is not essential, but it makes it easier to deal with PLC data in a project.

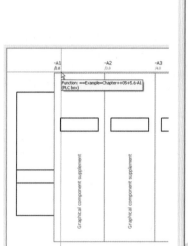
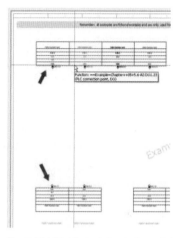

Fig. 5.132 Graphical overview (rack layout)

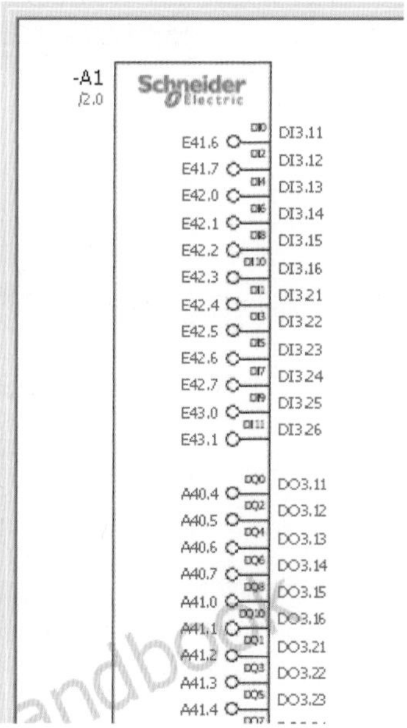

Fig. 5.133 Logical overview (references, etc.)

Fig. 5.134 Distributed view (schematic)

The various ways of editing are described in more detail on the following pages, whereby it is assumed that the PLC structure is the same as that described above.

This means that there is at least a finished logical PLC overview with correct entries (address assignments and connection point designations) and one or more inputs or outputs represented in a distributed manner in the schematic.

5.6.1 PLC / Write back connection point designations

The Write back connection point designations function in the **Project data PLC** menu writes the connection point designations from the PLC overview back to the PLC terminals that are represented as distributed in the schematic.

To use the function, PLC terminals are first drawn in the schematic or inserted from a macro. The PLC addresses are known and are already correctly entered into the schematic during project editing. The distributed representation of the PLC terminals is therefore finished:

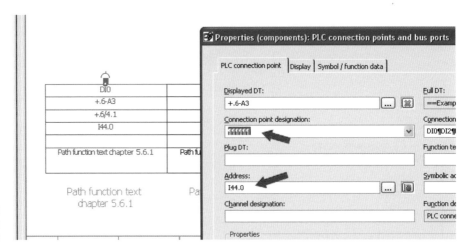

Fig. 5.135 Added distributed PLC data in schematic

When editing is complete, the view is changed to the logical overview page (PLC). Here the corresponding card(s) and address ranges are selected. Then you select the WRITE BACK CONNECTION POINT DESIGNATIONS item in the PROJECT DATA / PLC menu.

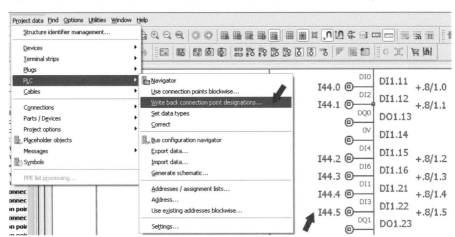

Fig. 5.136 Run the Write back connection point designations function

EPLAN opens the **Write back connection point designations** dialog. This dialog is simply confirmed with OK. The connection point designations from the logical PLC overview (of the selected PLC terminals) in the schematic are now written back to the PLC data that is in distributed view.

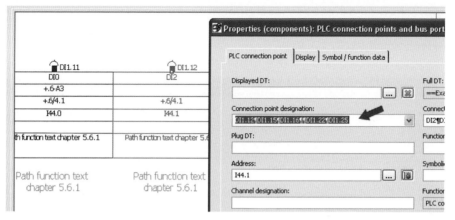

Fig. 5.137 Connection point designations written back

The result can be seen in the image. All connection point designations are correctly transferred to the PLC terminals. But this only works if the addresses have been entered correctly. Otherwise, EPLAN cannot recognize the relationship between the PLC terminal in the logical PLC overview and the PLC terminal in the distributed view.

5.6.2 PLC – Set data types

The SET DATA TYPES function is used to automatically update and write the corresponding data types to the PLC terminals if they are not already there.

EPLAN derives the data types from the functions. To do this, the PLC terminals are selected in the *logical PLC overview* or the *PLC terminals in the schematic* and the SET DATA TYPES function in the PROJECT DATA / PLC menu is called. EPLAN writes the data type into the properties of the PLC terminal.

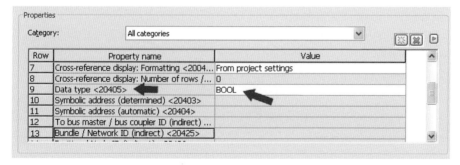

Fig. 5.138 Set data type

5.6.3 PLC / Addresses / assignment lists

In the PROJECT DATA / PLC menu, the ADDRESSES / ASSIGNMENT LISTS function opens the dialog of the same name (a table of all addresses used in the diagram).

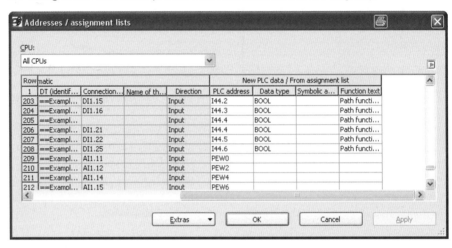

Fig. 5.139 Addresses / assignment lists dialog

In this dialog, the *PLC address*, *Data type*, *Symbolic address*, and the defined *Function text* properties can also be manually changed. EPLAN then automatically writes back the modified data to the PLC terminals and the modified function text to the manually placed path function text in the PLC terminal path.

At this point, if you remain at the function texts it is also possible to edit the function texts so that they can be written back to the schematic. Since this is done at a central place, it avoids tedious paging through the schematic in order to modify the placed path function text at the corresponding PLC terminals.

In addition to these manipulations, an *assignment list* can be exported from this dialog (for the downstream PLC software) or imported (from the PLC software).

To do this, click the EXTRAS button in the lower area of the dialog and EPLAN opens another menu.

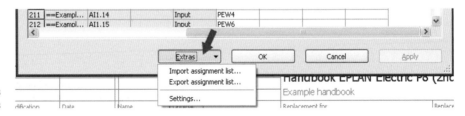

Fig. 5.140 The Extras button and its functions

This allows *assignment lists* to be *exported* or *imported*. Different settings may be required, depending on the PLC and its software.

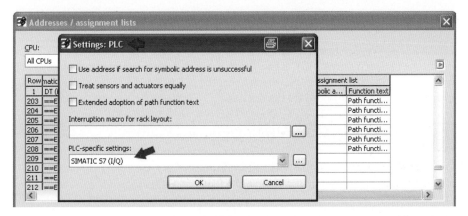

Fig. 5.141 PLC settings

EPLAN already contains predefined schemes for a range of PLC software for export and import, so that all you need to do is select the correct PLC software in the settings under *PLC-specific settings*.

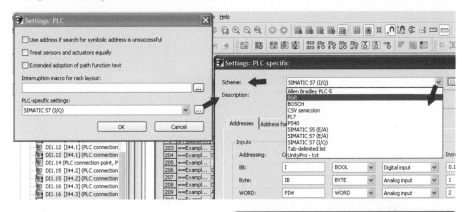

Fig. 5.142 Different PLC schemes

If all of the settings are properly selected, EPLAN can now generate the assignment list. To do this, the EXTRAS button in the EXPORT ASSIGNMENT LISTS entry in the **Addresses / assignment lists** dialog is selected, and the settings in **Export assignment lists** are checked once more and adjusted as necessary in the subsequent dialog. Select the correct option depending upon the available CPUs (there could be multiple controls in a project).

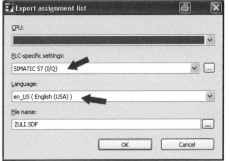

Fig. 5.143 Export assignment lists dialog

This dialog is then simply confirmed with OK. EPLAN generates the assignment list in the desired PLC software format and saves it in the specified directory.

5.6.4 PLC / Address

EPLAN allows simple addressing or readdressing of PLC terminals via the ADDRESS function in the PROJECT DATA / PLC menu. When doing this, the functions in the logical PLC overview and the functions in the schematic for the PLC terminals in distributed view are re-addressed or are addressed for the first time.

NOTE: In order for all PLC functions to be addressed, it is necessary to select the PLC card in the PLC navigator beforehand. This ensures that all of the functions, those of the logical PLC overview and the PLC functions represented in distributed view, are included and addressed.

The card to be addressed is selected in the PLC navigator. Then the ADDRESS function in the PROJECT DATA / PLC menu is called.

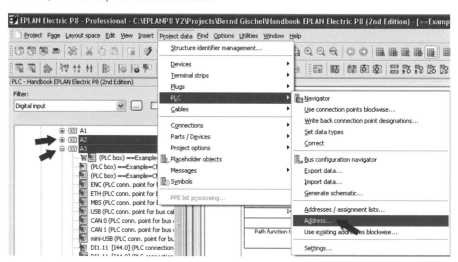

Fig. 5.144 Select the card to be addressed and call the addressing

EPLAN then opens the **Readdress PLC connection points** dialog.

Fig. 5.145 Define start addresses

In the fields Digital start address and Analog start address, the new address is entered *without* preceding signs. EPLAN obtains the preceding sign, meaning whether an entry begins with E or with I, according to the selected scheme, which is entered in the *PLC-specific settings* selection field.

It is possible, however, that you can enter a preceding sign in the start address entry field, e.g. an 'IX'. EPLAN would then use this during addressing.

I recommend leaving the *Preview of result* parameter switched on. After clicking OK the **Address PLC connection points / Preview of result** dialog is displayed so that you can check the results and use the CANCEL button to stop the process if necessary.

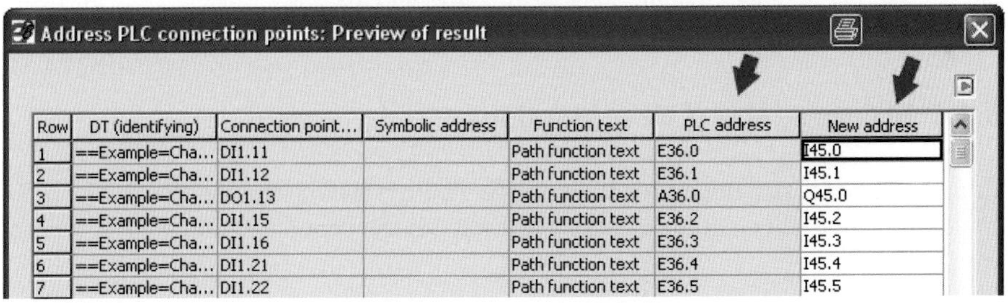

Row	DT (identifying)	Connection point...	Symbolic address	Function text	PLC address	New address
1	==Example=Cha...	DI1.11		Path function text	E36.0	I45.0
2	==Example=Cha...	DI1.12		Path function text	E36.1	I45.1
3	==Example=Cha...	DO1.13		Path function text	A36.0	Q45.0
4	==Example=Cha...	DI1.15		Path function text	E36.2	I45.2
5	==Example=Cha...	DI1.16		Path function text	E36.3	I45.3
6	==Example=Cha...	DI1.21		Path function text	E36.4	I45.4
7	==Example=Cha...	DI1.22		Path function text	E36.5	I45.5

Fig. 5.146 Check the new addressing

If the result is correct, then the **Address connection points / Preview of result** dialog is confirmed with the OK button and EPLAN readdresses the PLC functions of the selected cards in the entire project.

5.6.5 PLC navigator / New...

As in the other navigators, new devices with a function definition can also be created in the PLC navigator. The NEW... item in the popup menu is used for this.

EPLAN opens the **Function definitions** dialog. In the PLC navigator you can only select PLC function definitions. After applying the function definition by clicking the OK button, EPLAN opens the **Properties (components) PLC box** dialog. Here you must make the desired entries and save the properties. You can also immediately perform a *device selection* for the new PLC card.

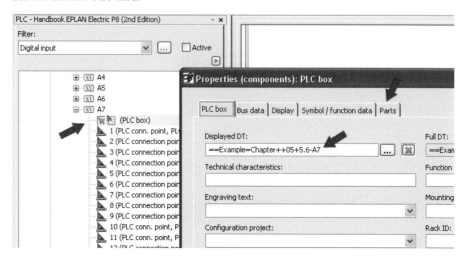

Fig. 5.147 New unplaced PLC card in the navigator

The new device was stored in the PLC navigator, with a function definition of PLC box and the selected part's function definitions, as an unplaced PLC box.

5.6.6 PLC navigator / New functions...

To add additional functions to devices in the PLC navigator, you use the NEW FUNCTIONS... function. This is started in the usual way via the popup menu in the PLC navigator.

EPLAN opens the **Generate functions** dialog. In the *Function definition* field, a PLC connection point, DI (for example) can be selected via the [...] button and the subsequent **Function definitions** dialog. For a card with 20 connection points, the numbering pattern is set to 1-20.

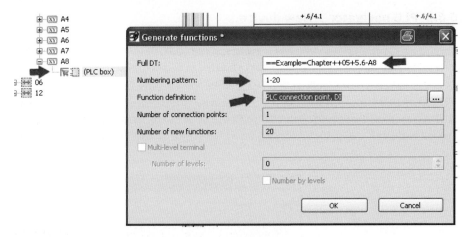

Fig. 5.148 Generating new PLC functions

These functions are generated when OK is clicked. This is also the approach for generating supply connection points for a card. This method creates a complete PLC card in the PLC navigator, with correct *PLC addresses* and their *connection point designations* ready to be placed.

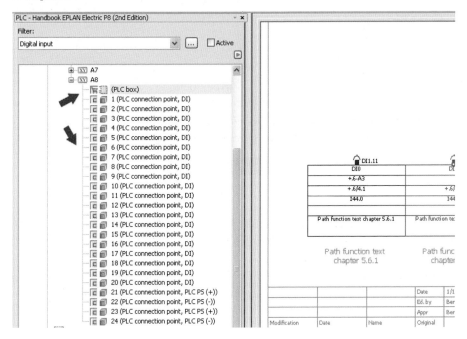

Fig. 5.149 New PLC functions

5.6.7 PLC navigator / New device...

In contrast to creating individual new functions, the NEW DEVICE... item in the popup menu creates a completely new device that already has default function definitions; in the PLC navigator this is, of course, a PLC function, such as an input card.

After NEW DEVICE... is selected in the popup menu of the PLC navigator, EPLAN opens the Parts management dialog. Only PLC parts can be selected. The selected device is now applied in the PLC navigator by clicking OK.

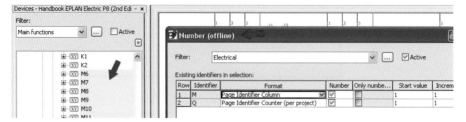

Fig. 5.150 New unplaced device in the navigator

EPLAN automatically assigns the next available device tag to the new device in the PLC navigator. The device is then ready to be used in the schematic or ready for its functions to be placed.

5.6.8 PLC navigator / View

The **PLC navigator** in EPLAN provides different *views* of the PLC data: the *DT-oriented* view, the *address-oriented* view, and the *channel-oriented* view.

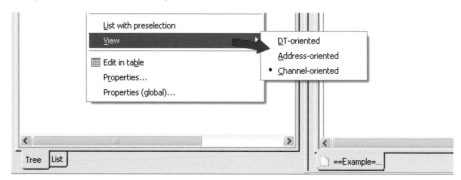

Fig. 5.151 View selection in popup menu

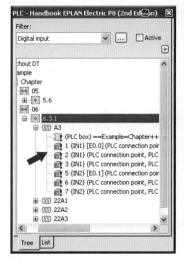
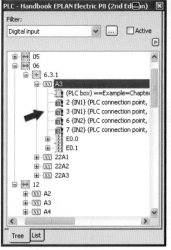
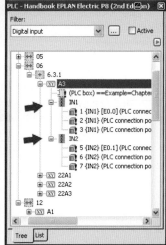

Fig. 5.152 DT-oriented view   Fig. 5.153 Address-oriented view   Fig. 5.154 Channel-oriented view

In the DT-oriented view, the PLC data is listed according to its device tag. In the address-oriented view, the PLC navigator organizes the PLC data by address. And lastly, the channel-oriented view shows the channel groups and all of their PLC functions.

The channel-oriented view has the advantage of enabling you to select and place all of the functions belonging to a channel at once. This ensures that you do not forget any of the functions in the channel group.

■ 5.7 Parts / Devices / Bill of materials

The **Bill of materials navigator** has a view of all devices with or without the part numbers that exist in a project. The bill of materials navigator is started via the PROJECT DATA / PARTS / DEVICES menu.

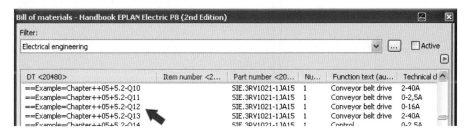

Fig. 5.155 The bill of materials navigator in list view

Just like the other navigators, the bill of materials navigator offers fundamentally different viewpoints of the project and its data.

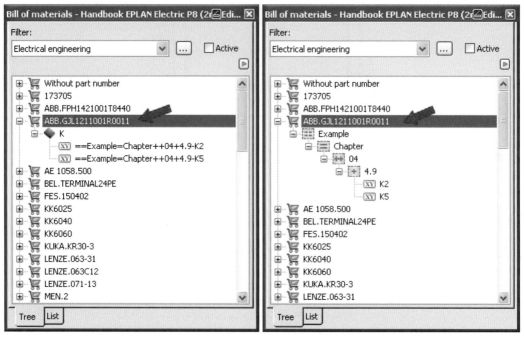

Fig. 5.156 Part number view (leading)

Fig. 5.157 Part number / Structure identifier view

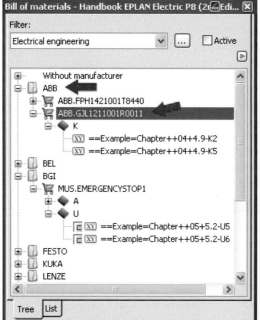

Fig. 5.158 View by manufacturer

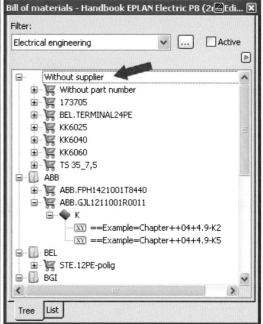

Fig. 5.159 View by supplier

The bill of materials navigator therefore offers a broad range of options for editing parts data throughout the project.

5.7.1 Popup menu functions

The bill of materials navigator offers different functions in its popup menu to make editing the project parts easier.

5.7.1.1 Add project part

If you select the ADD PROJECT PART menu item, EPLAN then opens *parts management*, and a part can be added to the project as a *project part*.

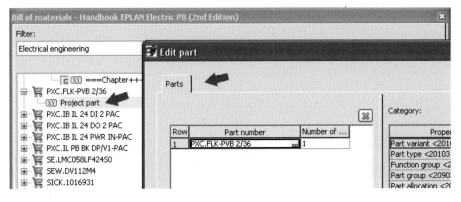

Fig. 5.160 Added project part

Project parts initially do not have device tags. These are normally parts that are shipped as spare or loose parts. Project parts can also be filtered or evaluated.

5.7.1.2 Add part

You can easily add additional parts to an existing device with the ADD PART menu item. To do this, select the device and call the ADD PART menu item.

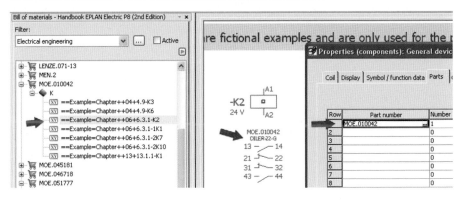

Fig. 5.161 Before running the Add part function

EPLAN starts *parts management* after running the ADD PART function. The desired part is chosen, selected and applied to the selected device by clicking OK.

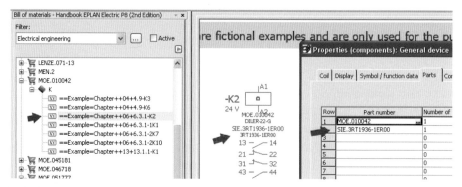

Fig. 5.162 A part was added.

This function of course only works with multiple selections. For example, you could add surge arresters for several contactors at once if you forgot them.

5.7.1.3 Place via...

The PLACE VIA... menu item contains additional menu items.

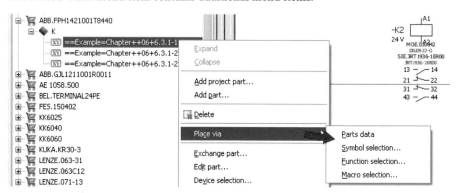

Fig. 5.163 Placement options

The menu item PLACING WITH PARTS DATA means that EPLAN places the *symbols* and *macros* stored with the part after placement on the schematic.

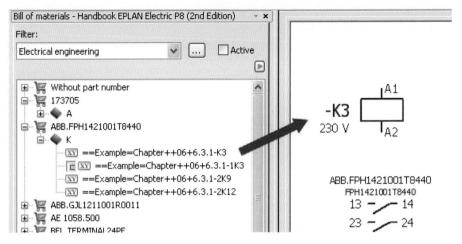

Fig. 5.164 Place via parts data

The SYMBOL DATA menu item PLACE allows EPLAN to open symbol selection before placement, and the corresponding or desired symbol must be selected. Then you can place this symbol on the schematic page and EPLAN will enter the parts data for this symbol.

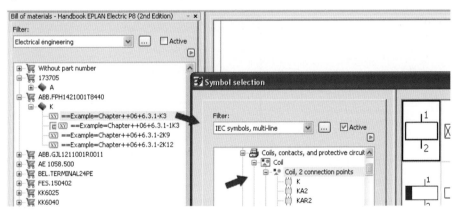

Fig. 5.165 Place via symbol data

The FUNCTION SELECTION menu item starts before placement. In the **Function definitions** dialog, the desired function definition is selected and applied before placing the part. Then the part can be placed. EPLAN then assigns the previously selected *function definition* to the part.

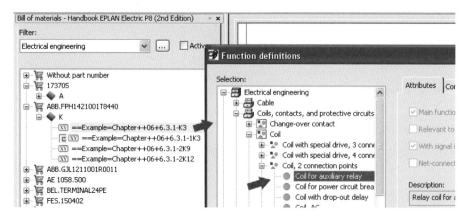

Fig. 5.166 Place via function data

The last menu item allows part placement via a MACRO SELECTION beforehand. To do this, EPLAN opens the **Select macro** dialog upon selecting the menu item. You can select the macro and then click the OPEN button to apply and place it.

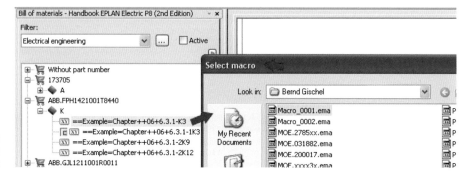

Fig. 5.167 Macro selection before placement

5.7.1.4 Exchange part

This menu item offers the option to exchange a part quickly and easily.

To do this, you select the part in the **bill of materials navigator** and select the EXCHANGE PART menu item. EPLAN then opens **parts management**, where the *part to be exchanged* is selected and applied with OK. EPLAN instantly exchanges this part with the selected one.

5.7.1.5 Device selection

This menu point is an interesting enhancement. For example, you can use this to perform *contactor selection* retroactively for multiple selected parts data (contactor coils). To do this, select the part and run the DEVICE SELECTION menu item.

EPLAN now allows you to perform device selection for every device, one by one. If for example device selection has not been performed for a device, then it suffices to press the ESC key. EPLAN then asks whether the entire action should be cancelled, or whether the selected device should be skipped.

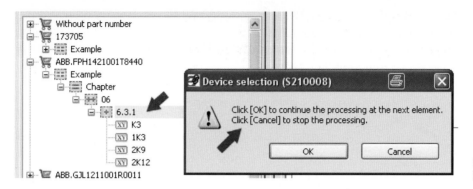

Fig. 5.168 Cancel option in device selection

5.7.1.6 Synchronize parts data

The **Synchronize parts data** menu item offers the option of performing master data synchronization for specific, selected parts with the parts management. This operation compares the parts data with the parts management and overwrites the saved parts data in the project if necessary.

■ 5.8 Assign function (navigators)

A number of navigators have the ASSIGN function in their popup menu.

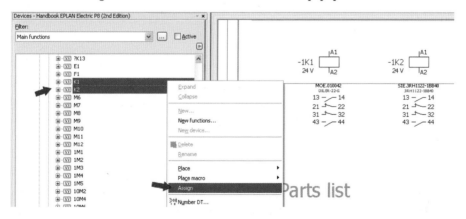

Fig. 5.169 The Assign menu item in the device navigator

The ASSIGN menu item allows the *properties of objects* selected in the navigator, e.g. a *device tag* or parts of a device such as a PLC connection point, to be assigned the properties of an existing object (e.g. a power contact).

It is also possible to select several objects in the navigator, execute the ASSIGN function, and then assign these properties to other objects by clicking them in the schematic.

A few devices are selected in the device navigator and then the ASSIGN function is executed. EPLAN now visibly hangs all of the properties for this device, such as for example the *device tag* or the *function definition*, on the cursor (the remaining unassigned device tags remain visible in the foreground).

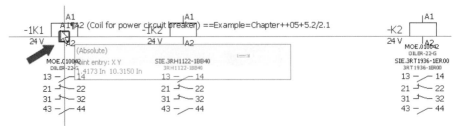

Fig. 5.170 The assigned properties hang on the cursor.

The individual power contacts in the schematic are then clicked one after another (always exactly those that are to receive a *device tag*).

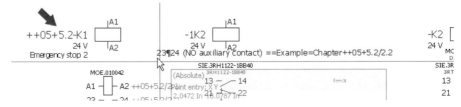

Fig. 5.171 Assigned device tags

EPLAN then transfers one DT after another to the power contacts without requiring this to be manually entered at each contact.

This is a very useful function that saves time and ensures error-free information transmission, and it is available in the following navigators (in addition to the device navigator): terminal strip navigator, plug navigator, PLC navigator, cable navigator, connection navigator and interruption point navigator.

6 Reports

This chapter deals with reports: Which ones are necessary, what settings are needed for which reports, how do I make the most effective use of templates and reports, and what are the advantages for the user? The versatile possibilities offered by external editing are also touched upon using simple, short examples.

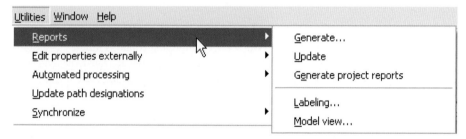

Fig. 6.1 Reports menu

The UTILITIES / REPORTS menu contains the functions for generating graphical reports such as terminal diagrams, parts lists or connection lists. The UTILITIES / REPORTS / LABELING menu item allows the user to export various reports such as cable overviews or summarized parts lists in (e.g.) Excel format.

EPLAN thus offers the user the two options for generating internal outputs (graphical outputs for project documentation) and for transferring the evaluated data into an external format (Excel, text format, etc.), for passing on the generated data from a schematic, for example.

Generation of graphical reports is only one option offered by EPLAN. Previously, you could externally edit the generated data via the *Labeling* function, but there was no way of synchronizing this with the EPLAN project, so an additional functional scope was added to EPLAN allowing properties to be exported, externally edited, and then re-imported back into EPLAN.

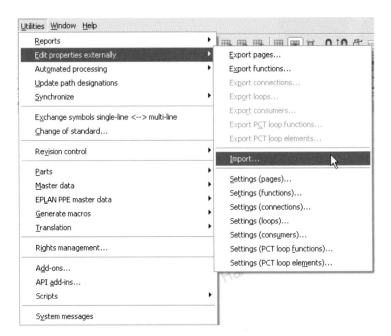

Fig. 6.2 The Edit properties externally menu

The UTILITIES / EDIT PROPERTIES EXTERNALLY menu item contains numerous options allowing you to more conveniently edit a wide range of properties in (e.g.) Excel and then import the externally edited data to the EPLAN project. This automatically changes the data in the project without you having to perform these changes in EPLAN.

6.1 What are reports?

Reports are selected project data that can output (e.g.) *a terminal diagram* in graphical form on newly generated project pages (internal report) or that can write the data into a *text file* (external report).

In addition, EPLAN also offers a special type of report. You can (e.g.) directly place a motor plug as a report onto the page containing the motor. These types of reports are *manually placed*, and versatile form design provides the user with a rich variety of custom uses.

Since EPLAN keeps all the data online and automatically updates the connection data before generating a report, the user does not need to update anything manually.

■ 6.2 Report types

As mentioned previously, EPLAN distinguishes between normal report pages, embedded reports, and frozen reports. Normal report pages are automatically newly created in the project depending on their settings, whereas embedded reports are manually placed directly on the page where the information (e.g. pin assignments, etc.) is to be displayed.

Frozen reports are automatically generated once and then „frozen" so that these reports can no longer be changed, neither automatically nor manually via an update. Frozen reports (the report pages) must be deleted and generated again if they need to be updated.

■ 6.3 Types of graphical reports

All graphical reports are based on a particular report type. Every form has a specific area of application of form properties. However, not all form properties can be used in every type of form. A number of specific form properties can, however, be used in a range of different forms.

This chapter lists all possible report types with a brief description for each one.

6.3.1 Report types (forms)

EPLAN manages different **report types** for the reports. Broadly speaking, the report types are distinguished by their file extension *.fnn, whereby nn represents a number.

- F01   Parts list
- F03   Device tag list
- F13   Terminal diagram
- F22   Plug diagram

 NOTE: Forms should **always** be edited using the internal EPLAN form editor because, when editing and saving forms, EPLAN also saves internal information in the form to allow later master data synchronization to be correctly performed.

6.3.1.1 Parts list *.f01

A parts list is the output of all parts without the creation of totals. The form corresponds to a normal single parts list.

Parts list

designation	part number	Quantity	Device Designation
Mini contactor	MOE.010042	1	==Example=Chapter++06+6.3.1
Auxiliary contactor	SIE.3RH1122-1BB40	1	==Example=Chapter++06+6.3.1
Auxiliary contactor	ABB.FPH1421001T9440	0	==Example=Chapter++06+6.3.1

Fig. 6.3 Sample parts list (file extension *.f01)

One special feature of the form should be noted. Parts with a quantity of "0" are also initially output by default. Unplaced parts are also listed in the standard forms. If this data is **not** to be output, then you can use a filter to limit the contents of the output (see also summarized parts list).

6.3.1.2 Summarized parts list *.f02

The summarized parts list is the output of all parts, but unlike the parts list, all of the same parts are output here as a total. This corresponds to a part quantity list.

Summarized parts list

supplier	Order number	Quantity	designation	Type number
MOE				
	010042	2	Mini contactor	DILER-22-G
SIEMEN				
	3RH1122-1BB40	0	Auxiliary contactor	3RH1122-1BB40
ABB				
	FPH1421001T9440	2	Auxiliary contactor	FPH1421001T9440

Fig. 6.4 Sample summarized parts list (file extension *.f02)

In this form, parts with a quantity of "0" are also output by default. In the standard forms supplied with EPLAN, unplaced parts are also listed by default. If this data is also not to be output in the summarized parts list, then you again have the option of using a filter to limit the data that is included in the output.

6.3.1.3 Device tag list *.f03

The device tag list allows you to output all the devices used in a project, along with their information.

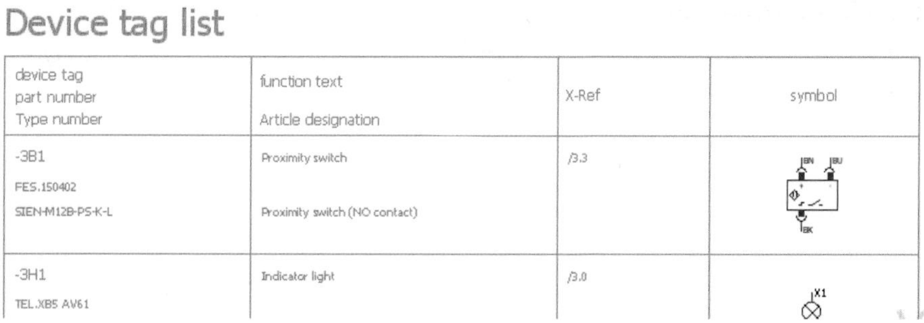

Device tag list

device tag part number Type number	function text Article designation	X-Ref	symbol
-3B1 FES.150402 SIEN-M12B-PS-K-L	Proximity switch Proximity switch (NO contact)	/3.3	
-3H1 TEL.XBS AV61	Indicator light	/3.0	

Fig. 6.5 Sample device tag list (file extension *.f03)

Whether or not a graphical representation of the corresponding component is to be displayed depends on the form used and / or the <13050 Component graphic> property. By default, unplaced devices are also output in forms, or you use the corresponding filter functions to output only the project data that is required or desired.

6.3.1.4 Forms documentation *.f04

Forms documentation provides an overview of which forms are stored and used in a project.

Fig. 6.6 Sample forms documentation (file extension *.f04)

This form has no additional filter possibilities, i.e. all forms stored in the project are output.

6.3.1.5 Device connection diagram *.f05

A device connection diagram allows the representation of a device with all its device connections and the internal / external devices connected to these.

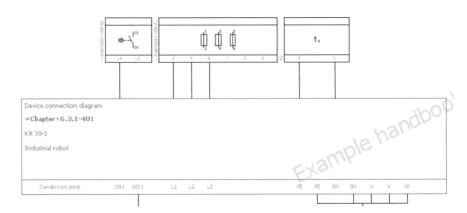

Fig. 6.7 Sample device connection diagram (file extension *.f05)

With this form and others, EPLAN offers the option of displaying the connected components over several connection levels on the internal and external pages.

 NOTE: In this form, internal connection points are not device-internal circuit diagrams etc.!

6.3.1.6 Table of contents *.f06

The table of contents is one of the typical uses for forms in EPLAN. The table of contents, also called a drawing directory, generates a list of the complete project documentation (the project pages).

Fig. 6.8 Sample table of contents (file extension *.f06)

Page	Page description
==ALL=DOC+DBL/1	Titel- / Deckblatt
==ALL=DOC+IHV/1	Table of contents
==ALL=DOC+IHV/1.a	Table of contents
==ALL=Example+Page/1	Parts list (Report)

As with other forms, the table of contents has a number of functions, such as filters, sortings, etc., available for controlling the graphical output to suit your needs.

6.3.1.7 Cable connection diagram *.f07

The cable connection diagram generates a form for a single cable (device) with all internal and external targets that are connected to the corresponding cable conductors.

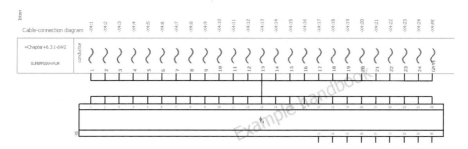

Fig. 6.9 Sample cable connection diagram (file extension *.f07)

The cable connection diagram belongs to the series of forms that can display the targets (external/internal) over several levels. Once again, predefined or user-defined filters can be used to control the output of the cable connection diagram and thus display only the required project cables.

6.3.1.8 Cable assignment diagram *.f08

The cable assignment diagram is the only form that cannot be generally entered for all cables in the *Output to pages* setting.

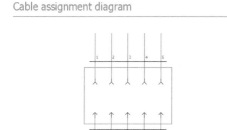

Fig. 6.10 Sample cable assignment diagram (file extension *.f08)

This is a special form that can be used to (e.g.) display a detailed representation of the internal structure of a cable together with a listing of the corresponding cables in the project (including various other information such as the cable device tag). Everything is displayed together on a single report page.

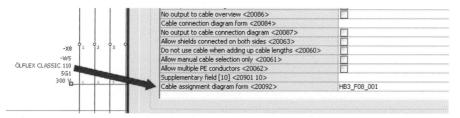

Fig. 6.11 Cable assignment diagram form entry on a cable

 Since different cable types are usually used in a project, it makes no sense to make a global entry for a cable assignment diagram or a global selection for all cables in the settings.

The cable assignment form can be selected directly in the cable properties (the property value would be <20092 Cable assignment diagram form>) from the selection list or saved with the part in parts management (Cable data tab).

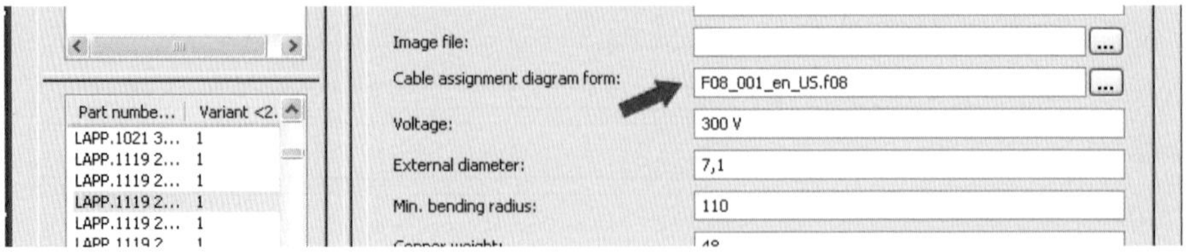

Fig. 6.12 Cable assignment diagram form entry in parts management

6.3.1.9 Cable diagram *.f09

The cable diagram report is the representation of a single cable with all its conductors and their information.

Cable diagram

Cable name	=Chapter+6.3.1-9W11		cable type		ÖLFLEX CLASSIC 110
function text	supply 24V DC		no. of conductors		12G
function text	Page / column	Target designation from	Connection point	conductor	Target designation to
supply 24V DC	=-Example++06/9.0	=-Example++06-X12	1	1	=-Example++06-X13
signal	=-Example++06/9.1	=-Example++06-X12	2	2	=-Example++06-X13

Fig. 6.13 Sample cable diagram (file extension *.f09)

Depending on the structure of the form used, the (e.g.) internal and external targets of the connected devices can also be displayed. You can control the output of the cable diagram, among others, via the filter settings. Without a filter setting, (e.g.) all cables used in the project are output.

6.3.1.10 Cable overview *.f10

By default, a cable overview contains a listing of all cables used in the project with all the desired information such as cable type, number of conductors, etc.

Cable overview

cable type	Cable description	device tag	Condctrs	cross-se
ÖLFLEX CLASSIC 100 H	Cable halogen free, 5G2,5	-W1	5G	
ÖLFLEX CLASSIC 110	Control line, 5G1.0	-W2	5G	
	Control line, 5G1.0	-W3	5G	
	Control line, 5G1.0	-W4	5G	

Fig. 6.14 Sample cable overview (file extension *.f10)

You can again control the output by setting filters or use sorting to (e.g.) display the cables by higher-level function or mounting location.

6.3.1.11 Terminal connection diagram *.f11

The terminal connection diagram belongs to the category of reports that can display several levels of connected internal and external targets at the corresponding terminal strip.

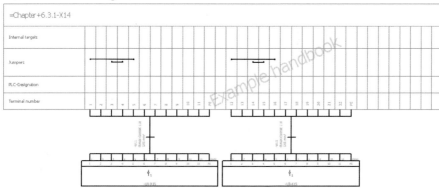

Fig. 6.15 Sample terminal connection diagram (file extension *.f11)

The number of levels to be displayed can be defined by the form structure and / or in the form properties. Once again there are extensive filtering and sorting functions in this form.

6.3.1.12 Terminal line-up diagram *.f12

The terminal line-up diagram generates a graphical report for each terminal strip with the associated terminal (strip) parts.

Fig. 6.16 Sample terminal line-up diagram (file extension *.f12)

The terminal line-up diagram can be graphically structured in a similar way to (e.g.) a terminal diagram, or you can use graphics to represent the parts and insert these visually into the form as properties.

You can also use the filter and sorting functions to selectively control the output in this report.

6.3.1.13 Terminal diagram *.f13

A terminal diagram outputs a single terminal strip (e.g.) with connected internal and external targets, including the cables used.

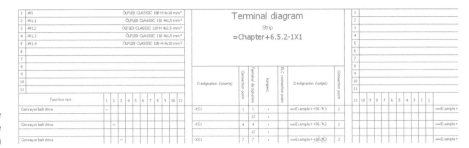

Fig. 6.17 Sample terminal diagram (file extension *.f13)

Terminal diagrams can be controlled via the numerous filter and sorting options. The structuring of the terminal diagrams is also very versatile.

6.3.1.14 Terminal strip overview *.f14

You can use the terminal strip overview form to obtain an overview of all terminal strips used in the project.

Terminal-strip overview

terminal strip	function text	Terminals			
		first	last	Total PE	Total N
		12	PE	4	0
-U1-X15	terminal 400V AC	1	PE	1	0
-U2-X15		1	PE	1	0
-U3-X15		1	PE	1	0
-X2	Power 690V AC	1	5	0	0

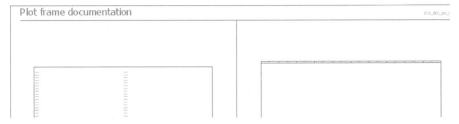

Fig. 6.18 Sample terminal strip overview (file extension *.f14)

In order to control the output of the desired terminal strips while editing the project, you can either enter the *<20857 No output to terminal strip / plug overview>* property into the terminal strip definition or limit the output later by using a filter.

As with many other forms, actions such as sorting using predefined or user-defined filters are also possible here.

6.3.1.15 Plot frame documentation *.f15

The plot frame documentation form is used to obtain an overview of all plot frames stored in the project.

Fig. 6.19 Sample plot frame documentation (file extension *.f15)

There are no filtering or sorting options available for plot frame documentation forms. This always produces a total output.

6.3.1.16 Potential overview *.f16

The potential overview report lists all potentials and signals used in the project, as well as further information, depending on the settings and structure of the form.

Potential overview HB3_F16_001

Name of potential	Potential value	Frequency	Potential type	Placement
0V DC			-	/17.3
24V DC	24V DC		+	/17.2
L1	400	50		++08+8.6.1/2.6
L1				++07+7.3.4/1.1

Fig. 6.20 Sample potential overview (file extension *.f16)

You can use the filter and sorting settings to control the output of the graphical report.

6.3.1.17 Revision overview *.f17

You use the revision overview form to obtain a graphical overview of the existing revisions and their corresponding entries such as the revision creator or the revision index.

0	1	2	3	4	5	6

Revision overview

Project name Revision control (change tracking)
Date 1/8/2011
revision Revision 3

Revision name Index	Revision comment (page) Revision description (page)	Page name Page description
Revision 3	Index 5 / - End of project	=A1+O1/3 Table of contents : =A1+O1/1 - =REPORT+REVUEB/1
Revision 3	Index 5 / - End of project	=A1+O1/3 Table of contents : =A1+O1/1 - =REPORT+REVUEB/1
Revision 3	Index 5 / - End of project	=A2+O2/1
Revision 3	Index 5 / - End of project	=A2+O2/2

Fig. 6.21 Sample revision overview (file extension *.f17)

The form can evaluate the respective page properties, e.g. the property *<11073 1 Reason for revision change>* or also the associated project properties such as the property *<10155 1 Revision name>*.

There are no additional filter or sorting possibilities for the revision overview, apart from the normal settings for graphical output.

6.3.1.18 Enclosure legend *.f18

A legend of the placed items (parts) (e.g.) on the mounting panel can also be included in a panel layout. The enclosure legend form in EPLAN allows this task to be automatically completed using the project data.

Enclosure legend

==Example=Chapter++06+6.3.1-M1		
item number	device tag	Type number
1	3H1	XB5 AV61
2	1K1	DILER-22-G
3	3Q1	3RV10 21-1JA15
4	3B1	SIEN-M12B-PS-K-L

Fig. 6.22 Sample enclosure legend (file extension *.f18)

An enclosure legend can be graphically generated as a type of window legend (embedded as a manual placement in a page) or as completely new report pages. Other filter and sorting options are available for graphical output using this form.

6.3.1.19 PLC diagram *.f19

In addition to the familiar (manual) logical PLC (card) overview, EPLAN can also automatically generate these in similar graphical forms, including additional information, using the existing PLC project data.

Fig. 6.23 Sample PLC diagram (file extension *.f19)

Data from the PLC modules, and other data such as rack or module number, path function texts, and automatically determined symbolic addresses are all taken into account.

You can again use numerous familiar filter and sorting options with PLC diagram forms.

6.3.1.20 PLC card overview *.f20

The PLC card overview provides a clear overview list of all PLC components used, such as power supply, CPU, or input cards.

PLC card overview

Device tag PLC	Workstation name	Is master	Rack
	Workstation type	Slave is appended to master	module
-22A1	Station 300 S7300	DP Master	2
-22A2	Station 300		04
-22A3	Station 300		0

Fig. 6.24 Sample PLC card overview (file extension *.f20)

You can, of course, display each PLC component on its own graphical report page of type PLC card overview. This depends on the structure of the PLC card overview form. Once more, you can also use filters and sorting to affect the graphical output in PLC card overview forms.

6.3.1.21 Pin connection diagram *.f21

Pin connection diagrams belong to the category of graphical reports that can manage several levels of internal and external targets and can output these graphically (depending on the form and settings used).

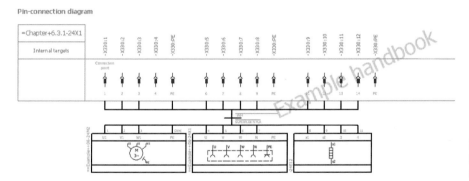

Fig. 6.25 Sample pin connection diagram (file extension *.f21)

A graphical output usually also includes a pin connection diagram. These can also be set to dynamic in the settings so that several pin connection diagrams can also be displayed on a single page (combined report). You can adjust the graphical output via filters or sorting settings.

6.3.1.22 Plug diagram *.f22

The plug diagram is a listing of all the internal and external targets connected to a plug with the maximum evaluation of one level. This is the normal evaluation of the standard forms.

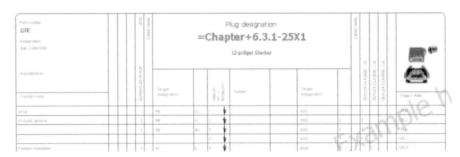

Fig. 6.26 Sample plug diagram (file extension *.f22)

The plug diagram can also be set to dynamic in the form properties so that several plugs are graphically output combined (consecutively) on a single page.

You can also use the familiar filter and sorting options to (e.g.) limit the output for the plug diagram.

6.3.1.23 Plug overview *.f23

The plug overview provides a listing of all plugs and / or sockets used in the project, depending on the structure of the plug overview form.

Plug overview

Plug designation	function text	pin			Page of plug connection diagram
		Normal	N	PE	
+2-X51		3	0	1	
-24X1	Hybrid cable	12	0	3	
-25X1		12	0	1	

Fig. 6.27 Sample plug overview (file extension *.f23)

You can filter and sort almost all forms using many different filter and sorting settings.

6.3.1.24 Structure identifier overview *.f24

EPLAN can manage the various structure identifiers such as functional assignment (==), higher-level function (=), or mounting location (+). You can assign additional descriptions to every structure identifier in EPLAN. You use the structure identifier overview to obtain an overview of all the structure identifiers assigned (or also unused) in the project.

Structure identifier overview   HB3_F24_001

Full designation	Structure description
Funktionale Zuordnung	
==ALL	General
==Example	Sample
==DOC	Info
==Example	Sample
==Chapter	chapter

Mounting location	
+7.3.4	subchapter number 7.3.4 in the book
+8	subchapter number 8 in the book
+8.6	subchapter number 8.6 in the book
+8.6.1	subchapter number 8.6.1 in the book

Fig. 6.28 Sample structure identifier overview (file extension *.f24)

No filters are possible for the output, but a range of sorting options are available.

6.3.1.25 Symbol overview *.f25

A symbol overview is used to display the symbol libraries used and / or saved in a project.

Fig. 6.29 Sample symbol overview (file extension *.f25)

Simple filtering and sorting options are available for the symbol overview.

6.3.1.26 Title page / cover sheet *.f26

The title page / cover sheet is usually the page with which a project begins. EPLAN allows the user to automatically generate this type of page.

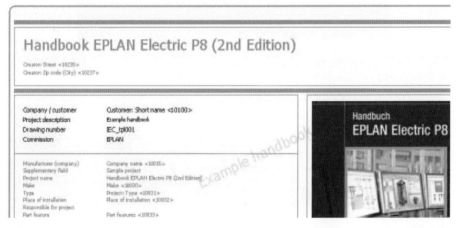

Fig. 6.30 Sample title page / cover sheet (file extension *.f26)

Various project or page properties can be assigned to the form. For example, you can output a title page / cover sheet for the entire project documentation and a cover sheet for each individual higher-level function. Additional filters unfortunately cannot be used or evaluated.

6.3.1.27 Connection list *.f27

The connection list form generates a report containing all the project connection data.

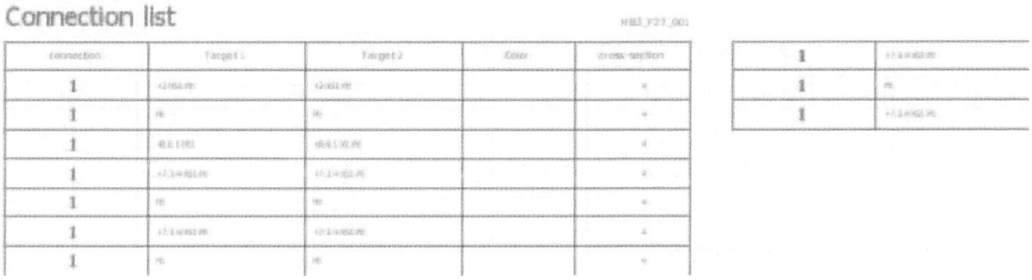

Fig. 6.31 Sample connection list (file extension *.f27)

The output can also be adjusted via existing or user-defined filters and sorting settings.

6.3.1.28 Graphic *.f28

The graphic form is not a „real" form in the sense of a report. I mention it here for the sake of completeness. Graphical „forms" have a file extension of *.f28 and can be entered into the page properties as „placeholder form graphics".

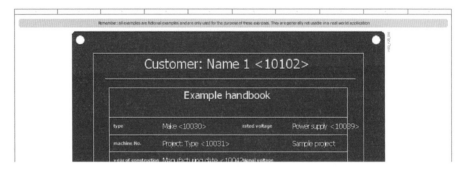

Fig. 6.32 Sample graphical „form" (file extension *.f28)

For example, you can enhance specific pages and fill them with page and / or project properties with specific information.

Since there are no report templates for a graphical „report", the form is entered directly in the page properties.

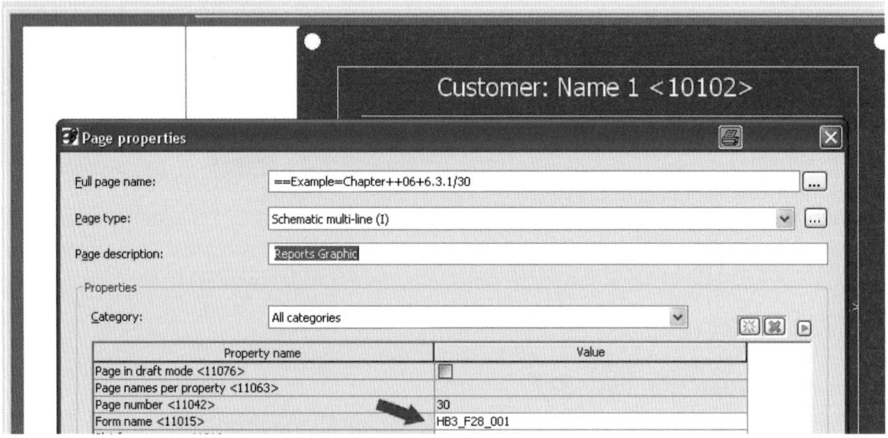

Fig. 6.33 Entering a form in the page properties

6.3.1.29 Project options overview *.f29

The project options overview report contains information on possible project options.

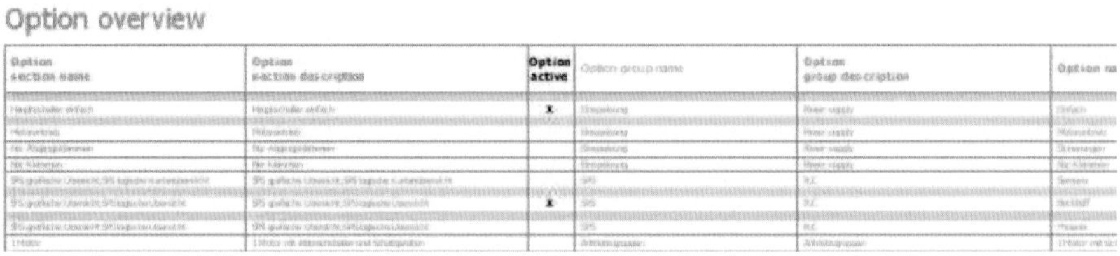

Fig. 6.34 Sample project options overview (file extension *.f29)

This can be information about the project options themselves or also the project option segments or via placeholder objects with the appropriate value sets. With this form, you can use a range of filter and sorting options for the graphical output.

6.3.1.30 Placeholder object overview *.f30

The placeholder object overview report contains information about the placeholder objects used and their different value sets.

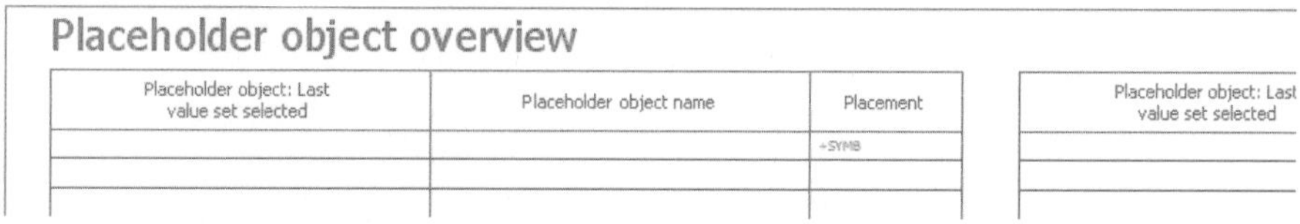

Fig. 6.35 Sample placeholder object overview (file extension *.f30)

You can control the output via filters and sorting options.

6.3.1.31 Manufacturer / supplier list *.f31

The manufacturer / supplier list report enables the specific listing of the manufacturers and / or suppliers used.

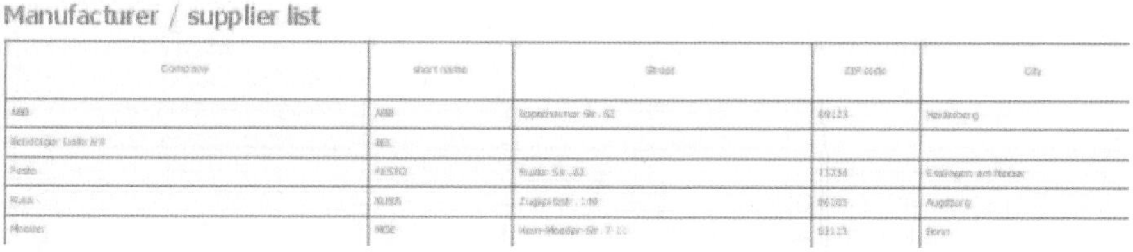

Fig. 6.36 Sample manufacturer / supplier list (file extension *.f31)

EPLAN's usual filter and sorting options are available here as well.

6.3.2 Special connection diagrams

In addition to the previous report types, EPLAN has special connection diagrams. These diagrams depict connection overviews between devices, independent of size and placement on the mounting panel.

EPLAN offers connection diagrams for the following reports: device connection points, terminal connection points, and pin connection points. A prerequisite for generating connection diagrams is that the devices must be placed on a mounting panel.

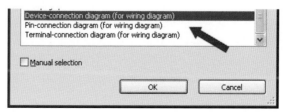

Fig. 6.37 Special connection diagrams

6.4 Settings (output options)

Before generating reports for a project, you will typically need to adjust certain settings to suit your requirements. These settings relate only to the output of data such as the form settings, or how EPLAN is to handle path function texts in the forms.

 NOTE: You do not need to perform report runs or the like, because EPLAN keeps all data online, and this is therefore always up to date. EPLAN also automatically updates the connection data before generating a report.

6.4.1 The Display / output project setting

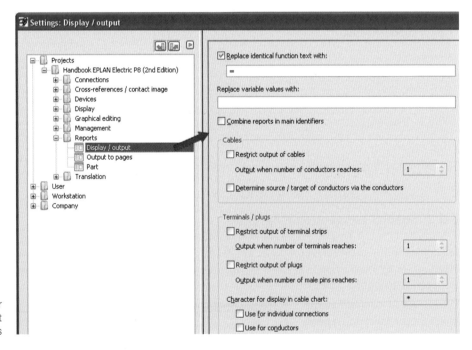

Fig. 6.38 Settings for the display and output of reports

These settings are used to define basic information for graphical evaluation of the forms. You can influence the way in which device tags are displayed in reports or from what number of data records particular reports should be output. You will find the 'Display / output' settings in the OPTIONS / SETTINGS / PROJECTS [PROJECT NAME] / REPORTS / DISPLAY / OUTPUT menu item.

Replace identical function text with – it is often the case that (e.g.) the function text would be repeated in the report form for a row of terminals. If this type of graphical output is not wanted, then here you can define one or more characters to represent the repetition of the function text. The equals sign "=" is often used for this.

Replace variable values with – this setting is only valid for summarized parts lists. It takes effect when placeholder objects with different value sets are used in a project. In the summarized parts list report, EPLAN can then replace the current placeholder text data with the data from this field (e.g. a reference to an extra page in the project). A precondition for this is that the *<13108> Replace variable values with text* property is activated in the form.

Combine reports in main identifiers – when this setting is activated, EPLAN lists the project data consecutively in the form in graphical reports. If the identifier changes, then it is immediately placed after the last different identifier in the graphical report without EPLAN generating a page break in the graphical output. When the setting is not active, EPLAN generates a page break in the graphical report when a main identifier changes.

Other settings for *terminals / plugs / cables* – this allows particular project data to be output only when it possesses a minimum number of data records. If (e.g.) a limit of five conductors is set for cables, then all cables with less than five conductors are not included in the graphical report. The default values are set to not active and they should usually remain so.

Character for display in cable chart – if the correct display of the cable conductors (or individual connections) in the terminal diagram and their conductor color property values is not important, then instead of the correct color designation, you can enter a value here to be entered instead of the conductor color in the graphical output.

For example, this could be an "X". Multiple characters can be entered, but in practice one character is better because this must also fit in the available space on the forms.

6.4.2 The Parts project setting

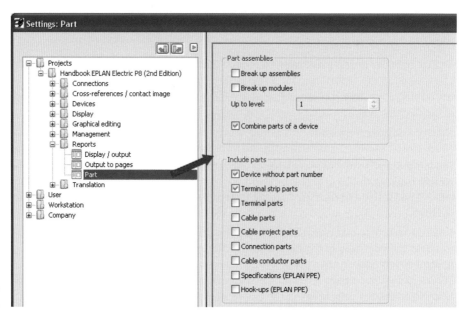

Fig. 6.39 Settings for the output of parts

This dialog contains settings that define how EPLAN is to handle parts when outputting graphical reports. You will find the settings for viewing parts in the OPTIONS / SETTINGS / PROJECTS [PROJECT NAME] / REPORTS / PARTS menu.

 TIP: These settings are mainly intended for evaluating forms related to **parts** project data. These would be parts lists or summarized parts lists, for example.

Part assemblies / Break up assemblies & modules – you can use the level setting here to define the level to which EPLAN should break up assemblies (and assemblies within assemblies) and modules when generating graphical reports.

Combine parts of a device – this means that (e.g.) the terminals of a terminal strip are combined in a parts list, so that not every terminal part is listed individually in the parts list.

Include parts – you can influence the graphical output for each type of part here by activating the individual options such as terminal parts or cable parts.

6.4.3 The Output to pages project setting

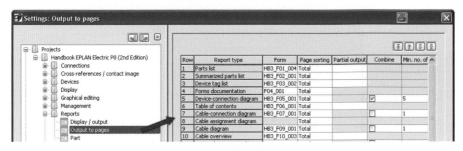

Fig. 6.40 Settings for reports

Forms are a fundamental element in outputting reports. They are always set on a project-specific basis via OPTIONS / SETTINGS / PROJECTS [PROJECT NAME] / REPORTS / OUTPUT TO PAGES or adopted from the templates or basic projects as default settings for the project. These settings define a number of global default conditions for graphical output of the project data.

6.4.3.1 The Report type column

The **Report type** column cannot be changed in this table. EPLAN provides different report types as default. This does not mean that all these reports must be generated for a given project. The user can decide whether (e.g.) only one terminal diagram should to be output for the schematic or the full palette of all possible reports.

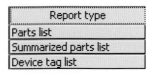

Fig. 6.41 Report types

6.4.3.2 The Form column

The desired form can be set in the Form column. To do this, you click the FORM row and then click the BROWSE button in the drop-down list that appears. The **Select form** dialog opens. Now you can select another form from the directory and press the OPEN button to load it into the *Form* column.

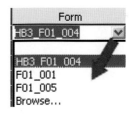

Fig. 6.42 Forms

 TIP: The form is automatically stored in the project, if it does not already exist in the project master data. Initially it makes no difference where the new form was selected from.

6.4.3.3 The Page sorting column

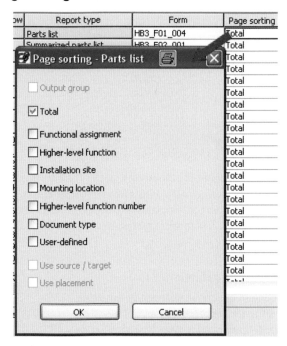

Fig. 6.43 Page sorting

The **Page sorting** column plays a decisive role in defining how the graphical output pages are later sorted into the existing page structure of the project. Selecting *Total* causes all reports to be summarized under the selected total identifier. If you wish to (e.g.) output all terminal diagrams by mounting location, then you should select the *Mounting location* entry.

 NOTE: Not all types of page sorting are available for all report types. For example, output of forms documentation by higher-level function and mounting location is not possible because it does not make sense.

6.4.3.4 The Partial output column

Partial output is an interesting setting. For reports where this is possible, as well as a type of main form such as a total cover sheet, a partial cover sheet per higher-level function can also be automatically generated. For this to work properly, the page sorting must also be set to (e.g.) *Total + Higher-level function*. With this setting, EPLAN then generates a (total) cover sheet and a (partial) cover sheet for each individual higher-level function. We explicitly point out here that the structure and graphical content of these forms can be different.

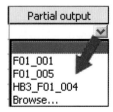

Fig. 6.44 Partial output

6.4.3.5 The Combine column

The **Combine** function is ideal for (e.g.) placing several terminal strips having only a few terminals on *one* graphical output page (limited only by the available space on the output page; then EPLAN generates a new page as before).

The examples in the following images should help to explain this. In the following image, the Combine option was *not* selected. EPLAN generates one page of the terminal diagram report type for each terminal strip.

Fig. 6.45 Combining

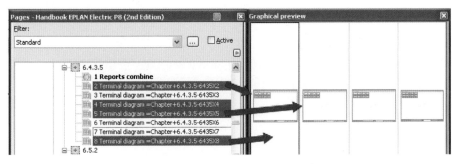

Fig. 6.46 Combine option not activated

In the following image, the Combine option was set to *active*. EPLAN now generates one or more report pages of the terminal diagram type for several terminal strips, limited only by the size of the plot frame and the set maximum number of data rows that EPLAN should display on a report page.

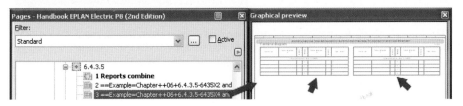

Fig. 6.47 Combine option activated

 NOTE: Only dynamic forms can be used with the **Combine** option. Static forms **cannot** be used with the Combine setting!

6.4.3.6 The Min. no. of report rows on report page column

The *Min. no. of rows on report page* setting specifies a particular number of data sets to be output before EPLAN generates a page break.

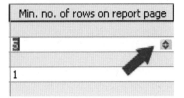

Fig. 6.48 Minimum number of rows

TIP: This setting only makes sense when used together with the Combine option. When the Combine option is not used, then changing the minimum number of rows has no effect on the graphical output. ∎

The usual value is 1. This then places no limitations on the graphical output pages, and the project data is output consecutively.

If the *minimum number* is now changed to 10, then at least ten terminals of a terminal strip must be displayed on a single graphical output page before EPLAN can force a new page break.

6.4.3.7 The Subpage column

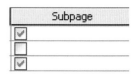

Fig. 6.49 Subpages

This setting defines whether or not EPLAN should generate subpages for the graphical output pages. Report pages must not only be generated using consecutive integers in EPLAN, in some cases the actual page number for the report should remain the same. This means that *subpages* must be created for the graphical output pages.

6.4.3.8 The Character column (for subpages)

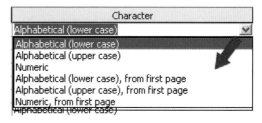

Fig. 6.50 Characters

This property only makes sense when used together with the Subpages property. The *Character* setting defines the format of the subpage. There are multiple selections available, but all subpages are generally separated from the main page with a point.

Alphabetical (lower case) → Report pages begin with **20; 20.a; 20.b; 20;c**

Alphabetical (upper case) → Report pages begin with **15; 15.A; 15.B; 15;C**

Numeric → Report pages begin with **301; 301.1; 301.2; 301.3**

Alphabetical (lower case), from first page → Report pages begin with **5.a; 5.b; 5;c**

Alphabetical (upper case), from first page → Report pages begin with **7.A; 7.B; 7;C**

Numeric, from first page → Report pages begin with **43.1; 43.2; 43;3**

6.4.3.9 The Blank pages column

Fig. 6.51 Blank pages

EPLAN can automatically maintain a certain spacing between the report pages and the actual schematic pages, the so-called *blank pages*. This setting defines the size of this spacing.

The last schematic page is 2. The blank pages setting is set to 5. 2+5=7; the report pages then begin with page number 8.

6.4.3.10 The Round column

The *Round* setting relates to the setting for blank pages.

In the example, we round to 10. If the last schematic page 12 has a spacing of 5 (setting for blank pages results in 17), then the setting Round = 10 will cause the first report page to have a page number of 20.

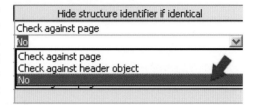

Fig. 6.52 Rounding

6.4.3.11 The Hide structure identifier if identical column

This setting checks to see if structure identifiers should be included in the graphical output or suppressed when certain conditions are satisfied. To put it briefly, this allows "abbreviation" of device DTs in the reports, meaning a check for corresponding identical entries.

Fig. 6.53 Identical?

There are three possible settings:

- *Check against page* – this causes the structure identifier of the DT to be checked against the structure identifier of the report page. If EPLAN finds a match here then the DT is shortened by this matching value in the report.
- *Check against header object* – this setting is intended for function-related reports such as terminal or cable diagrams. In this case, EPLAN checks the output DT against the header object of the report page (e.g. a terminal strip definition or a cable definition) and abbreviates the output DT by this structure identifier if it is the same.
- *No* – this setting always causes the complete structure identifier to be output.

6.4.3.12 The Synchronize column

Synchronize is an important setting for keeping project master data up to date, i.e. synchronized with the system master data. The Synchronize setting in EPLAN means that forms having synchronizing active (check box set) are automatically synchronized with the system master data when the project is opened.

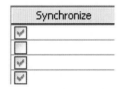

Fig. 6.54 Synchronize

 A precondition for this is that the **Synchronize project master data when opening** setting is set in the OPTIONS / SETTINGS / PROJECTS [PROJECT NAME] / MANAGEMENT / GENERAL menu.

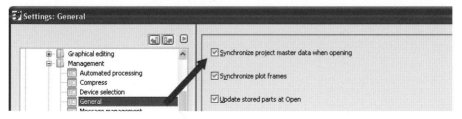

Fig. 6.55 Allow automatic synchronization of master data

When the *Synchronize* check box is not set in the column, then it will not be synchronized despite the active project setting! You should always consider this for each project. You should not synchronize if the project master data must retain its original edited status. This setting should therefore not be activated.

6.4.3.13 The Next form column

Next forms are forms that display additional information to the information in the main report (e.g. parts data).

For example, a cable assignment diagram can be followed by a summarized parts list. To do this, the main form (in this case the cable assignment diagram) must be created with a form insertion point. Only then does EPLAN know that it must generate a corresponding next form containing the subsequent data (parts data).

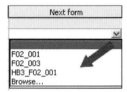

Fig. 6.56 Next form

 NOTE: At present, only the summarized parts list is available as a next form!

6.5 Generate reports

This chapter deals with the generation of reports and the most important points of the various ways for doing this.

EPLAN can generate reports with or without templates. Without templates when a quick report is needed or no templates are available. With templates when these are available, or have been created by the user, or have been imported from other projects.

You do not need to use templates for generating reports in EPLAN. They enable project editing when no reporting structure has been defined and reports are output as desired.

6.5.1 The Reports dialog

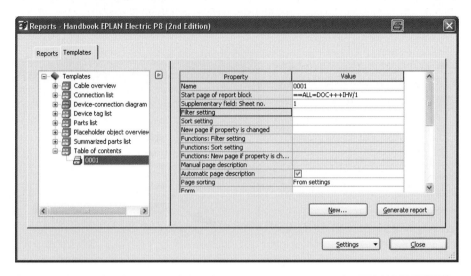

Fig. 6.57 The Reports dialog

To generate graphical reports while editing a project, you use the EPLAN UTILITIES / REPORTS / GENERATE menu item to open the **Reports** dialog.

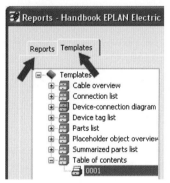

Fig. 6.58 Left area with tabs   Fig. 6.59 Right area with additional information

The dialog is divided primarily into the *Reports* and *Templates* tabs, as well as the right part, which contains the report data, e.g. from which page the report should be generated, which filter and sorting settings should be used, and much more.

The *Reports* tab is usually empty before reports are generated from the project data for the first time, i.e. without any entries, and the *Templates* tab is also usually empty.

If the project was copied with the *Copy with reports* option, then the old reports will, of course, be visible in the *Reports* tab. Templates may also already be present.

After graphical reports, e.g. a parts list, have been generated from the project data for the first time, then the *Reports* tab will contain an overview of all reports in the project. Normal graphical outputs (each graphical cal output on a new page) are then shown in the Pages folder and are visually distinguished by a small graphical symbol.

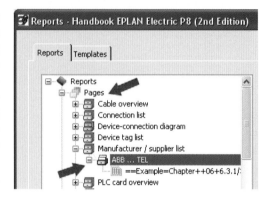

The graphical symbols are distinguished as follows:

Fig. 6.60 Pages tree

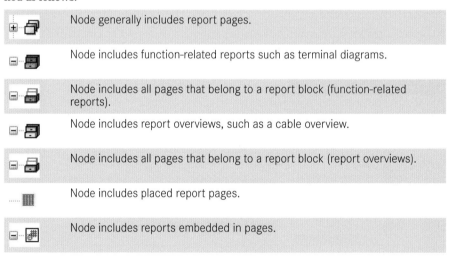

	Node generally includes report pages.
	Node includes function-related reports such as terminal diagrams.
	Node includes all pages that belong to a report block (function-related reports).
	Node includes report overviews, such as a cable overview.
	Node includes all pages that belong to a report block (report overviews).
	Node includes placed report pages.
	Node includes reports embedded in pages.

6.5.2 Generate reports without templates

In general, you can directly generate reports in EPLAN. You do not need to create any templates for this. To generate a report, you simply begin by clicking the NEW... button.

6.5.2.1 Output of project data for graphical output on new report pages

EPLAN opens the **Select report** dialog. Now you select the desired report (report type) in the window. *Only one* type of report can be selected each time this dialog is called up. You can only generate several reports at once by using suitable templates. After selecting the report, e.g. a terminal diagram, you can now set the remaining options in the **Select report** dialog.

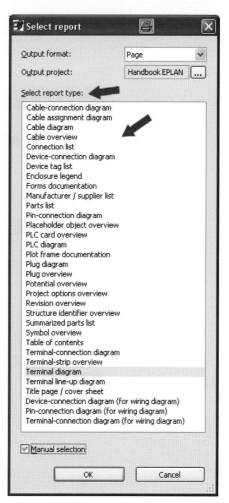

Fig. 6.62 Output format

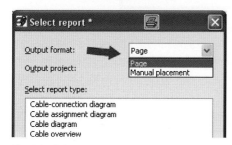

Fig. 6.61 Select report

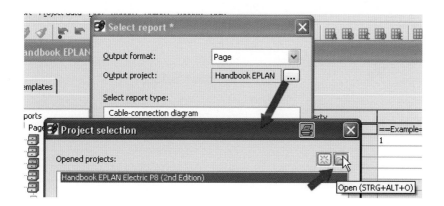

Fig. 6.63 Project selection

There are two options available for *output format*: output in pages (to be generated by EPLAN) or output as manual placement (in an existing page). The *output project* is normally the current project. If this is not what you want, then EPLAN can also write the reports to another project.

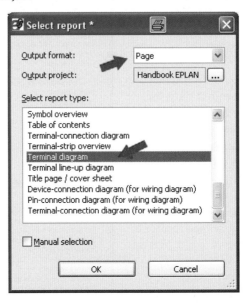

Fig. 6.64 Select report

When you click the OK button (we will ignore the **Manual selection** check box at this point, because all of the terminal strips are to be output), EPLAN allows you to define the sorting into the project via particular predefined filter and / or sorting schemes, or your own filter and sorting schemes (the possibilities here may be limited by the particular report type).

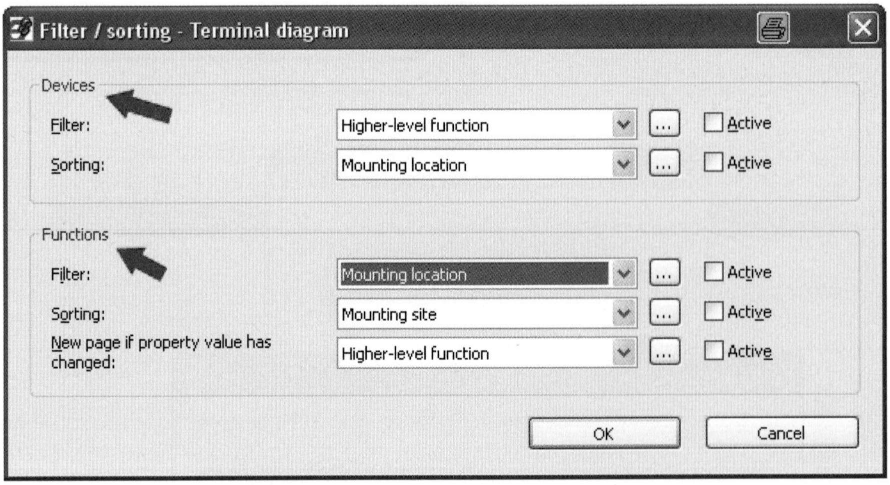

Fig. 6.65 Filter and sorting options for a report

If EPLAN is to use a filter, then the *Active* check box must be set! Only then is a filter active. If the option is grayed out, then no filters can be used with this area of graphical output. If the check boxes are not set, then the set filters have no effect.

Fig. 6.66 Select output

EPLAN then displays the **[Report type] (Total)** dialog. Here you can define default values for sorting such as the selection of *structure identifiers* or specific entries in the page properties, e.g. *automatic page description*. EPLAN then automatically creates the page properties from the form properties, whereby something must, of course, be assigned to the *<13019 Format for automatic page description>* form property. The remaining dialog fields contain the usual entries and will not be further explained here. When you click OK, EPLAN generates the reports.

> **NOTE**: If one of the specified structure identifiers is not yet present in the structure identifier database, then after the graphical report is generated, the **Place identifiers** dialog is displayed. The identifiers are manually sorted as requested here.

Fig. 6.67 Generated output of terminal diagram report

After leaving the dialog by clicking the OK button, EPLAN now generates the graphical report pages and sorts them into the project according to your settings. For example, the terminal diagram report is generated as graphical output and inserted into the project as a report page.

6.5.2.2 Manual selection of specific project data for graphical output with manual placement

When reports are generated as described above, then all project data is always output. This means that if the project contains ten terminal strips, then all of these ten terminal strips are output to the project as graphical report pages (depending on the settings in the terminal strip definition, e.g. *<20851 No output to terminal / plug diagram>*).

In the following, data is selected via manual selection, but, in contrast to the previous chapter, not on new report pages; instead, the data is manually placed on existing schematic pages.

 NOTE: Before you use the **Manual placement** function, the schematic page where the report will finally be placed must be open and active. You can no longer scroll between the pages once the report has been generated.

You should also note that a manual placement works in principle with all form sizes (in this case the graphical extent of the form). However, the form should, of course, suit the corresponding schematic page, both visually and in the sense of available space.

Fig. 6.68 Select report

The **Manual selection** option allows you to define which particular project data is to be output, e.g. specific plug strips. To do this, you select the corresponding report and activate the **Manual selection** option. In addition, the data should be placed on an active schematic page. To do this, you set the output format to **Manual placement**. Then you can confirm the dialog by clicking OK.

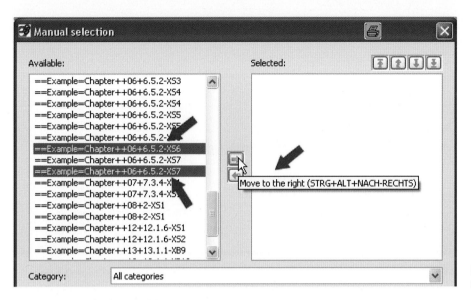

Fig. 6.69 Manual selection of project data

The **Manual selection** dialog is then displayed. The left side of the dialog shows all data that are *available* for the selected report.

You use the buttons with the blue arrows to add the desired data to the *Selected* field at the right of the dialog. The ⇨ button inserts data from left to right. The lower button ⇦ (which is only available when there is data in the right field) removes the selected data from the selection.

You select the data via Ctrl (or Shift) + left mouse button, or you can use the usual Windows functions such as Ctrl + A (select all). To select data, the cursor must be located either in the left (*Available*) field or the right (*Selected*) field of the dialog.

Once all desired data has been selected, it can also be sorted as necessary with the SORT button. EPLAN sorts the data with this function based on the sorting of the existing structure identifiers.

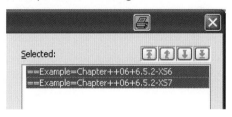

Fig. 6.70 Selected data

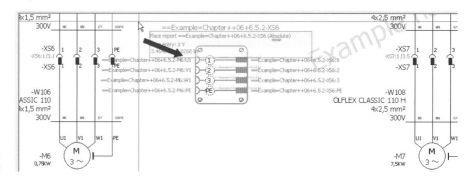

Fig. 6.71 Filter and sorting options for manually selected functions

After clicking OK, EPLAN then displays the familiar **Filter / Sorting [Report type]** dialog. Filters and sorting cannot be selected for devices because these are already manually selected. Filters and sorting can be used for the functions themselves, though (data such as plug pin or terminal).

Fig. 6.72 Manually place report

After clicking OK, EPLAN closes all dialogs. The reports hang on the cursor and can now be placed on the active schematic page.

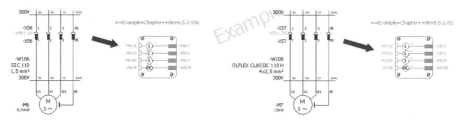

Fig. 6.73 All reports are manually placed.

 NOTE: The **Manual selection** option cannot be used with all report types. This is not allowed for technical programming reasons.

6.5.3 Popup menus in the Reports tab

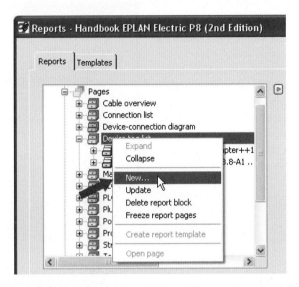

Fig. 6.74 Popup menu

There are a number of other functions in addition to the previously described functions, such as creating or manually placing reports. Reports should, of course, be able to be deleted, or quickly updated from within the dialog (e.g. after changing a setting).

These and other functions are located in the right-click popup menu (click a field in the *Reports* tab with the left mouse button and then click the right mouse button or open the popup menu using the button).

In addition to the usual EXPAND and COLLAPSE functions for the report tree, the menu contains the following functions.NEW – creates a new report. This is the same function as the NEW... button in the **Reports** dialog.

UPDATE – this function updates one or more reports. This is useful when (e.g.) you have changed the report sorting and wish to quickly check the results where the graphical output pages were placed.

DELETE REPORT BLOCK – this removes the reports (i.e. the graphical output pages) from the project. Here too, you can delete one or more reports at the same time.

FREEZE REPORT PAGES – this "freezes" the selected reports and removes them from the list of reports in the dialog. EPLAN displays a safety warning before executing this function. This allows you to cancel the action before it is executed.

CREATE REPORT TEMPLATE – if a report is suitable for creating new reports, then a template can be generated by clicking this menu item. The template can then be found in the **Templates** tab.

OPEN PAGE – this is a useful function allowing the user to directly navigate to the selected report in the project.

6.5.4 Generate reports with templates

What are *templates*?

You could describe this term in another way. These are special *defaults* that define how, where, and to what extent EPLAN is to generate the graphical output of various project data. The procedure for creating a template is the same as that for creating pure reports, except that no reports are generated when the template is finished.

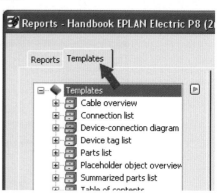

Fig. 6.75 The Templates tab

TIP: The following restrictions apply to the use of templates to generate reports. The *output format* is fixed in the *page* selection. You also cannot define a *manual selection*.

How are templates created? The "how" involves a number of different settings for a template.

You open the **Reports** dialog via the UTILITIES / REPORT / GENERATE menu and switch to the **Templates** tab. You then click the NEW... button to display the **Select report** dialog. You now select the desired *report type* here. This selection is confirmed with OK.

Fig. 6.76 Select report type

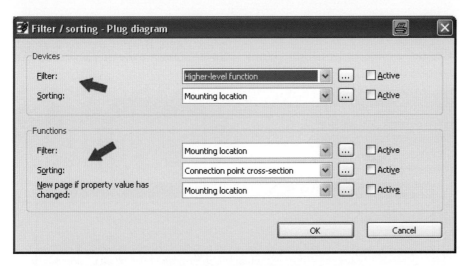

Fig. 6.77 The Filter / sorting dialog

As usual in EPLAN, you can use predefined filter and sorting options, or set up your own, in the **Filter / sorting [Report type]** dialog. However, you must always make sure that the check boxes are correctly set, otherwise the filter and sorting options will not be active.

 NOTE: Depending on the report type, the **Filter / sorting** dialog may look different, or may not display a few options at all!

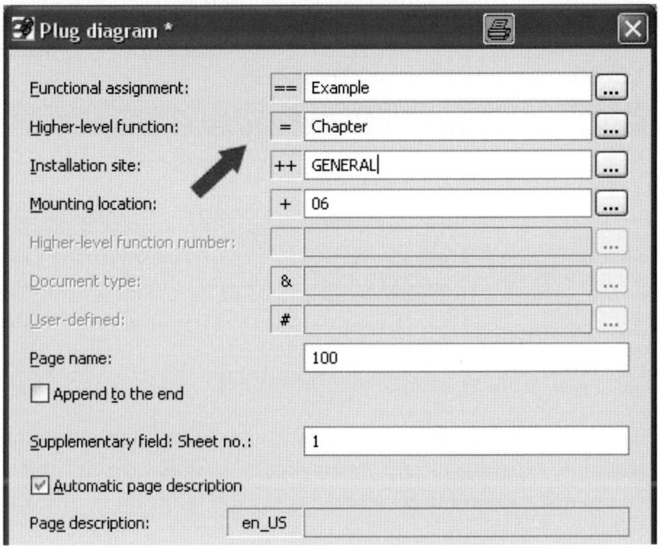

Fig. 6.78 Setting the structure of report pages

After the data filter and sorting have been selected, there is the *Selection of structure identifier* dialog for the output of report pages. The structure identifiers can be directly entered into the data entry fields or, if they already exist, they can be selected and applied from a table via the ⊟ selection button.

No further entries are needed. Clicking OK saves the settings in the template and EPLAN closes the dialog.

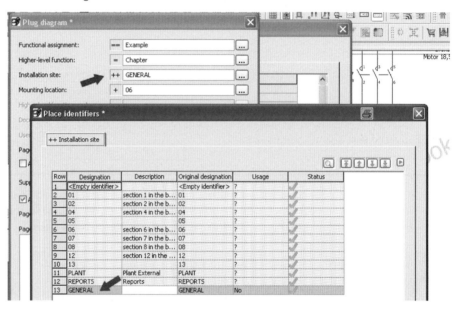

Fig. 6.79 Select missing structure identifiers

If structure identifiers were defined that are not yet in the structure identifier database, then the **Place identifiers** dialog is displayed so that the identifiers can be sorted into the project.

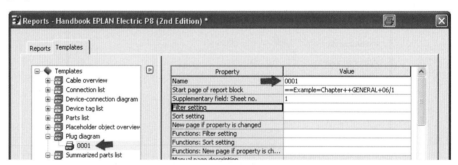

Fig. 6.80 Template addition complete

EPLAN now adds the template and its settings to the **Templates** tab. In order to keep templates functionally separate, EPLAN automatically assigns a certain structure to them when they are created (according to report type). The template is given the standard name 0001.

 TIP: If filters were selected previously, EPLAN assigns the filter's name to the template!

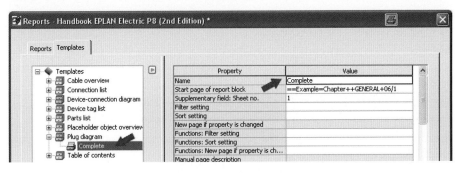

Fig. 6.81 Template name changed

Because there can be several templates for the generation of report types, it is possible to manually change the names of templates later. To do this, click in the PROPERTY – NAME: VALUE line and simply enter another name.

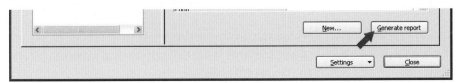

Fig. 6.82 Generate this template report

This would complete the generation of a template. If you select this template and click the GENERATE REPORT button, EPLAN will generate the report pages and sort them in the predefined structure of the project.

This and other information can be seen at the *right side* of the *Templates* tab. The table shows additional properties and their associated values. In contrast to the **Reports** tab (which also has this table), here you can change the corresponding values of basic properties.

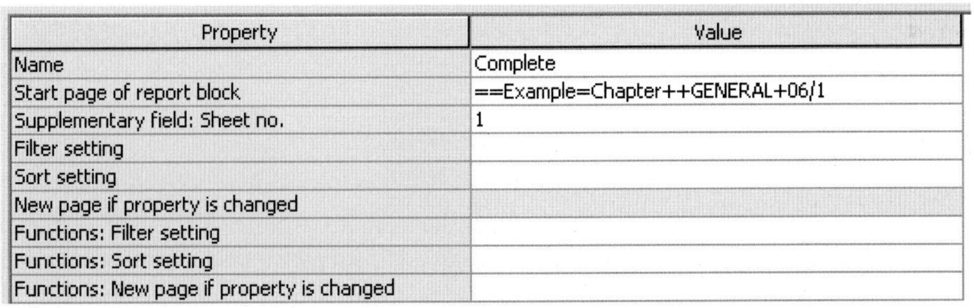

Property	Value
Name	Complete
Start page of report block	==Example=Chapter++GENERAL+06/1
Supplementary field: Sheet no.	1
Filter setting	
Sort setting	
New page if property is changed	
Functions: Filter setting	
Functions: Sort setting	
Functions: New page if property is changed	

Fig. 6.83 Basic properties for structure and output

Name – this contains the name of the template. It can be manually changed to suit your personal requirements.

Start page of report block – starting page of the graphical report. Can also be subsequently changed. To do this, click the field and then click the [...] selection button that appears. The familiar dialog for entering the structure identifier is then displayed.

Supplementary field: Sheet no. – additional information that appears in the page properties of the report pages.

Filter setting – this field is filled with the selected filter settings. Can also be changed by clicking the field and then clicking the [...] selection button and assigning a new filter scheme.

Sort setting – can be subsequently changed in exactly the same way as the Filter setting property.

New page if property is changed – possible default value that causes an automatic page break when the value of a particular property changes. Cannot usually be changed.

Functions: Filter / Sort setting and *New page if property is changed* – these filters operate directly on the device functions, e.g. on a terminal.

Manual page description	
Automatic page description	✓
Page sorting	From settings
Form	
Partial output	

Fig. 6.84 Properties that affect the report page

Manual page description – can be subsequently changed to any value.

Automatic page description – can be activated. When this property is activated, the *Manual page description* property can no longer be edited.

Page sorting – usually empty because the basic value is defined in the default setting (Output to pages setting). However, this value can still be changed.

Form – here you can enter a different form to that specified in the default settings (Output to pages setting). This form then has priority over the global settings.

Partial output – can be changed but only for certain report types that support partial output, e.g. the table of contents.

Break up assemblies	From settings
Break up modules	From settings
Device without part number	From settings
Terminal strip parts	From settings
Terminal parts	From settings
Cable parts	From settings
Cable project parts	From settings
Connection parts	From settings
Cable conductor parts	From settings

Fig. 6.85 Properties that affect parts output

Break up assemblies and *Break up modules*; *Device without part number*; other additional settings for *terminals, cables, project* and *connection parts* – these settings can be changed later but only affect some of the part report types.

Project filter	FACTORY_PRINT_OK
Template active	✓

Fig. 6.86 Special report settings

Project filter – the setting of a project filter enables you to generate reports that are linked to a condition. You can generate conditional reports independently of certain project properties.

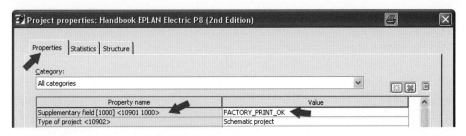

Fig. 6.87 Select project property and its value

This means that if a defined project property's value matches the filter value, a report will be generated.

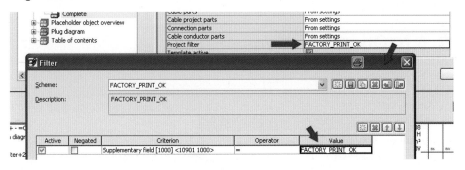

Fig. 6.88 Selecting a filter for the desired project property and its value

Template active – this setting cannot be edited; it is automatically used or not used by EPLAN. The check box depends on whether the project filter was fulfilled or not. When the condition is fulfilled, the check box is set; when the condition is not fulfilled, EPLAN removes the check. This means that this report will not be generated.

6.5.4.1 Export and import of templates

Once you have created useful templates for graphical reports in a project, EPLAN can also use these for other projects. EPLAN provides *export and import functions* for this. However, this applies only to templates. Existing reports cannot be exported or imported.

To export or import templates, you open the **Reports** dialog via the UTILITIES / REPORTS / GENERATE menu. If not immediately displayed, then switch to the *Templates* tab. You access the Export and Import functions via the popup menu (accessible via the button or the right mouse button).

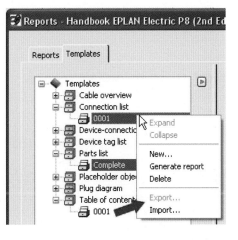

Fig. 6.89 Export / import of templates

EPLAN allows the user to decide which templates should be exported or imported. If the tree is selected at the highest level, for example, then all templates below this are exported together. You can also export individual templates. To do this, you expand the template tree, select the desired template only, and run the EXPORT function. EPLAN then displays the **Export report templates** dialog. Select a file name for the template and then save it by clicking the SAVE button. The template is then exported.

Importing existing templates functions in exactly the reverse direction. You open the **Reports** dialog and start the IMPORT function (right mouse button or popup menu). EPLAN opens the **Import report templates** dialog. Here you select the desired template and open it via the OPEN button. EPLAN now inserts the new template into the project.

■ 6.6 Other functions

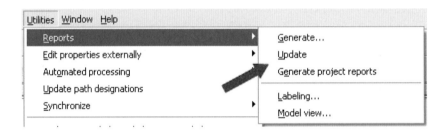

Fig. 6.90 Generate project (templates) reports

You will usually not just constantly create new reports and develop pretty templates for reports. Reports with particular report types are usually not completely finished until the end of project editing and then only updated when project data changes, such as terminal designations or a different cable type. EPLAN offers two further functions that allow this to be performed quickly.

6.6.1 Generate project reports

The GENERATE PROJECT REPORTS function is accessed via the UTILITIES / REPORTS menu. If you select this function, all of the reports (that exist as report templates) are evaluated completely. For large projects, this can take a while.

 NOTE: Once the menu item has been selected, EPLAN begins project evaluation immediately (without an intermediate query dialog)!

You can, however, use the CANCEL button in the **Generate project reports** dialog to immediately exit the function. The reports then remain in their original state or are returned to their original state.

6.6.2 Update

An additional function that EPLAN offers is the UPDATE function for *existing* reports. This function is also available in the UTILITIES / REPORTS menu.

In contrast to the *Generate project reports* function, the *Update* function always only affects the currently open report or a report selected in the page navigator (a page, but can also be a manually placed report) in the project. If you run the *Update* function, the corresponding report is updated.

If the Update function is used on a schematic page, EPLAN displays a message stating that you are not on a report page or no report page is selected in the page overview.

6.6.3 Settings for automatic updates

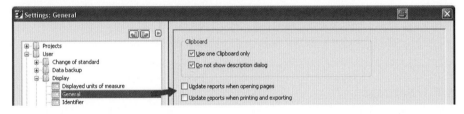

Fig. 6.91 Settings for automatically updating reports

If this is not enough for you, then you can also determine a user setting defining whether or not report pages should be automatically updated when opened, printed, or exported. This is not a project setting but rather a user-specific setting. It can be found under OPTIONS / SETTINGS / USER / DISPLAY / GENERAL in the two settings **Update reports when opening pages** and **Update reports when printing and exporting**.

However, you should note that this type of constant updating slows down project editing, e.g. when scrolling through project pages. Opening report pages also takes longer when this user setting is active than when it is not, because the reports must always be brought up-to-date.

You should therefore carefully consider whether or not the report pages should be constantly kept up-to-date. In my opinion, this is not absolutely necessary, but EPLAN offers this setting. This is up to the user to decide.

6.7 Labeling

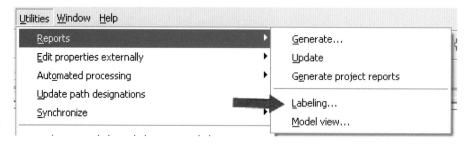

Fig. 6.92 Labeling

The second way to generate a report from project data is via the functions under the UTILITIES / REPORTS / LABELING menu item.

This function does not generate graphical report pages but rather writes the project data to external files. These external files can then be passed on and subsequently edited, e.g. terminal diagrams in Excel format or files for printing labels.

6.7.1 Settings

This chapter deals with the basic settings for outputting labeling. When the LABELING function is started, EPLAN displays the **Output labeling** dialog.

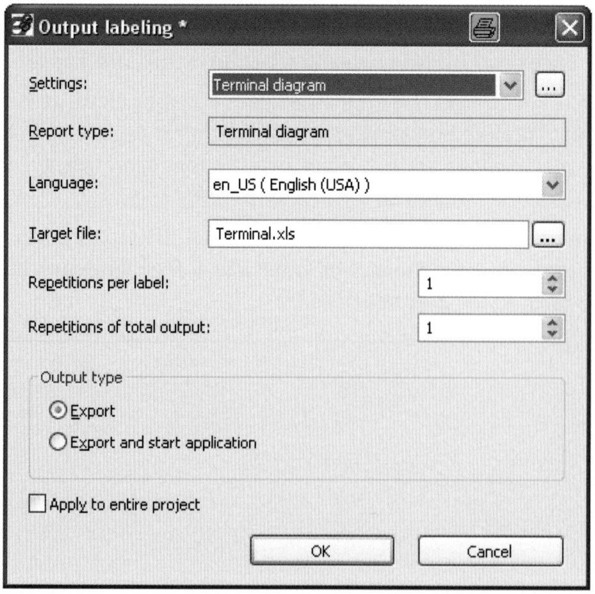

Fig. 6.93 Output labeling

This dialog offers the following settings options.

Settings – here you can select which data are to be output. You can filter and / or sort the output.

Report type – this field merely displays which report type forms the basis of the set labeling scheme.

Language - this setting controls the language output of the labeling file. Only the languages set in the project are available here. EPLAN offers two ways of selecting the language: Either a single language, or all project languages can be output together.

Target file – this is where you enter the name of the output file, according to the selected scheme, for the labeling data to be generated.

Value for repetitions – all values greater than 1 increase the output of the labeling data by the value specified here. The default value here is 1 (output the data once).

Output type – the generated file is exported, or you can cause the exported file to be subsequently opened with a specified editor so that you can check or edit the contents.

> **TIP:** The editor to be used is defined by an entry in the user settings. This can be found via the OPTIONS / SETTINGS / USER / INTERFACES / LABELING menu item under the OPTIONS button.

Apply to entire project – regardless of which data is selected, this setting always causes the entire project's data to be written!

6.7.1.1 Preparatory settings

The most important area in the **Output labeling** dialog are the *settings*: Which data should be output here and under which relationships? This is defined via predefined schemes or your own schemes.

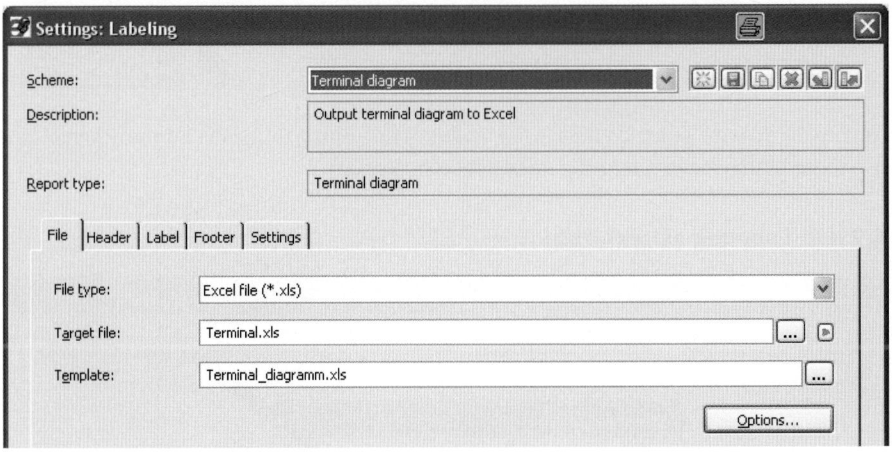

Fig. 6.94 The Labeling dialog with the scheme and the settings

To add a setting, you click the selection button next to the **Settings** field. The **Settings / Labeling** dialog is displayed.

The dialog is basically divided into two areas: the area with information on the scheme itself (name of the scheme, description, and the report type), and the lower area with the tabs for various data and settings such as file, header, label, and footer, and further output settings such as filter, sorting, etc.

The file type to be created is important when creating a labeling file. Further settings, e.g. which program to use for opening the generated file, depend on this value.

EPLAN offers several different file types here.

Fig. 6.95 Selecting the file type

Excel file *.xls – as the name indicates, this file can be opened with Excel. Whereas a text file is usually output without formatting, the Excel format can be used to (e.g.) output cable diagrams formatted in an Excel template. This produces report files that look similar to those generated in the graphical report pages. You can use the OPTIONS button to also output column designations.

Text file *.txt – this data type generates a normal text file. It can be edited with a conventional text editor. The OPTIONS button allows further settings to be defined for a text file, e.g. the ANSI output character set or separators for multi-language output.

XML file *xml – the last possible data type generates an XML file that can be opened and edited with an appropriate XML editor. No other options are available for this file type.

When outputting a labeling file, the correct file format should be chosen before generating the output. The *.txt file format is usually the right one for outputting project data such as cable device tags (e.g. for labeling cable shields). In practice, you should always check what data format can be edited by the downstream systems.

Fig. 6.96 Toolbar for editing the selected scheme

Next to the *Scheme* selection field is the toolbar with editing functions for the scheme. You use these buttons to create, edit, and delete schemes, etc.

6.7.1.2 Labeling as text output

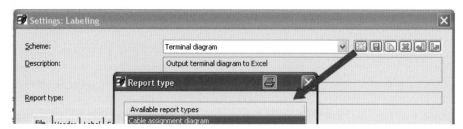

You create a new scheme via the ⬚ button. EPLAN first asks for the *report type*. You should enter a suitably descriptive name and description in the subsequent dialog. One advantage of this is that every user gains an immediate impression of what the scheme produces. Another advantage is that this scheme name is also used for exporting the scheme. This saves duplicate work later.

Fig. 6.97 Select report type

Once the header of the scheme has been defined, you can fill the scheme with the required format elements. Format elements in EPLAN represent the *available properties* of the project, pages, and devices. Format elements are thus nothing more than *properties*. The file type is set to *.txt. You can now use the *Header*, *Label* and *Footer* tabs to assign *format elements* to the scheme.

The tabs are all structured in a similar way. To the left is the field with the *available format elements* (the actual properties) and the right area contains the overview of currently selected format elements.

You use the ⬚ button to „copy" selected elements from the left field to the right field.

Depending on the property selected, you can open an additional dialog to make a more detailed selection of one of the properties.

You can delete elements from the right field by using the ⬚ toolbar. You must, of course, first select the element you wish to delete.

NOTE: Clicking the ⬚ button immediately removes the element. No confirmation prompt is displayed!

To assign a format element to the scheme, you switch to the *Label* tab. EPLAN offers a large amount of possible project data (properties) that can be used for labeling.

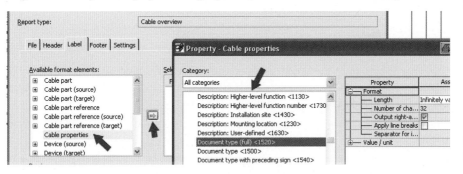

Fig. 6.98 Selection of desired properties

The cable device tag should be output in a cable overview. The cable device tag belongs to the cable properties group, since the DT is a property of a symbol (terminal, cable, etc.). You select the format element in the left field with the left mouse button. You now use the ⬚ button to move the *Cable property* format element from the left field (*Available format elements*) to the right field (*Selected format elements*).

EPLAN then opens the **Property – Cable properties** dialog. All available cable properties are listed here.

You can limit this overview of the properties by selecting a more suitable value in the **Category** field, e.g. *devices*. This hides all properties that do not belong to this category. This greatly simplifies the overview.

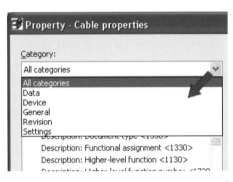

To allow the cable device tag to be output later in the labeling file, the *<20006 DT (full)>* property must be selected from the list and applied by clicking OK.

Fig. 6.99 Categories

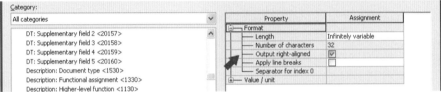

Fig. 6.100 Possible assignments for selected properties

The additional options in the right side of the dialog should be set as required. This requires no further explanation. After applying the property, EPLAN closes the dialog. Property 20006 is now located in the right field with the selected format elements and with the correct settings (unlimited variable length). These entries can be changed at any later time. You use the toolbar above the field for this. This applies to all available tabs in the Labeling dialog.

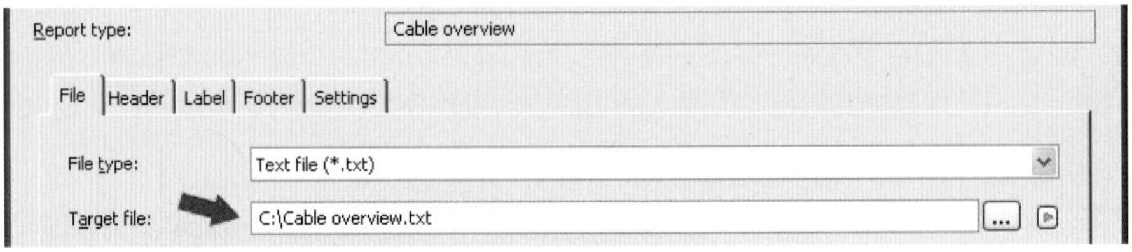

Fig. 6.101 Select target file

Now only the name of the target file is missing. This will complete all entries and the scheme can now be saved by clicking the [H] button. You can now exit the settings dialog via the OK button. EPLAN returns to the **Labeling** dialog.

The labeling file can now be generated. You must now only set the output type to suit your wishes.

If the file is only going to be exported, then the *Export* setting is sufficient. If the file is to be subsequently edited, you should set the Output type setting to *Export and start application*.

In the example, the result appears as follows (when opened in a text editor):

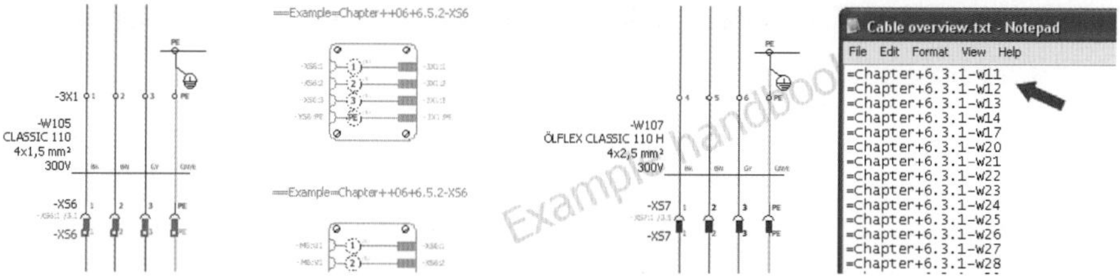

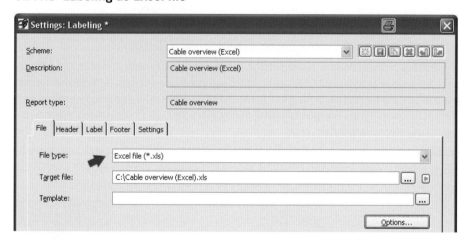

Fig. 6.102 Generated text file with information

6.7.1.3 Labeling as Excel file

Fig. 6.103 Scheme for output to Excel

In addition to text output, EPLAN also allows output in the Excel format *.xls. This allows you to replicate the graphical output in Excel if necessary. The preparatory procedure is similar to that for outputting a text file. This means that you again generate a scheme containing assigned format elements.

However, one point must be noted when generating Excel files. For an Excel output, a template **can** be stored in the scheme. This template is created in Excel with variables before the data is output and then later assigned to the scheme. EPLAN later writes the project data (as specified in the scheme) into these variables.

The following example is explained **without** an Excel template, since an Excel template is not required.

 NOTE: You **do not need to, but you can,** use an Excel template!

You move the data from the pool of available properties to the list of *selected format elements* in the familiar way.

Fig. 6.104 The Options button

The OPTIONS button on the *File* tab is important for output to Excel. If the *Include column header in output* setting is activated, the column headings are designated in Excel with the names of the corresponding format elements (here the full DT).

Once these conditions have been met, the data has been correctly stored in the scheme, and the scheme is saved, EPLAN can start the output of project data in the Excel format. As you already know, this is done via the UTILITIES / REPORTS / LABELING menu item. This is followed by the **Output labeling** dialog. Here you select the desired scheme and set the corresponding output type.

You now confirm the output by clicking OK. EPLAN generates the data and saves it in the Excel format. When you open the Excel file, in our example you can see the transferred cable overview with all the data that was previously assigned to the scheme.

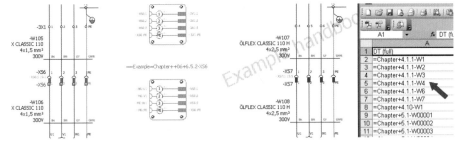

Fig. 6.105 Generated Excel file with the data from EPLAN

6.8 Edit properties externally

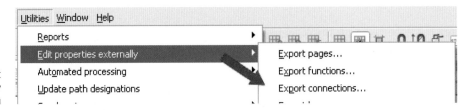

Fig. 6.106 The Edit properties externally menu

In addition to generating reports in the form of graphical output pages or files, EPLAN provides another powerful tool with the *Edit properties externally* function.

NOTE: One important thing should always be noted. Whatever you wish to externally edit must always first be highlighted (selected) in EPLAN.

If you wish to edit page properties then you must select the desired pages or select the entire project (project name) in the page navigator. This also selects all pages. If you wish to edit only particular terminals (functions) on a page then you must select precisely these terminals.

These functions can be described very simply. They allow project data to be transferred to Excel, edited there, and then transferred back to EPLAN.

NOTE: At the moment, Excel is the only spreadsheet program available for external editing!

Data that is easy to rename, exchange or enhance with the functions in Excel is then synchronized with the existing project data. This avoids subsequent editing in EPLAN; i.e. this is a very powerful function.

However, we should explicitly mention here that there is no UNDO function after transferring the data back! Once the data has been synchronized with the schematic, there is no way back. You should therefore always be very careful when using this powerful function.

If you are not sure, then you should work with a copy of the current project. Here you can check the results and adjust the settings for the external properties as necessary.

Fig. 6.107 File types for external editing

Using Excel is, of course, not the only way of doing this. EPLAN offers a number of other file types for output that are listed below.

- Excel file (*.xls)
- Tab-delimited Unicode file (*.txt)
- Tab-delimited text
- XML file (*.edc)

However, the following examples will only deal with output to Excel. Exporting of page data will be used to illustrate the creation of a scheme, the export to Excel, and the import into EPLAN.

The other possibilities of exporting functions or connections are performed in an identical manner. The only differences are usually just the data and / or the property values.

6.8.1 Export data

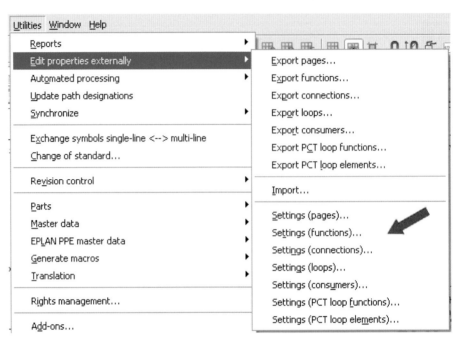

Fig. 6.108 Settings for external editing

A number of preconditions must be met before EPLAN can transfer data to Excel. These are determined in the settings under EDIT PROPERTIES EXTERNALLY / SETTINGS (pages, functions, connections).

EPLAN offers several ways of transferring project data to Excel. This relates to the following project data (extract):

Pages – all data relating to pages. For example, this relates to the page names, the page supplementary fields, or other properties assigned to a page, e.g. a form.

Functions – functions are device tags, cable comments, connection point designations, and everything relating to devices, symbols, or the associated parts. The range of possible data is, of course, much larger, but this is sufficient as an example.

Connections – everything relating to connection data.

 TIP: Basically, most properties are suitable for exporting. However, not all properties can then be externally edited. You should make a note of this!

This is because EPLAN needs certain important information in order to write back the modified data at the correct position in the schematic. If this data basis is denied to EPLAN, then it has no idea where the data should be placed and may end up generating incorrect data.

Some properties are locked for editing to prevent this. You can still change these externally, but EPLAN simply ignores these changes when reading back the data tables.

6.8.1.1 Settings – sample pages

EPLAN allows a large amount of data for pages and their descriptions. Most of these can be conveniently edited in the page navigator (page overview).

This is even easier when you create an export to Excel for the pages and their properties. You can then use many Excel functions that do not exist, or are very difficult to implement, in EPLAN. To use this feature you must first create a scheme in EPLAN with the associated properties. An Excel template is also useful, but not absolutely necessary.

For example, without an Excel template you get the following result:

Fig. 6.109 Exported data without Excel template

The problem with this representation is that all cells look the same here and can also be edited. However, as already mentioned, not all edited values are re-imported into EPLAN. To make it clear that this is not caused by errors in the program but rather errors in the application, it is a good idea to use a formatted Excel template.

A formatted Excel template provides an overview of which properties have changed and which properties are not able to be written back to EPLAN for programming reasons.

An example of a formatted Excel template may appear as follows:

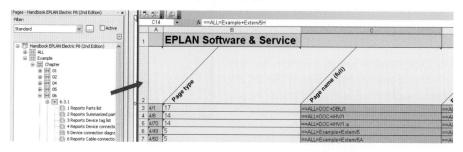

Fig. 6.110 Exported data with Excel template

The easiest way to create a template is to modify one of the templates provided in the EPLAN master data. You apply or use the usual Excel editing functions for this.

Only the following points are important for the template. There is a *header* and a *data area*. The header area is marked with the **#H#** identifier and the data area with the **###** identifier. The number of properties transferred by EPLAN is irrelevant for these templates. The rest of the template can be structured in any desired manner.

To allow EPLAN and Excel to select the different areas (which properties are editable or not), the Excel template has a second *Format* worksheet.

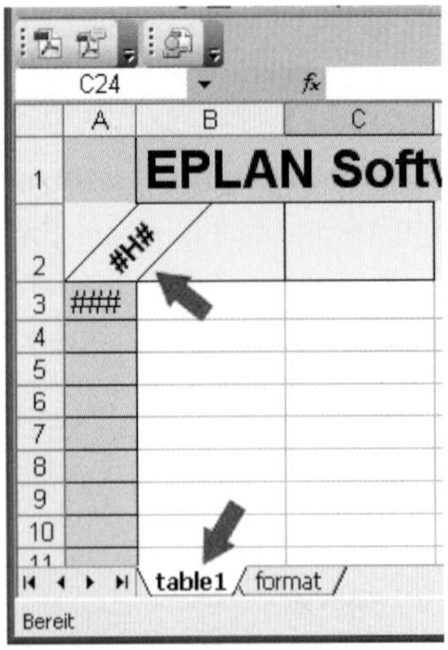

This worksheet has several entries. These entries should not be deleted, nor should the text be modified. However, the format of the cells (background color, font, text size, etc.) can be changed to suit your needs. Changing the cell format affects the later visual appearance of the worksheet.

Fig. 6.111 Worksheet table 1

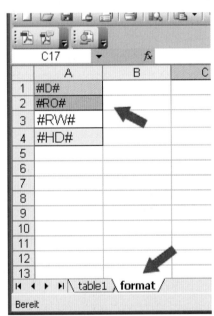

For the sake of completeness, the meanings of these entries are listed here.

- #H# – Header area
- ### – Data area
- #ID# – identifier for a property.
- #RO# – property has the read-only format and cannot be changed, or EPLAN ignores these values when reading back the data.
- #RW# – property has the read-write format and the values can be changed as desired. However, illogical values are not adopted by EPLAN.
- #HD# – Automatically added header data

The Excel template is now complete. You now save this in any desired directory. The EPLAN templates directory is an ideal place. We still don't have the EPLAN scheme allowing the properties to be transferred to Excel.

Fig. 6.112 The Format worksheet

To do this, you use the UTILITIES / EDIT PROPERTIES EXTERNALLY / SETTINGS (PAGES) menu to open the **Settings / External editing** dialog.

Fig. 6.113 Create new scheme

Click the [icon] button to create a new scheme. The **New scheme** dialog appears. Here you enter a descriptive name and a description for the scheme and apply these settings by clicking OK. You can now fill the new scheme with the desired values.

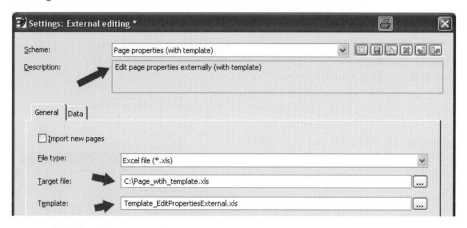

Fig. 6.114 Settings in the General tab

In the *File type* selection field, you select the Excel file (*.xls) value. You can enter any value into the *Target file* field. The *Template* field is important. Here, you should select the Excel template that you just created for the scheme. This completes the entries in the **General** tab.

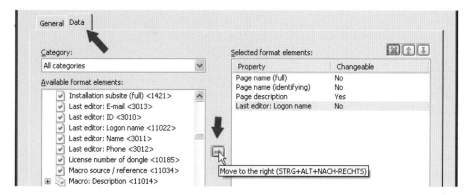

Fig. 6.115 Settings in the Data tab

To transfer the properties to Excel, you must now select the desired properties in the **Data** tab. This is done in a similar way to selecting properties for labeling data. As you can see in the image for this example, a number of properties have been added to the *Selected format elements* field.

This completes all of the required steps. The scheme is now saved, and the dialog can be confirmed by clicking OK.

6.8.1.2 Export data – sample pages

To be able to export page properties of a project with this scheme so that you can externally edit them, you call up the **page navigator** and select the desired project with the left mouse button or individually select the desired pages in the page navigator.

NOTE: If no pages are selected, then the EDIT PROPERTIES EXTERNALLY / EXPORT PAGES function cannot be executed!

You then use the UTILITIES / EDIT PROPERTIES EXTERNALLY / EXPORT PAGES menu item to open the **Edit pages externally** dialog. You must select and adopt the required *scheme* in the **Scheme** selection field. Select the desired option in the **Language** selection field and then select either a single language or all of the languages set in the project.

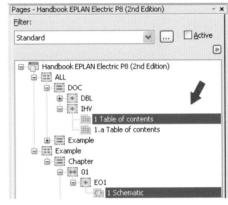

Fig. 6.116 Selected pages

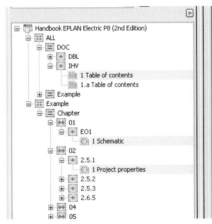

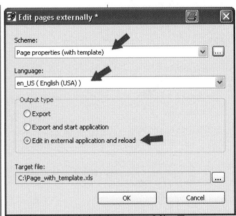

Fig. 6.117 Export settings

The **Output type** options are the most important point for outputting the data. The *Export* output type generates the data and just creates the output file. The *Export and start application* option generates the output file, and then opens this in the specified application (in this case Excel), where you can then edit and save the output file. This does not update any data in EPLAN. The saved output file can later be read back into the EPLAN project via UTILITIES / EDIT PROPERTIES EXTERNALLY / IMPORT. This option is useful when the data is to be further edited, e.g. by a customer.

The last option, *Edit in external application and reload*, includes a transfer of the data from EPLAN to Excel. It can now be edited there. After editing, the file can be saved and Excel can be exited. This is followed by a confirmation prompt asking if the data should actually be imported. If you confirm the dialog by clicking Yes, then EPLAN synchronizes the data from the Excel file with the project data.

To export the data, edit it in Excel, and then re-synchronize it with the EPLAN project, you should select the last output type. The target file (where the data should be written to) is specified in the scheme.

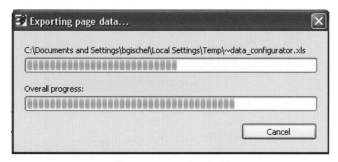

Fig. 6.118 EPLAN exports the data

Clicking the OK button generates the output file and then starts Excel. The loaded data can now be edited or changed.

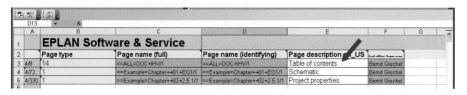

Fig. 6.119 Editing data in Excel

After making all your changes, you save the Excel file (click the 💾 button once or press CTRL + S) and can now close Excel.

Fig. 6.120 Prompt asking whether the changed data should be imported

This is followed by a confirmation prompt asking if you really want to import the changed data. When you confirm this by clicking YES, then all modified data is written back to the project and is now available there.

Fig. 6.121 Changed page description

NOTE: It is not possible to undo this action (importing changed data)!

6.8.2 Import

Up to now, we have exported data, edited the data externally, and then written it back to the EPLAN project. Immediately writing back the data may sometimes not be desired.

In this case, in the **Edit pages externally** dialog, you should set the output type to *Export* or *Export and start application*. This then only transfers the data to Excel and either immediately stores it in the target file, or writes it to the application where it can then be edited further and subsequently saved.

In this case, the project editing for the aforementioned cases can also be exited. The exported data can now be edited as desired, and then read back into the project at a later time.

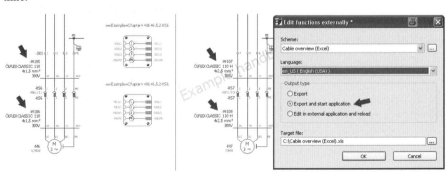

Fig. 6.122 Data sets selected for export (here cables)

This will be explained with an example. Cable function texts are meant to be edited subsequently or entered for a cable. To do this, you either select all cables in the cable navigator (by selecting the project) or select only the required cables on a page.

Using a scheme that contains the properties you want to change, the functions are now exported and then saved in an Excel file. This file can now be edited by an employee who (e.g.) doesn't even have EPLAN installed.

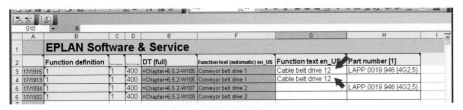

Fig. 6.123 The external file is edited.

The cable function texts are then entered or modified in the Excel file. After editing, the saved Excel file can now be read back (imported) into EPLAN.

To do this, the UTILITIES / EDIT PROPERTIES EXTERNALLY / IMPORT menu item is used to select the Excel file to be imported. The **Open** dialog is displayed. Here, you use the normal Windows functions to select and open the file using the OPEN button.

Fig. 6.124 Selecting the saved file

EPLAN now reads the file **immediately and without additional queries** and writes the modified data back to the project. This completes the import!

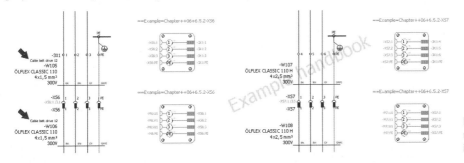

Fig. 6.125 Externally modified data after the import

 NOTE: No confirmation prompt is displayed at this point. EPLAN imports this data into the project immediately after opening the file! There is no UNDO function at this point! You should note this!

7 Management tasks in EPLAN

A large amount of different types of data must be managed in EPLAN. For this, EPLAN has functions that can handle these various, and sometimes different, tasks. This chapter will explain some of these functions.

Fig. 7.1 Structure identifier management

In EPLAN, you must not only manage all types of structure identifiers, but also project messages occurring either online while editing projects, or offline as a result of project checks. You must also organize and manage parts data and the various layers of a project. Thus the collective term "management".

■ 7.1 Structure identifier management

As soon as a project has a different page structure than **Sequential numbering**, the different structure identifiers in the project will eventually be used somewhere.

 Every possible identifier block in a project is assigned a specific identifier that may consist of a string of letters and/or numbers. These identifiers are known as structure identifiers.

The structure identifier management is called up via the PROJECT DATA / STRUCTURE IDENTIFIER MANAGEMENT menu item. EPLAN opens the **Identifiers - [Project name]** dialog.

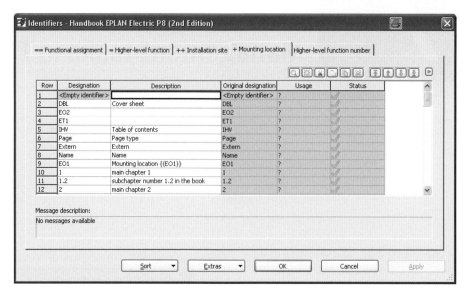

Fig. 7.2 The control center of structure identifier management

The possible structure identifiers such as **higher-level function**, the **installation site** or the **mounting location** are managed together in EPLAN. The sequence of these structure identifiers and the sequence of the pages in a project are closely related. The page structure is initially sorted in the page navigator in the same way as the structure identifiers.

Several mounting locations are entered in the structure identifier management (not all mounting locations need to have been used previously in the page structure).

Fig. 7.3 Sequence of the mounting location

Thus, the page structure in the page navigator is now sorted according to the defaults in the structure identifier management as follows:

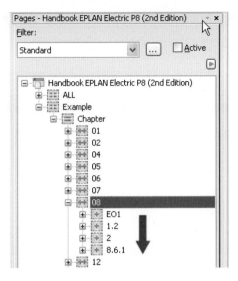

Fig. 7.4 Sequence in the page navigator

For example, if you want to sort mounting location ++EO1 after mounting location ++8.6.1 in the page structure, this sequence will be changed in the structure identifier management.

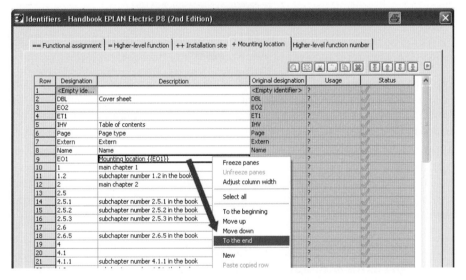

Fig. 7.5 Change sequence of structure identifiers

This is done, for example, by selecting the row and then calling the popup menu via the right mouse button. Here, you will see functions such as TO THE BEGINNING (will be moved there) or DOWN (moved by a row) and many others.

Following a resorting and a click on the OK button at the bottom of the dialog, EPLAN will prompt you whether the changes are to be adopted.

Fig. 7.6 Prompt whether changes to project data are to be adopted

Then, the changes are adopted and the page structure is adjusted in the page navigator.

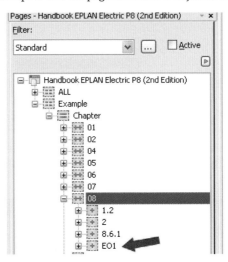

Fig. 7.7 Changed sequence

This is only one of many functions in structure identifier management. Among other features, it is also possible to (e.g.) copy structure identifiers from an Excel table into the structure identifier management. This saves the time-consuming process of manually entering the data into EPLAN.

A detailed description can be entered for every structure identifier. This means that reports or graphical generation of structure identifier overviews can be sensibly filled with exactly this information and the data need not be entered manually. The graphical output and the information contained therein depend, of course, on the structure of the forms used.

7.1.1 Tabs in the structure identifier management

The available tabs correspond to the page structure entered that was selected at the time the project was created.

 NOTE: Every possible identifier block in a project is assigned a specific identifier that may consist of a string of letters and/or numbers. These identifiers are known as structure identifiers.

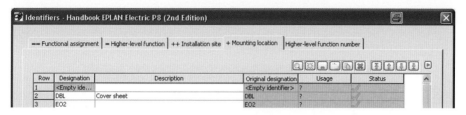

Fig. 7.8 Tabs in the structure identifier management

EPLAN can manage up to seven structure identifiers in one project, the so-called identifier blocks. The following structure identifiers are possible:

- **Functional assignment** – superior to the higher-level function identifier block. The following preceding signs are possible for functional assignment in EPLAN: "==", "#" or "≠".
- **Higher-level function** – the higher-level function identifier block is (e.g.) used for the identifiers of system parts. A possible preceding sign in EPLAN is "=".
- **Installation site** – the installation site identifier describes the physical site of devices or system parts. EPLAN uses "++" as the identifier.
- **Mounting location** – the mounting location describes the place where devices are mounted / installed. For example, pushbuttons would have a mounting location of "control panel". EPLAN allows "+" as a preceding sign for this.
- **Higher-level function number** – further structuring of devices. Is used (e.g.) for fluid devices. The preceding sign can usually be freely chosen, but should be compatible with the other identifier blocks.
- **Document type** – identifier block for the so-called KKS (power station identifier system). EPLAN allows a preceding sign of "&". A space character can also be used for this.
- **User-defined** – this identifier block can be freely chosen. Any character can be used as the preceding sign. However, the preceding sign used should, of course, not conflict with the existing identifiers.

Any number of identifiers can be entered into an **Identifier block**. New identifiers are (or can be, depending on the setting) sorted into the structure identifier management via a query dialog during graphical editing. Whether or not the **Identifiers** dialog is used during editing is defined in a user-specific setting. This setting is accessible via OPTIONS / SETTINGS / USER / DISPLAY / IDENTIFIER.

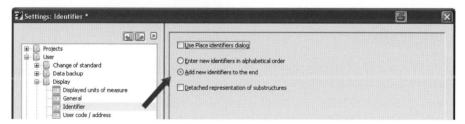

Fig. 7.9 Settings for Place identifiers

If the parameter **Use Place identifiers dialog** is switched on, then when an identifier is entered that is not already in the structure identifier management, EPLAN automatically displays the **Place identifiers** dialog.

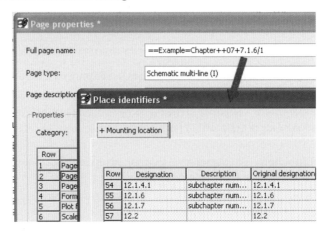

Fig. 7.10 Place identifiers dialog

Depending on the setting of the two options, **Enter new identifiers in alphabetical order** or **Add new identifiers to the end**, the focus in the **Place identifiers** dialog will be different.

 TIP: Every possible identifier block in a project is assigned a specific identifier that may consist of a string of letters and/or numbers. These identifiers are known as structure identifiers.

Before closing the **Place identifiers** dialog, EPLAN allows the user to reposition the new identifier in the table using the graphical buttons (see next chapter for explanation).

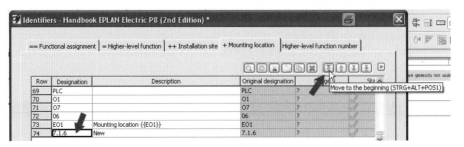

Fig. 7.11 Buttons in the structure identifier management

To do this, you select the row header and then use the graphical buttons to move the identifier to the desired position.

If the USE PLACE IDENTIFIERS DIALOG parameter in the setting **Options / Settings / User / Display / Identifier** is not used, then EPLAN *does not* open the **Place identifier** dialog during project editing, and sorts the identifiers automatically without a query, according to the settings in the options, **Enter new identifiers in alphabetical order** or **Add new identifiers to the end**.

In order to assign **Descriptions** to these automatically sorted identifiers, the structure identifier management must be started and the descriptions entered later.

7.1.2 The graphical buttons

When working on a project, it is not always possible to ensure that only desired identifiers are entered into the identifier structure. Various types of copy and macro actions usually create unwanted entries in the structure identifier management.

For this situation, EPLAN allows convenient subsequent editing of the structure identifiers via functions such as Move or Delete in the graphical toolbar:

These (and more) functions are also available via the right-click popup menu. The graphical buttons have the following functions:

	Opens the **Replace** dialog. Allows the user to perform extensive find and replace actions. This usually replaces the manual editing of identifiers.
	Inserts a new row before the currently selected row.
	Cuts the selected cell(s) and allows this (these) selected cell(s) to be inserted at a different place in the **Identifiers** dialog.
	Paste function. Inserts copied (or cut) cells at the specified location.
	Copies selected rows to the clipboard.
	Deletes selected rows.
	Moves the selected row to the beginning.
	Moves the selected row one position upwards.
	Moves the selected row one position downwards.
	Moves the selected row to the end.

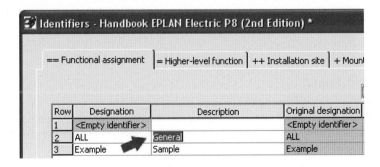

Fig. 7.12 Direct editing using F2

To edit an identifier, click on the corresponding field and edit the entry. Or you can press the F2 key to edit entries.

Either option opens up different right-click popup menus. But in either popup menu you can use similar functions, such as COPY, REMOVE or TRANSLATE.

Single selection **Direct selection**

Fig. 7.13 Selected row

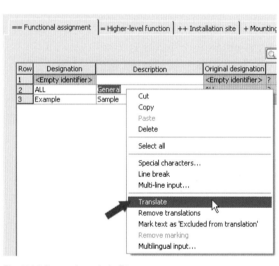

Fig. 7.14 Row selected via F2

 NOTE: The right-click popup menus can also be accessed via the button.

To translate column entries, select them for translation and translate them using the CTRL + L shortcut key or the right-click popup menu. If EPLAN finds multiple matching entries in the dictionary, then it displays the **Found words** dialog. You can then select the most suitable entry, and then adopt this by clicking OK.

7.1.3 Sort menu

The **Sort** menu provides a few automated sorting options for identifiers. Regardless of the sorting used, it is initially a temporary sorting and will not be adopted until the **Identifiers** dialog is closed via OK or APPLY.

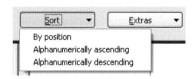

Fig. 7.15 Sort button

Using the SORT menu and the BY POSITION option, EPLAN sorts the identifiers according to the internal item numbers.

The ALPHANUMERICALLY ASCENDING option sorts the identifiers according to their character strings (numbers take precedence over letters).

The ALPHANUMERICALLY DESCENDING option sorts the identifiers in descending order.

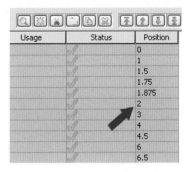

Fig. 7.16 Item numbers

The identifiers do not generally need to be selected in order to arrange them using the SORT option. The cursor must only be in a field. You should note that when the identifiers are automatically sorted, the sorting of the page structure follows that of the identifiers.

 NOTE: A new sorting always applies to all identifier types collectively! It is not possible to sort individual identifiers automatically!

7.1.4 Extras menu

The EXTRAS button displays an additional menu. This menu contains items allowing unnecessary entries to be cleaned from the database and also offers a FIND function.

Fig. 7.17 Extras button

The CHECK USAGE option works as follows: When this function is executed, EPLAN checks all identifiers to see if they are used and exist in the project or if they are only entries in the identifier database.

These can be entries that (e.g.) were created from previous copy operations and which can now be deleted. EPLAN designates such identifiers that are no longer used with a **No** in the **Usage** column. These identifiers can be safely removed from the database via the graphical [DELETE] button.

Row	Designation	Description	Original designation	Usage	Status	Position
1	<Empty identifier>		<Empty identifier>	*	✓	10
2	AL2		AL2	No	✓	11
3	DOC	Info	DOC	3	✓	12
4	Example	Sample	Example	58		13
5	Chapter	chapter	Chapter	4084	✓	14
6	ALG		ALG	No	✓	15
7	51		51	No	✓	16

Fig. 7.18 Check usage

The CLEAN UP ALL option automatically removes all entries from the database that are no longer required.

NOTE: Once the function is started, no additional queries are displayed, the entries are immediately checked for usage, and all unnecessary identifiers are immediately removed.

Since the system does not know when (e.g.) identifiers for higher-level functions or mounting locations have been inserted for later use, these identifiers are also removed. You should therefore be careful when using this function.

If these changes are applied with a click on the OK button, these additional (unused) entries will have been removed definitively from the database!

The FIND option displays the **Replace** dialog. You can use this to locate entries in the database and replace them with other strings.

Find and replace allows for a simple way of replacing terms (or partial terms) with other terms. Here too you should work carefully because no Undo function is available. Replaced then means replaced!

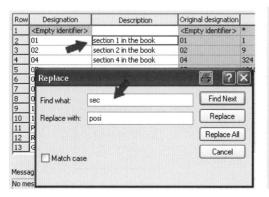
Fig. 7.19 Before the replacement

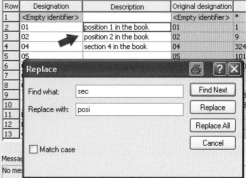
Fig. 7.20 After the replacement

 NOTE: An unwanted replace action can only be undone if the **Identifiers** dialog is exited via the CANCEL button.

Fig. 7.21 More buttons

In addition to the previously described SORT and EXTRAS buttons, the dialog also has OK, CANCEL and APPLY buttons.

You use the OK button to exit the **Identifiers** dialog and, depending on the actions you have performed, you may need to confirm your changes in a subsequent dialog.

The CANCEL button is very important, because it allows you to discard all changes made, and the identifier editing has no Undo function.

This APPLY button adopts the changes in the identifier management without closing the **Identifiers** dialog. When performing comprehensive actions such as renaming or deleting of unnecessary identifiers, you should first save the contents of the **Identifiers** dialog via the APPLY button before performing the replacement action. If something goes wrong with the replacement action, you can then exit the Identifiers dialog with exactly the status it had before performing the replacement action without saving. This means that all the previous work done will be lost.

 TIP: A kind of „interim save" (APPLY button) is necessary when editing the identifier structure so as not to lose all changes!

7.2 Message management

Why does the message management exist? EPLAN is an online system, and the message management is a perfect tool for immediately checking (i.e., online) whether the action you have just performed was correct (depending on the project check settings).

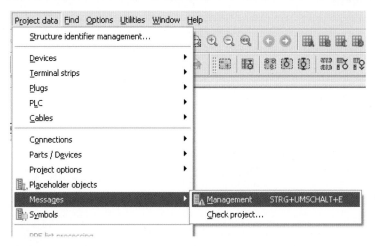

Fig. 7.22 Message management

Message management is started as an extra dialog via the PROJECT DATA / MESSAGES / MANAGEMENT menu, or via shortcut key, such as CTRL + SHIFT + E.

The **Message management** dialog can be placed anywhere on the desktop and can be displayed / hidden via a (user-defined) keyboard shortcut (e.g.) CTRL + SHIFT + E.

You do not need to keep this dialog open when editing projects. However, if you wish to receive immediate notification of faulty data entries (depending on the types of checks set), then you should keep the message management open.

>
> **TIP**: The greater the number of messages in message management, the slower project and/or graphical editing may be. Therefore, you should hide the dialog unless you need it.

When project editing is almost finished and the project is to be checked for faulty entries, the dialog can be displayed for extra information, since all **Errors, Warnings**, and **Notes** generated by the corresponding **Check run** are listed here.

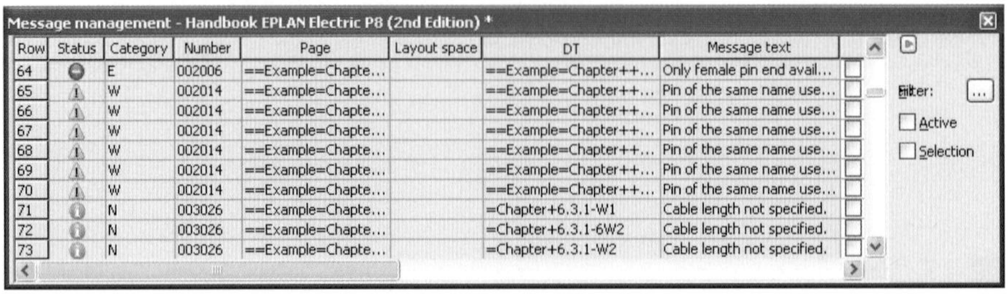

Fig. 7.23 Message management

When you start editing a project in EPLAN, various actions, such as inserting entire pages from other projects or inserting existing macros into the project, can result in unwanted data in the various databases.

For example, unwanted device combinations are created or new structure identifiers are added to the project. You cannot really avoid the creation of such data that do not really belong to the project without being constantly „interrupted" in your actual project editing. To make sure that you do not forget to change this unwanted data, EPLAN provides an online monitoring feature, or you can use manual check runs to check the project using a specified scheme (offline) and generate messages in the message database.

7.2.1 The visual appearance of message management

The dialog supports the user with the following features: Display of the message type, the message priority, and further information such as a brief message description or the jump point to the faulty position in the project.

Small colored icons make it easy to visually distinguish between the different types of messages.

●	Error	For example: „Duplicate connection point designation"
⚠	Warning	For example: „Connection point description missing"
ⓘ	Note	For example: „A placement lies outside the range"

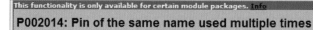

Fig. 7.24 Note on message in online help

If you position the cursor in a message row and then press the F1 key, EPLAN displays the online help for the corresponding message number, lists the possible cause(s), and suggests one or more solutions to the selected message.

7.2.2 Project checks

EPLAN generally distinguishes between **offline checks** and **online/offline checks** as well as the **Prevent errors** check type and also allows you to "remove" checks entirely from the check run, i.e. switch off the check: the *No* check type

Fig. 7.25 Types of check

However, this feature should only be used occasionally. Otherwise, you run the risk of certain true project editing errors not being discovered (e.g., duplicate DT) until the actual system is built, which will then present a much bigger problem.

The user decides what type of check routines to use. The **online check routine** is the best option, but is also the most computing-intensive because all the data and resulting messages must be continually checked and updated. This, of course, has a negative effect on the performance during project editing.

The **offline check routine** is the "second choice", but I prefer to use this one. Nobody manages to constantly correct all the online messages (errors, warnings or notes) while editing a project!

Check runs are started via the PROJECT DATA / MESSAGES / CHECK PROJECT menu or directly via the CHECK PROJECT function in the opened message management via the right-click popup menu.

7.2.3 Message classes and message categories

EPLAN allows the user to define the categories that messages belong to. This has the great advantage of allowing company-specific check runs to be developed that exactly match the way in which projects are developed within the company.

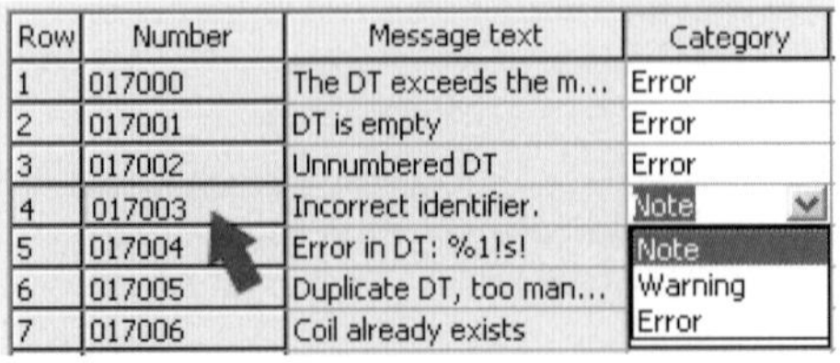

Row	Number	Message text	Category
1	017000	The DT exceeds the m...	Error
2	017001	DT is empty	Error
3	017002	Unnumbered DT	Error
4	017003	Incorrect identifier.	Note
5	017004	Error in DT: %1!s!	Note
6	017005	Duplicate DT, too man...	Warning
7	017006	Coil already exists	Error

Fig. 7.26 Check run message 017003

For example, not all messages generated or supplied by EPLAN are true error messages. The message <017003 Incorrect identifier> is an example of that. It can be an error message, a warning, or a note, depending on the project editing sequence. You can create your own schemes for check runs in the message management so that these messages are adjusted to exactly suit your project editing process.

To do this, you use the PROJECT DATA / MESSAGES / CHECK PROJECT menu to open the **Check project** dialog.

In this dialog, you can select a scheme in the **Settings** field for the subsequent check run.

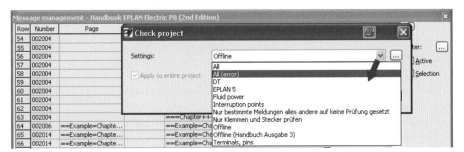

Fig. 7.27 Check project dialog

To add your own scheme to this selection, you click the selection button [...]. The **Settings / Messages and checks** dialog is displayed. To create a new scheme, click the [NEW] button in the upper toolbar.

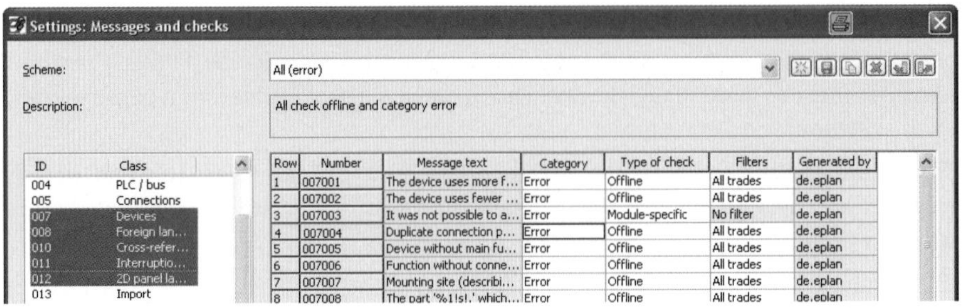

Fig. 7.28 Dialog for creating test schemes

Fig. 7.29 Create new scheme

EPLAN opens the **Create new scheme** dialog. Here you assign a suitably descriptive **Name** and **Description**. When you click the OK button, the scheme is saved and automatically entered into the **Scheme** selection field of the **Settings / Messages and checks** dialog.

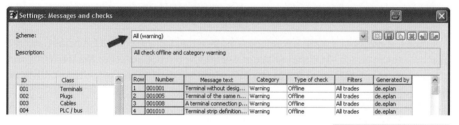

Fig. 7.30 Example of a new scheme

The individual messages can now be set and edited according to the desired specifications.

Every message belongs to a message class. There are currently 25 message classes. They range from the message class 001 Terminals to message class 999 External. The message classes cannot be extended and are permanently defined by EPLAN.

This means that every message begins with a specific three digit number. This defines the message class. In our example, the message <017003 Incorrect identifier> is assigned the DT message class. The rest of the message number is a three-digit number.

Fig. 7.31 Message classes

Message text
A pin connection point has too many targets: %1!s!. Maximum num...
Pin without targets.

Fig. 7.32 Message text

Every message number has a corresponding message text. As with the message number, this message text is defined by EPLAN and cannot be changed.

However, to return to their own scheme, users can set the category and type of check for this message.

EPLAN divides messages into three categories. There are **Notes**, **Warnings**, and **Errors**. The user must now decide on one of these three categories for the selected message. Should this be a note, a warning, or an error?

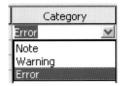

Fig. 7.33 Category

Once the category for the message has been defined, a check type must be decided. There are several check types in EPLAN.

Fig. 7.34 Types of check

The **No** type of check does not carry out any checks for this message either automatically or manually. This will be clarified with a small example.

Fig. 7.35 Type of check set to „No"

Number	Message text	Category	Type of check
001001	Terminal without designation.	Error	No

A terminal is positioned in the schematic. This terminal is not given a designation, and the check for the message **Terminal without designation** has been set to **No**. If a check run is now carried out, there will not be any check run message **Terminal without designation**.

Fig. 7.36 Message management after completed check run

The check type **Offline** only generates messages when a check run is manually started.

Row	Number	Message text	Category	Type of check
1	001001	Terminal without designation.	Error	Offline

Fig. 7.37 Type of check set to „Offline"

Again, a terminal is positioned in the schematic. This terminal is not given a designation either, and the check for the message **Terminal without designation** has been set to **Offline**. EPLAN does not yet check this terminal during placement. Only once the check run has been started manually (offline) does EPLAN generate the check run message **Terminal without designation**.

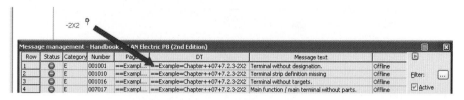

Fig. 7.38 Type of check set to „Offline"

The check type **Online / Offline** generates messages online and immediately enters them into the opened message management. Completed messages are not removed from the message database until the next check run.

Row	Number	Message text	Category	Type of check
1	001001	Terminal without designation.	Error	Online / offline

Fig. 7.39 Type of check set to „Online/Offline"

Another terminal is positioned in the schematic. This terminal is not given a designation, and the check for the message **Terminal without designation** has been set to **Online/Offline**. In contrast to the previous check runs, EPLAN checks whether a designation has been entered already during the placement of the terminal. Since in our example no designation has been assigned, EPLAN generates immediately (online) the corresponding check run message **Terminal without designation**.

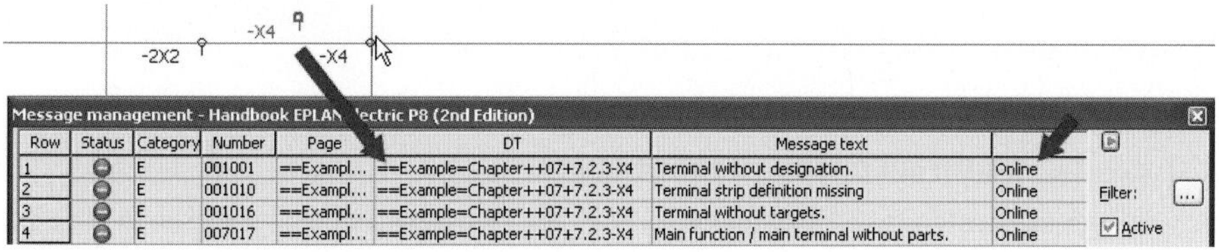

Fig. 7.40 Type of check set to „Online/Offline"; EPLAN reports the error immediately.

The **Prevent errors** type of check means that EPLAN checks the input during an action, for example, when inserting devices, and, if this results in errors, shows a message immediately and then "prevents" the action.

Row	Number	Message text	Category	Type of check
1	001001	Terminal without designation.	Error	Prevent errors

Fig. 7.41 Type of check set to „Prevent errors"

Again, a terminal is inserted and placed without a designation. During the placement, EPLAN checks whether this action violates any check and immediately reports the error in the **Prevent errors** dialog.

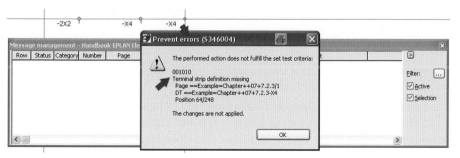

Fig. 7.42 EPLAN prevents the error on the basis of the type of check defined.

Fig. 7.43 Terminal can be repositioned.

EPLAN terminates this action, and the terminal can be repositioned and relabeled. If the default requirements are met (terminal designation exists), the terminal can be placed.

> **NOTE:** The **Prevent errors** type of check, of course, is very strict and should, in my personal estimation, not be used during an ongoing project editing. But when, for example, schematics are revised, this is a very good opportunity to prevent errors immediately while revising schematics, instead of processing such errors again later on as part of a check run!

The **Module-specific check** type of check also generates messages during certain automated actions, but cannot be influenced any further.

7.2.3.1 Filters for all check run messages

Apart from the previous settings for types of check or categories, it is possible to define a separate filter for each check run message.

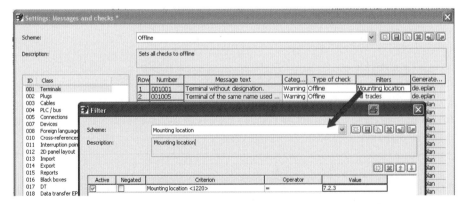

Fig. 7.45 Filter options for each check run message

These filters allow you to limit certain checks to specific areas or properties. In other words, checks can be targeted quite specifically.

The **Terminal without designation** check is set to the **Offline** type of check, but is to be checked only in a specific mounting location. This will exclude from the check all other faulty terminals that do not match the defined filter criterion.

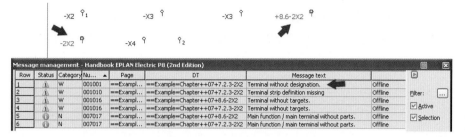

Fig. 7.46 Only checks defined by filter will be carried out.

In the example, the terminal without a designation, whose mounting location was different than the one defined in the filter, was ignored by the check run! The terminal whose mounting location matches the filter, however, was captured by the check run, and EPLAN generated a corresponding check run message.

 TIP: This feature, i.e., placing filters on individual check run messages, should be used very sparingly. The check run could be slowed down considerably as a result, because EPLAN would always have to evaluate the check run message as well as the filters!

7.2.4 Filters in the message management

As described in the last chapter, EPLAN messages are categorized into specific message classes. All message classes and areas (electrical engineering, fluid, or e.g. errors) can be switched on and off via filters.

7.2.4.1 Filter active setting

As described in the last chapter, EPLAN messages are categorized into specific message classes. All message classes and areas (electrical engineering, fluid, or e.g. errors) can be switched on and off via filters.

To do this you use the ⋯ button to open the **Filter / Messages** dialog.

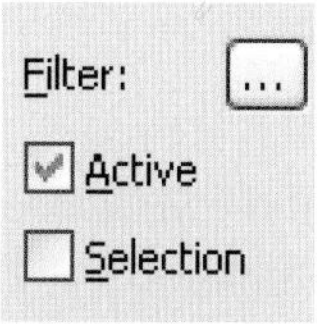

Fig. 7.47 Filter active

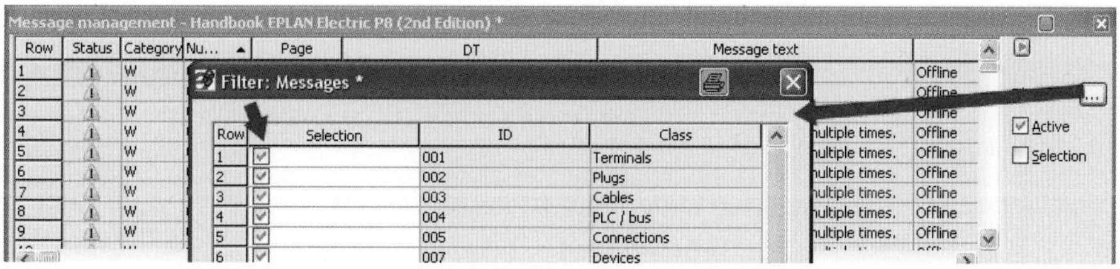

Fig. 7.48 Filter settings for the selection of message classes

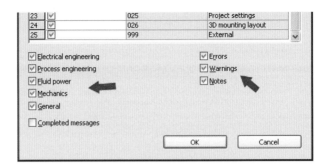

Fig. 7.49 Filter settings for the areas

In this dialog, by deselecting the **message classes** (terminals, plugs, cables, etc.), as well as the **areas** (electrical engineering, process engineering, etc.), it is possible to filter different data. You finish your entries and close the dialog by clicking the OK button. To activate the filter for the message management display, you must set the checkmark in the **Active** check box below the filter.

A typical example of a filtered display would be (e.g.) removal of the completed messages from the message management until the next check run.

7.2.4.2 Filter selection setting

If this check box is set, then only messages from selected objects are displayed in the message management. All other messages from other objects etc. are hidden. This avoids tedious searching for messages relating to particular objects.

Fig. 7.50 Filter selection

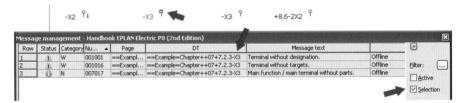

Fig. 7.51 Selection has been activated.

If the SELECTION selection box is active, then only the messages relating to the selected objects are displayed.

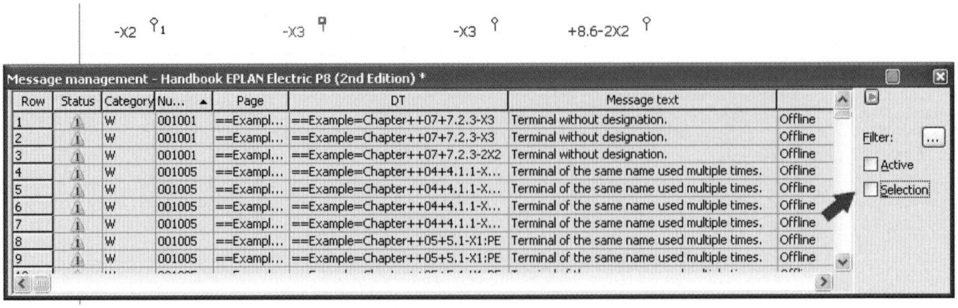

Fig. 7.52 Selection has not been activated.

If the SELECTION selection box is not active, message management displays all messages. A selection of objects does not play any role here.

7.2.5 Various ways of editing messages

This chapter does not deal with the correction of generated error, warning, or note messages. I refer you instead to online help, which explains problems and related solutions for check run messages on a much wider scale.

It focuses more on the ways in which messages can be localized in the schematic, and what tools EPLAN provides for this. The general functions for displaying, for example, Freeze panes or Adjust column width as well as copy functions will not be explained any further here.

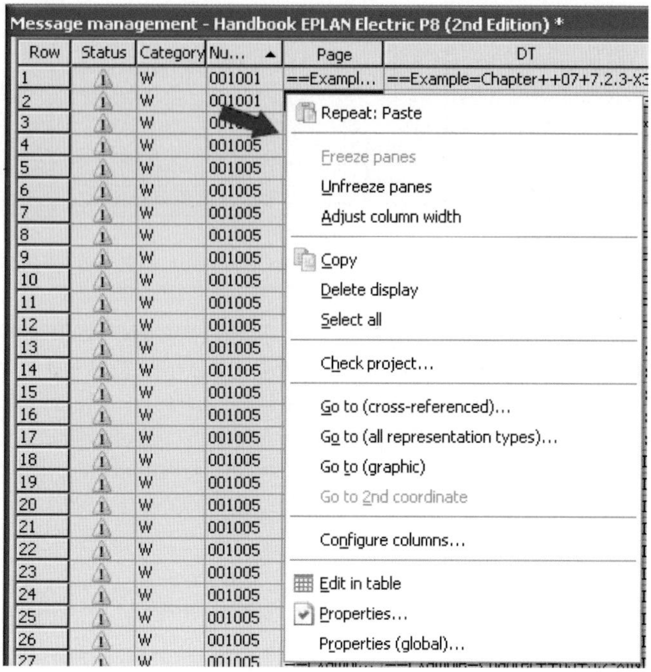

Fig. 7.53 Popup menu

The right-click popup menu provides several options for easily finding errors. The key ones are described here.

DELETE DISPLAY – this option deletes all messages from the message management display. The messages are still present, but are no longer displayed. They re-appear after the next check run.

CHECK PROJECT - this function carries out the set check run; depending on the selection, this is done for the entire project or only an individual page.

GO TO (CROSS-REFERENCED) - with this function, EPLAN enters all cross-referenced objects in the **Go to** list and then opens the **Find** dialog with the **Go to** tab.

GO TO (ALL REPRESENTATION TYPES) - this function determines all representation types (schematic, report pages, etc.) of the function and also enters them in the **Go to** list. From here, you can navigate directly to the desired representation.

GO TO (GRAPHIC) - this function allows you to navigate to the object in question directly from message management.

GO TO 2ND COORDINATE - for example, if there are two duplicate terminals, you can use this function to jump directly to the other terminal.

CONFIGURE COLUMNS – this function allows the number of columns displayed in the message management to be increased or decreased.

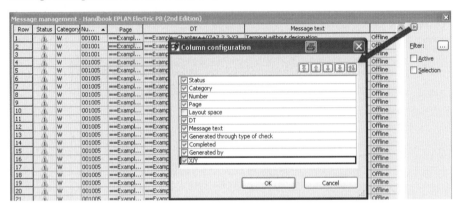

Fig. 7.54 Configure columns

When this function is executed, EPLAN displays the **Column configuration** dialog. Here you can remove or add columns, or change the sequence of the columns.

 TIP: To obtain a preview of the incorrect objects, the GRAPHICAL PREVIEW should be enabled in the VIEW menu. Ideally, this preview should be used with a multi-screen system.

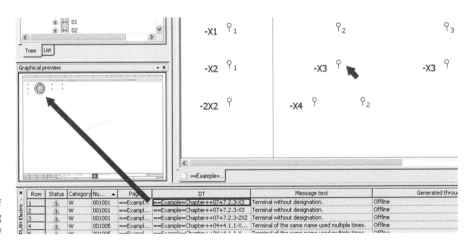

Fig. 7.55 Preview of error locations using graphical preview

7.3 Layer management

What are layers? Layers are an essential element originating from mechanical engineering (CAD). Information of the same type (e.g., dimensions) is placed on the same layer. For example, the form, color, font size or other formats of this layer can later be changed at a central place easily and without errors.

This is a great advantage, because otherwise every property that is not located in a layer must be manually edited. The layer management makes such editing actions a "piece of cake".

EPLAN has adopted and extended this well-known idea from mechanical engineering into the CAE area. You start the layer management via the OPTIONS / LAYER MANAGEMENT menu. The **Layer management** dialog is displayed. The layer management is a project-specific setting, and you can recognize this from the window title bar that displays the currently active project.

The layer management has the following structure. The dialog has windows at the left and right sides. The left window contains a tree with the superior layer designations (nodes and any subnodes and then the layer designation EPLANxyz), and the right window contains the associated sub-entries (the actual layer information).

Fig. 7.56 Layer management

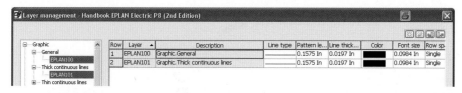

Fig. 7.57 Layer management dialog

7.3.1 Standard layers

EPLAN is supplied with a number of standard layers. They begin with the name EPLAN followed by a three-digit number.

NOTE: Original EPLAN layers cannot be removed (deleted).

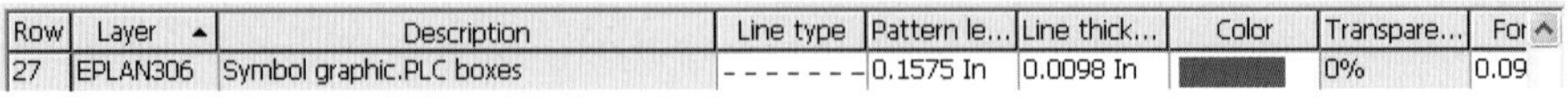

Fig. 7.58 Standard EPLAN layer

Row	Layer	Description	Line type	Pattern le...	Line thick...	Color	Transpare...	For
27	EPLAN306	Symbol graphic.PLC boxes	- - - - - -	0.1575 In	0.0098 In		0%	0.09

7.3.2 Export and import layers

All layers can be modified and/or adapted to the requirements, including the original EPLAN layers. However, you should first export all layers before making any changes to the standard layers. You then have a functioning backup copy of the previous layer settings.

EPLAN layers are exported in the familiar manner. Before starting the export, you must first select all desired layers using the usual Windows functions. The upper right of the dialog contains the graphical buttons.

You use the button to start the export. EPLAN opens the **Layer export** dialog. The file name contains a default value set by EPLAN, which is the name of the project. When you confirm the dialog by clicking SAVE, the layers are exported and saved.

Fig. 7.59 Layer export dialog

Layer import is done in a similar way. Clicking the button causes EPLAN to open the **Layer import** dialog. Here you select the desired layer configuration file and import it into the opened EPLAN project by clicking the OPEN button.

Fig. 7.60 Layer import

NOTE: All layers not already present in the project are imported and created as new layers. If layers of the same name already exist in the project, then the data in the import file always has priority. This means that layers in the project may be updated.

7.3.3 Create and delete your own layers

In addition to the standard layers provided with EPLAN, you can, of course, create your own layers. You should be somewhat systematic in doing so, otherwise you will end up with a "proliferated chaos". Layers can only be (directly) created in existing nodes such as graphics or property placements, but new (superior) nodes cannot be directly created.

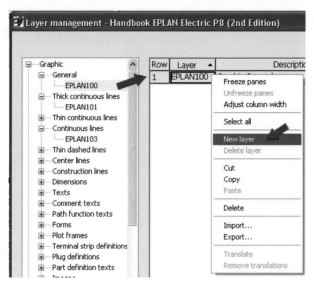

Fig. 7.61 Create a new layer

You create new layers from the **Layer management** dialog by right-clicking and selecting the NEW LAYER popup menu item (in the right area of the layer management dialog), or by using the [icon] button. To do this, you use the mouse to select the layer in which the desired layer should be created.

Now you call up the NEW LAYER function. EPLAN generates a new layer with the name NEW_LAYER_1.

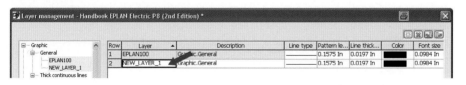

Fig. 7.62 New user-created layer

This layer can now be edited. The key columns here are **Layer** and **Description**.

The **Layer** column bears the name of the layer. When you create your own layers, you should give them sensible names. One possibility is to name the layers after the company and to include a range of numbers, e.g. COMPANY1000.

The **Description** column describes the layer itself. This description can be viewed later at the time of assigning layers of individual objects in the project. You should therefore enter a descriptive text here that allows the function to be assigned to a layer based on the description.

TIP: Subsequently, the layer will be sorted into the left-hand area of layer management on the basis of the description!

The layer is now changed from NEW_LAYER_1 to HB3-0100. The layer description is changed to Manual.Graphic.Special. This way, EPLAN sorts the layer after being saved as follows.

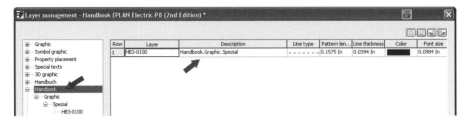

Fig. 7.63 Newly sorted layer HB3-0100

Properties like line type, pattern length, line thickness, etc. are self-explanatory and are stored or selected in accordance with the layer requirements.

Fig. 7.64 Other layer settings options

7.3.4 Uses of layers

What can you use all these layers for? You can use your own layers (e.g.) to provide extra project information for the workshop. This could be cabling notes or deadlines. The possible uses are limitless.

As described above, a new layer (HB3-0101) was created with the description **Manual. Graphic.Texts.Workshopinformation**. This layer is to display further information for workshop production in the schematic. When the workshop has finished using this information, it should then be easy to hide this information for the entire project.

Fig. 7.65 Workshop layer

A schematic page with corresponding entries is then opened in the project. Now, special information for the workshop is placed on the page as text using the EPLAN default settings for texts.

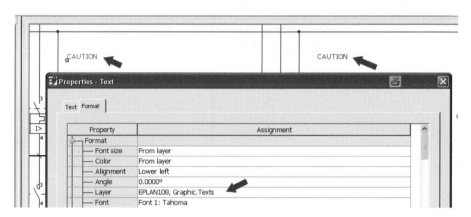

Fig. 7.66 Text with EPLAN default setting

Then, these texts are selected and placed on the newly created layer **Manual.Graphic.Texts.Workshopinformation**.

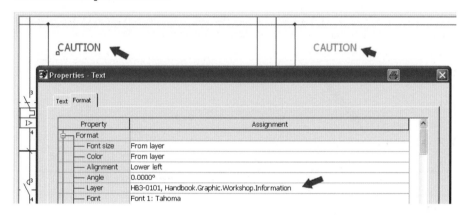

Fig. 7.67 Texts placed on the new layer

EPLAN handles the settings that pertain to this layer and formats the texts accordingly. Via layer management, this workshop information can now be hidden, for example, prior to printing.

Row	Layer	Description	Color	Font size	Row spacing	Paragraph spacing	Text box	Visible	Print	Locked
1	HB3-0100	Handbook.Graphic.Spezial	■	0.0984 In	Single	0.0000 In	No	✓	✓	□
2	HB3-0101	Handbook.Graphic.Workshop.Information	■	0.1378 In	Single	0.0000 In	No	✓	□	□

Fig. 7.68 Exclude layer from printing

This is only a small example of what is possible with the layer management. The user has almost unlimited possibilities here. The biggest advantage of layers is that it is easy to globally change the format, representation, etc. of the respective element on this layer with just a few mouse clicks.

7.3.4.1 Remove unnecessary layers from the project

To remove layers that have been rendered superfluous, for example, layers that were stored in the project and subsequently changed as a result of copy actions or the insertion of DXF/DWG files, you can use the compress function of the project.

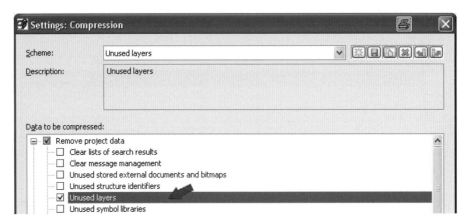

Fig. 7.69 Compression scheme to remove unnecessary layers

After the scheme is launched, EPLAN removes the unnecessary layers and then displays a summary of the layers removed.

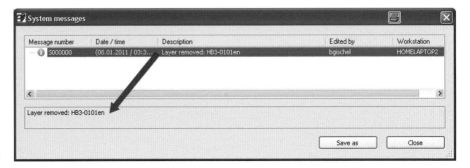

Fig. 7.70 System message

7.4 Parts management

All devices in EPLAN are managed with all their technical and commercial data, such as technical characteristics, dimensions (width, height, depth) or prices, in **Parts management**.

However, not only the device-specific data is managed in the parts management, but also the corresponding function definitions, symbols or also symbol macros can be stored for every device.

The parts management is called up via the UTILITIES / PARTS / MANAGEMENT menu.

7.4 Parts management

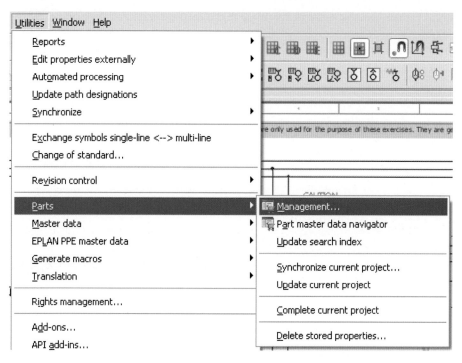

Fig. 7.71 Parts management menu item

The actual parts database used is defined via SETTINGS / OPTIONS / SETTINGS / USER / MANAGEMENT / PARTS MANAGEMENT. You can also run the parts management from an SQL Server database instead of an Access database.

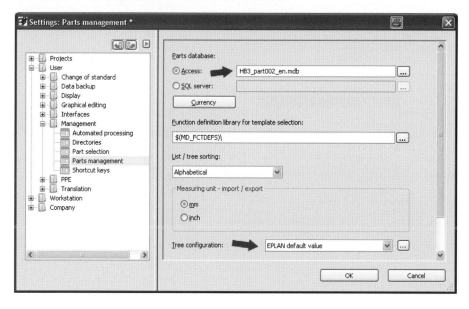

Fig. 7.72 Parts management user settings

In these **User-specific settings**, the **View** in the parts management (tree configuration) can also be set. There are various options to adjust the tree configuration view to your own specifications by using your own schemes.

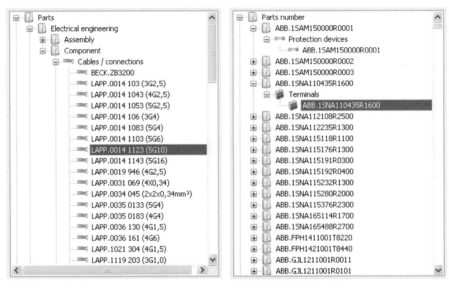

Fig. 7.73 Standard view EPLAN

Fig. 7.74 View by part

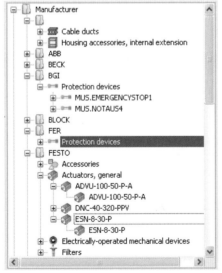

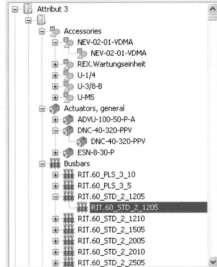

Fig. 7.75 View by manufacturer

Fig. 7.76 View by attribute 3

7.4.1 Structure of the parts management

Describing parts management in all its details would probably require a separate book. As a result, this chapter will be limited to the most important information that is of interest and useful to your practical work.

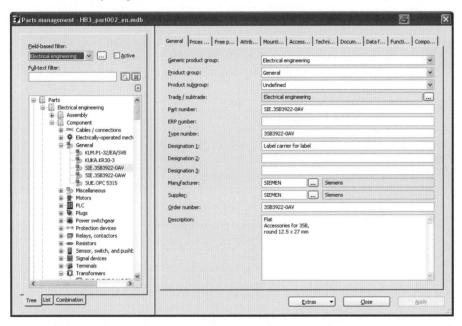

Fig. 7.77 Overview of the parts management dialog

The **Parts management** dialog basically consists of an area to the left - an overview of all parts, customers and manufacturers/suppliers - and an area to the right containing detailed information. Both parts are separated by "splitters" and, therefore, scalable to (almost) any size.

The area to the right always relates to one (or more) selected parts in the area to the left – the parts or customers and/or manufacturers/suppliers. The area to the right, i.e. the parts data, is further divided into several tabs containing the various data relating to the part.

7.4.2 Tabs in the parts management

Each part has several tabs. Various technical and commercial data for the part are defined in these tabs.

The existing tabs of the part Siemens SIE.3LD2 504-0TK53 will serve as our example here.

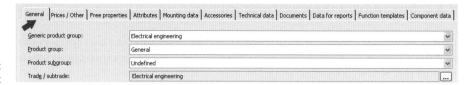

Fig. 7.78 The different tabs of a part

NOTE: The tabs shown here *may* differ from part to part. For example, there are differences for terminals, plugs, PLC, etc. You must take this into consideration while editing!

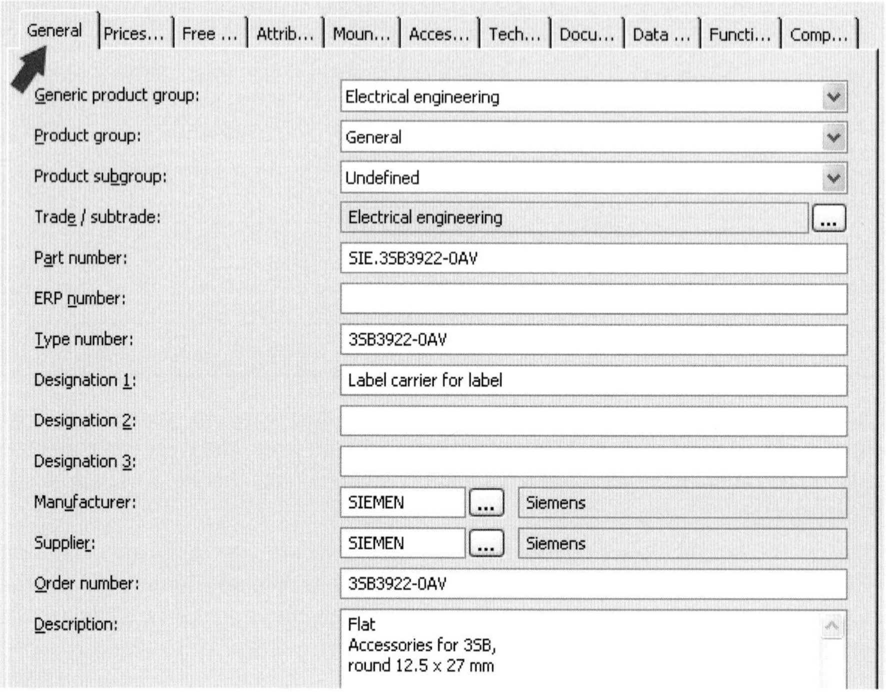

Fig. 7.79 General tab

The **General** tab contains general information on the part, such as the part number, ERP number, *product group* (e.g., a contactor or the *trade*). Also, you will find information here such as the *part number*, *ERP number*, *order number*, the *designations* of the device or the *manufacturer*.

NOTE: The part is identified by its part number and the ERP number. This means that they have to be unambiguous and may be used in parts management only once!

Most of these fields are pure input fields; where this is impossible, for example, for the generic product group, there are fixed selection lists. The Manufacturer and Supplier fields, if the manufacturer and/or supplier have been created in parts management as a record, can be selected from the following Manufacturer and/or Supplier dialog via the button. EPLAN then adopts this entry and automatically fills in the full name of the manufacturer or supplier.

> **NOTE:** These selection lists - generic product group, product group and product subgroup - cannot be added on to by the user! Only EPLAN can do that.

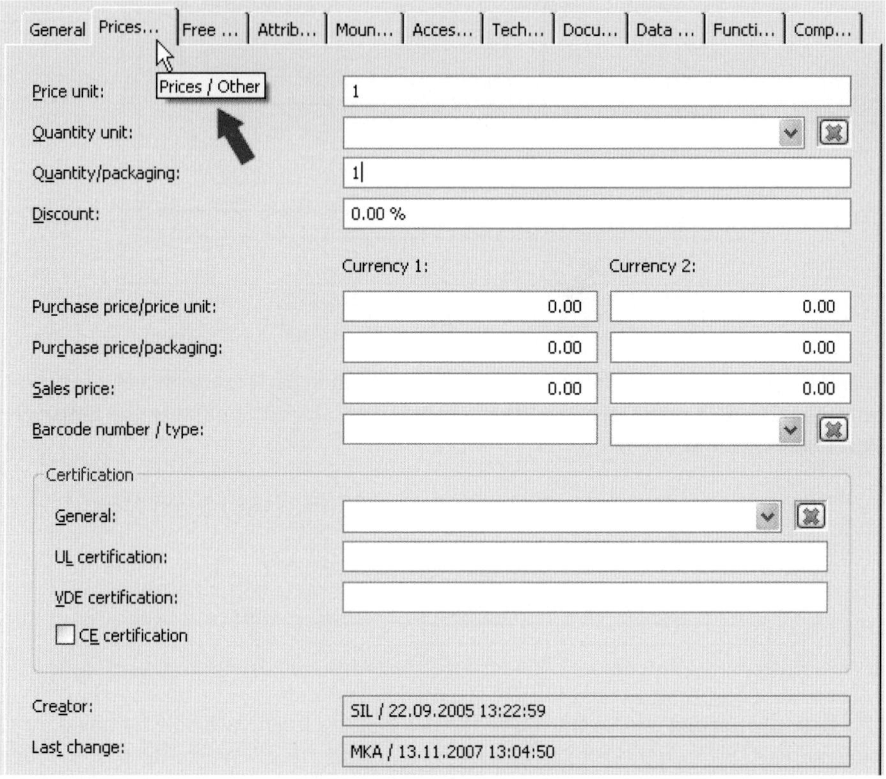

Fig. 7.80 Prices / Other tab

The **Price / Other** tab contains commercial and organizational entries like *Quantity unit* or *Package sizes*, *Prices* or also linked documents for *Certification*.

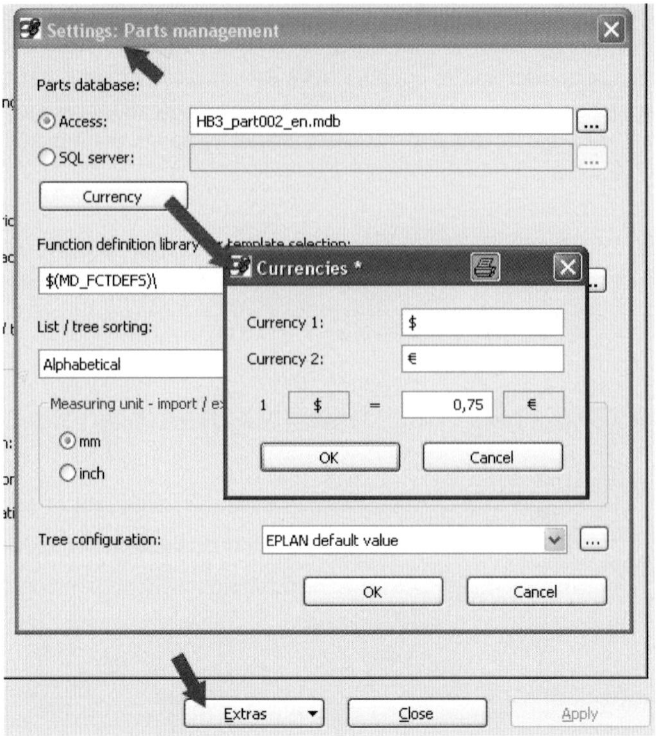

Fig. 7.81 Definition: Currency

You define the currency by clicking the [Extras ▼] button in parts management, then selecting the SETTINGS menu item, and then clicking the [Currency] button in the **Settings / Parts management** dialog.

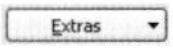

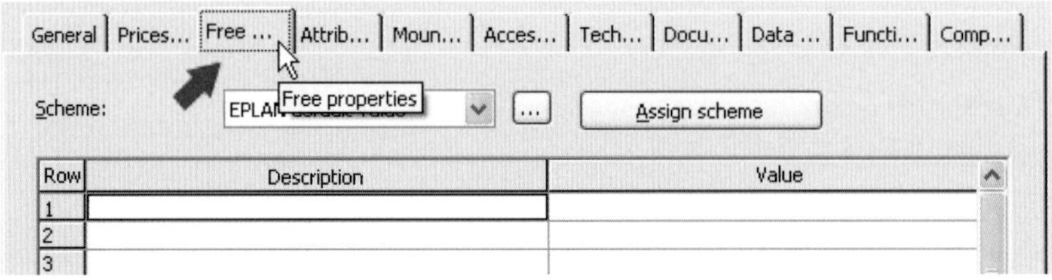

Fig. 7.82 Free properties tab

The **Free properties** tab covers 100 properties that are freely available for any desired purpose. Each of these free properties consists of a *Property*, a *Value* for the property, and a *Unit*.

All fields can be freely used and can be used (e.g.) for information that is not available as a standard field for the part. Block properties can be used with the free properties to display necessary or additional information at a specific symbol in the schematic (with stored parts).

You can create your own schemes with free assignment of the free properties and assign parts to these schemes without having to manually enter the contents of the free properties for every part.

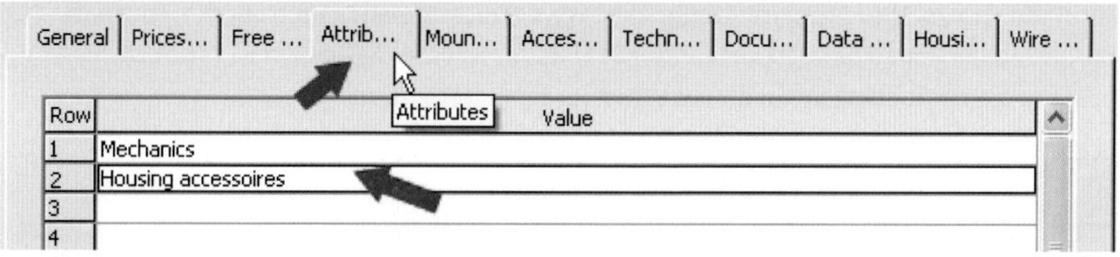

Fig. 7.83 Attributes tab

The **Attributes** tab allows storage of additional information (in the *Value* column) that is not available as standard in the parts management. For example, these entries can be used for filtering purposes or as a sorting characteristic for the tree configuration. The entries themselves are limited to 200 characters.

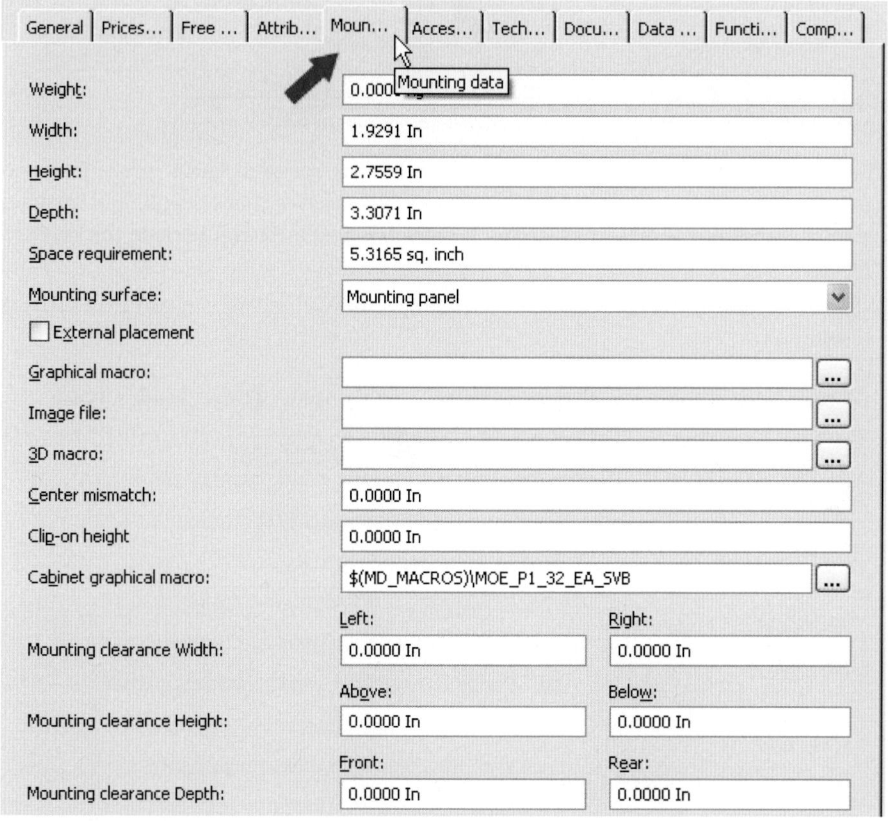

Fig. 7.84 Mounting data tab

The **Mounting data** tab contains all the information that is necessary to use the part, for example, for a mounting panel layout.

It is recommended to maintain meticulously this information, such as *Width*, *Height* or *Depth* (especially important for the 3D mounting layout (EPLAN Pro Panel)) for the part and to update any missing data. A stored image file can also be useful, because this information is evaluated at different locations in EPLAN.

In other words, this helps facilitate the work tremendously, such as the equipment of mounting panels or data reports like parts list or device tag lists, etc., because these necessary data can be processed automatically from parts management.

 NOTE: The fields provided here for *Graphical macro*, *3D macro* or *Cabinet graphical macro* are not intended for EPLAN Electric P8. A macro that can be used optimally in connection with EPLAN Electric P8 should be stored on the **Technical data** tab in the *Macro* field.

Fig. 7.85 Accessories tab

The **Accessories** tab allows you to assign specific accessories to "main" parts. These accessories, too, are stored as normal parts. Via the *Accessories* check box, it then becomes an actual accessory and can be selected for and assigned to "main" parts in the *Accessories* tab.

Selected accessories can be marked as *Required* or, if the identifier is missing, as optional accessories.

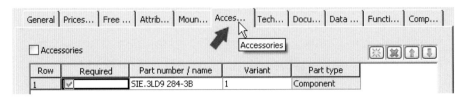

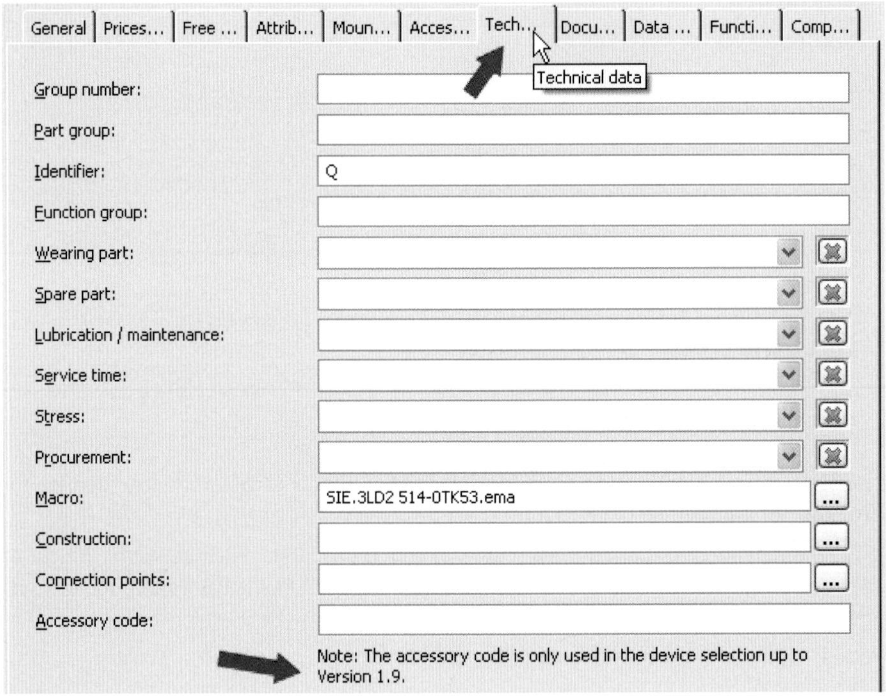

Fig. 7.86 Technical data tab

The **Technical data** tab contains part data such as *service time* or whether the part is a *wearing part*. Most fields are pure input fields, such as the *Group number* or *Function group* fields.

 NOTE: The *Identifier* field is *not* used for the automatic assignment of identifiers during the placement of a device. Instead, you can use this field, for example, to set a filter to the identifier Q in parts management. ∎

The *Wearing part*, *Spare part*, *Lubrication / maintenance*, *Service time*, *Stress* and *Procurement* fields are also input fields. However, every entry is recognized by EPLAN and written to a selection list (internal) so that, over time, this selection list contains many entries that have already been selected (and can be selected again).

The remaining fields of this tab, such as *Macro*, *Construction*, *Connection points*, can be selected via the [...] selection button. The *Macro* field can contain the corresponding window macro for the project. Using the familiar macro functions with the different representation types (e.g., part placement, overview or other representation types), this macro can be set up in such a way that it can be put to multiple uses during project editing.

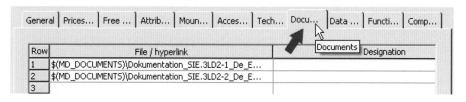

Fig. 7.87 Documents tab

The **Documents** tab can hold up to 20 different documents. These can be user instructions, manuals, etc. - plus an option of storing them with the respective part in several languages.

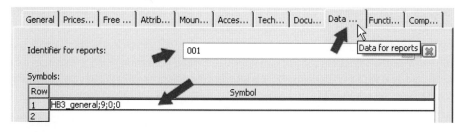

Fig. 7.88 Data for reports tab

It makes sense to use the **Data for reports** tab in combination with the device tag list report type and conditional forms (so-called subforms for dynamic forms).

This way, certain graphical and more complex requirements regarding the output of parts in a device tag list can be displayed in a specific manner.

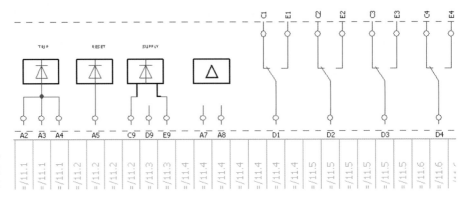

Fig. 7.89 Example view of a complex graphical report of a part

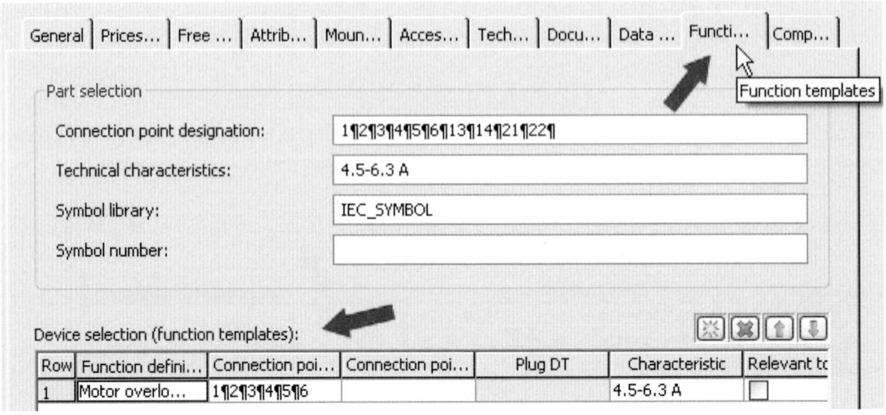

Fig. 7.90 Function templates tab

The **Function template** tab contains key information on the actual part and for the subsequent device selection of the part.

The *Function templates* for the device selection of a part are defined in this tab. I.e. here you define whether this part is a lamp or an auxiliary contact.

EPLAN needs function templates in order to compare the devices already used in the project with the devices in the parts management when performing device selection (not to be confused with parts selection). This allows selection of exactly the devices matching the items in the schematic.

Device selection (function templates):

Row	Function definition	Connection point designation	Connection point description	Plug DT	Ch
1	Motor overload s...	1¶2¶3¶4¶5¶6			4.5-

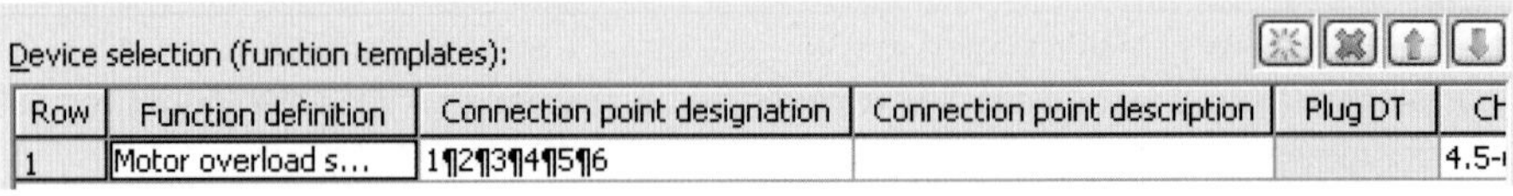

Fig. 7.91 The first columns of the function template

Device selection (function templates):

Row	Characteristic	Relevant to safety	Intrinsically safe	Symbol	Symbol macro
1	4.5-6.3 A	☐	☐	...	

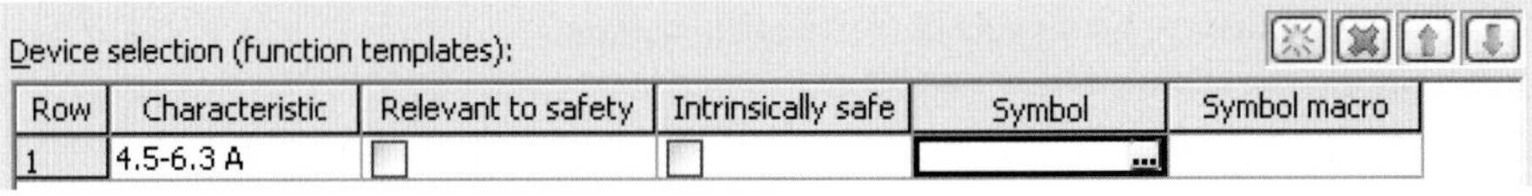

Fig. 7.92 The last columns of the function template

Entries such as the *Function definition*, the *Connection point designation*, the *Connection point description* or the *Characteristic* allow comparison criteria to be defined. For example, when selecting a device, a pushbutton that was defined in the parts management with a function definition of pushbutton and connection point designations of 7 and 8 can only be assigned to an item in the schematic that has exactly these properties, assuming that the device selection settings have been set for this.

For each function definition, you can define, in addition to the previous values, also specific *symbols* or *symbol macros* that EPLAN is then to use for graphical project editing should the standard symbol not fit.

But you can also define for each function definition whether the function is *relevant to safety* or *intrinsically safe*.

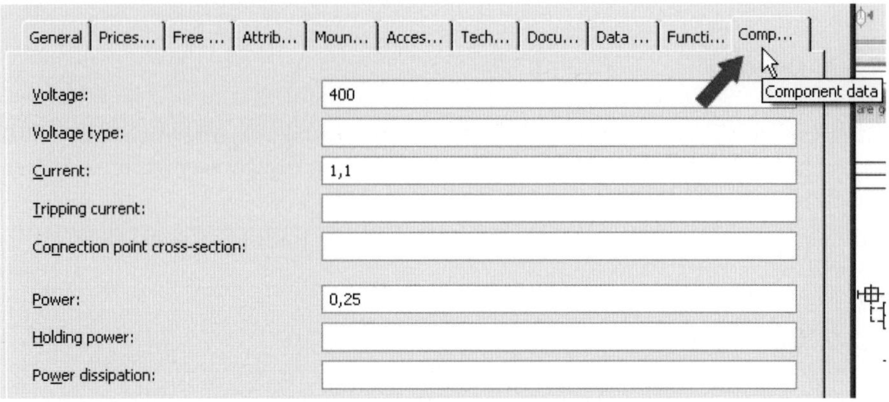

Fig. 7.93 Component data tab

The last tab, **Component data**, contains further technical data for the part, which can also be evaluated - for the purposes of reports, calculations or simply as an additional representation in the schematic on the device.

■ 7.5 Revision control

Apart from the project creation proper, EPLAN Electric P8 also provides a management tool that allows you to capture, evaluate or document revisions (e.g., changes to the site, end customers, etc.) in this project at the end of project editing.

This task - of capturing, managing and documenting revisions - is handled by revision control in EPLAN Electric P8. EPLAN distinguishes between two fundamental procedures for executing revision control.

Revision based on a **project comparison** and revision based on **change tracking**.

Fig. 7.94 Revision control with property comparison

Fig. 7.95 Revision control with change tracking

7.5.1 General

A revision in EPLAN Electric P8 is essentially defined as follows: A revision is an edited project in which the various changes are marked and then tracked via revision. A revised project is stored with the file extension *.ell.

1. This chapter will explain revision control using change tracking as an example.

7.5.2 Generate new revision

Before a revision, and thus a change tracking of modified data, can be executed, you must first activate it (the revision) for the project.

To do so, in the UTILITIES / REVISION CONTROL / CHANGE TRACKING menu, start the GENERATE REVISION menu item.

Fig. 7.96 Generate a revision

Fig. 7.97 Enter revision data

EPLAN opens the **Generate revision** dialog. Here you must enter a *Revision name*. The *Comment* on the revision is optional. The *User* and *Date* cannot be changed and are set by EPLAN automatically.

These data always apply to a complete revision. They are a kind of global "header" for the revision data. These data in the **Generate revision** dialog are part of the project properties.

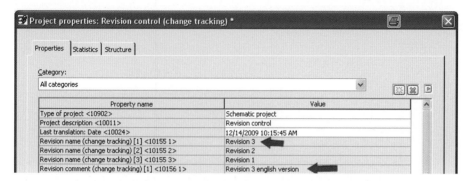

Fig. 7.98 Project property names of revision entries

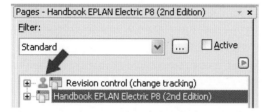

Fig. 7.99 Modified icon

For example, these global revision data can be integrated with the plot frame. If all data have been entered, you can close the **Generate revision** dialog by clicking on OK. EPLAN now generates the revision and, for the purposes of the visual check, changes the project icon in the page navigator.

7.5.3 Execute changes

After the revision has been created, the changes can be entered in the schematics.

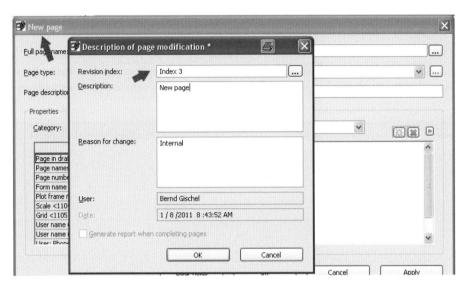

Fig. 7.100 Index query following a change

Following changes, EPLAN queries a revision index. To do so, EPLAN opens the **Description of page modification** dialog. Here the *Revision index* is defined as well as a *Description*, which can also be optional. Again, the user and date of the modification are retrieved by EPLAN from the system and set.

For the revision index, of course, you should choose one system to avoid chaos and to prevent the numbering of the revision index from becoming mixed up upon completion of the project.

It would be a good idea to integrate the revision index, and possibly its description, with the plot frame. This would show immediately, for example, in case of a printout, that a page has been modified.

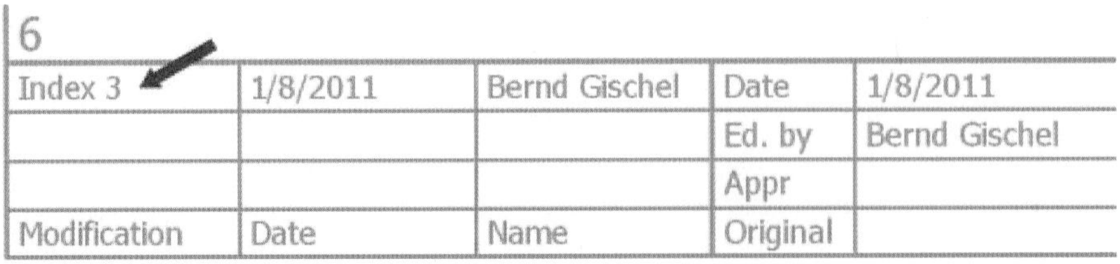

Fig. 7.101 Adding the revision index to the plot frame

The properties of the revision index and their additional data such as the description are pure page properties.

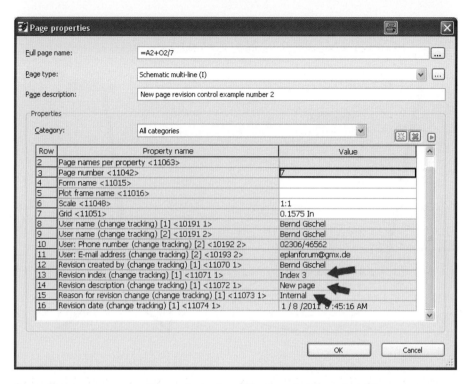

Fig. 7.102 The revision index in the page properties

Generally, you can now make any desired modifications to the project. Changes on a page are marked with the **Watermark** (the text of the watermark can be modified in the settings) across the plot frame, and in the title bar of the window you will also see the word "**Draft**". This serves as a visual indication that this page with the modifications has not been completed yet.

EPLAN allows you to place revision markers for every modification. As well, you can also use visually different revision markers for different types of changes.

Fig. 7.103 Graphical markers for various modifications

So, EPLAN differentiates between *Added*, *Changed*, *New in report* and *Deleted*. Depending on your requirements, you can adjust these graphical markers for each project differently.

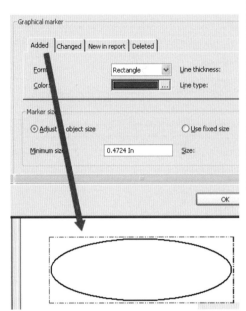

Fig. 7.104 Graphical marker for added objects

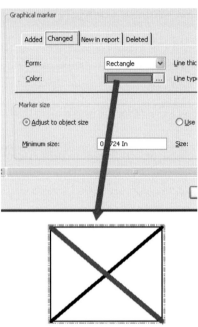

Fig. 7.105 Graphical marker for changed objects

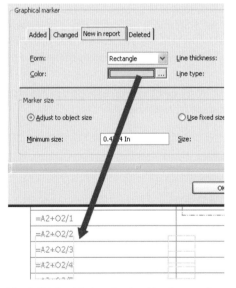

Fig. 7.106 Graphical marker for „New in report"

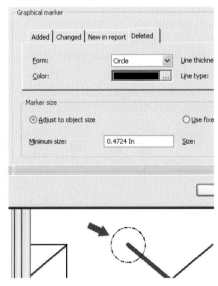

Fig. 7.107 Graphical marker for deleted objects

7.5.4 Complete page(s)

After the modifications have been incorporated into the project, the page or pages should be completed. To do so, in the UTILITIES / REVISION CONTROL / CHANGE TRACKING menu, start the COMPLETE PAGES menu item.

Fig. 7.108 Complete pages

EPLAN opens the **Description of page modification** dialog. Here you must now enter the revision index. But it is also possible to switch to the **Select revision index** dialog via the More button. From there, select, apply or modify an existing revision index.

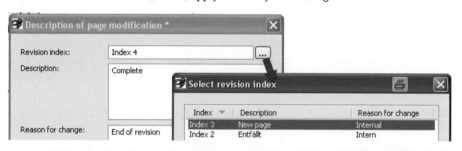

Fig. 7.109 Define description of page modification

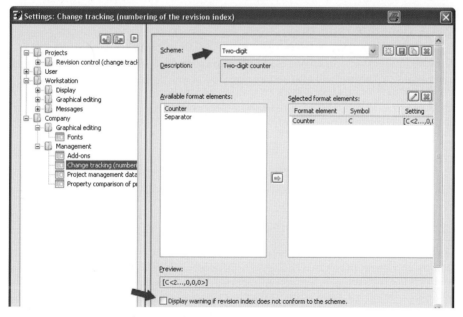

Fig. 7.110 Define numbering of the revision index

If a numbering scheme has been defined for the revision index in the settings (OPTIONS / UTILITIES / COMPANY / MANAGEMENT / CHANGE TRACKING (NUMBERING SCHEME FOR REVISION INDEX)), a revision index will be recommended automatically based on the selected scheme.

NOTE: As long as the project itself has not been completed, any completed page can be modified again. In that case, you merely have to complete the page again!

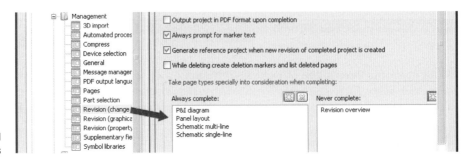

Fig. 7.111 Special handling of page types

The **Always complete pages** option means that these pages are always assigned a revision index even if no modifications have been made. But EPLAN can assign an automatic revision index only if a scheme for this has been defined in the settings.

Never complete pages has the opposite effect. These pages are ignored for revisions even if modifications have been made.

7.5.5 Generate reports

Once all modifications have been made, you can generate and/or update the reports. Special revision overviews will give you an overview of the changes made to the project.

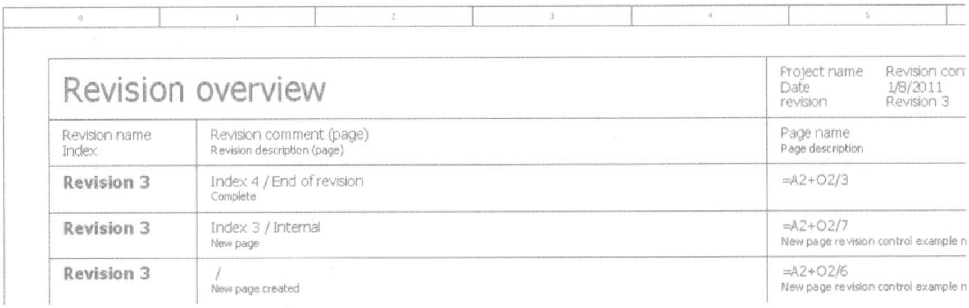

Fig. 7.112 Revision overview report

A special menu item allows you to list even the pages deleted from the project. You can call up this menu item via UTILITIES / REVISION CONTROL / CHANGE TRACKING / DELETED PAGES. When this function is started, EPLAN displays the **Deleted pages** dialog. This contains information about when and how pages have been deleted (e.g., when reports were updated), plus the data you entered, such as a reason for change.

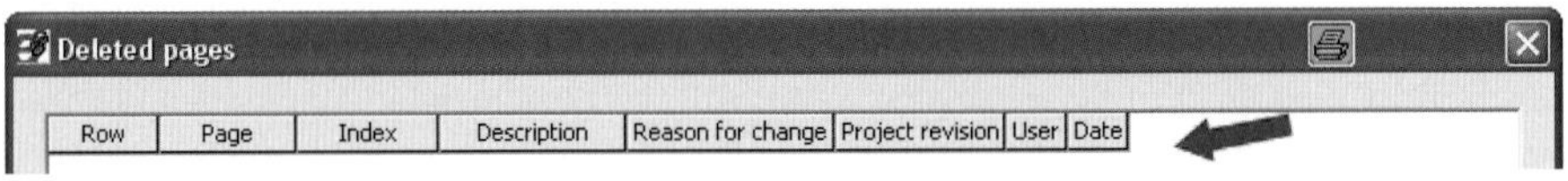

Fig. 7.113 Deleted pages dialog

7.5.6 Complete a project

Fig. 7.114 Complete project

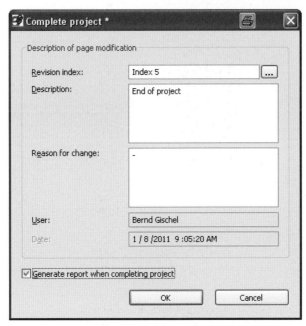

Fig. 7.115 Complete project

If all modifications have been finished, the reports generated and updated, and the pages completed, you are ready to complete the project itself. To do so, in the UTILITIES / REVISION CONTROL menu, start the COMPLETE PROJECT menu item. EPLAN opens the **Complete project** dialog. Here, you enter the already familiar data like revision index or reason for change.

If you select the *Generate report when completing project* option, the following happens. EPLAN updates the project, evaluates all reports and completes each page, i.e., it removes the entry "Draft" and/or the watermark. All revision indices that EPLAN assigns for this automatically now show the most recent revision index defined.

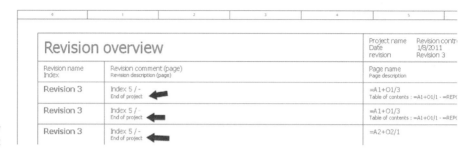

Fig. 7.116 Automatically completed project

A completed project is represented in the page navigator by a corresponding icon.

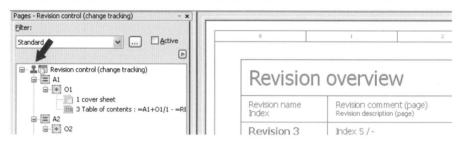

Fig. 7.117 Icon for a completed project

8 Export, import, print

EPLAN offers a range of export and import functions for the entire project or also individual pages. Export or import is a relatively simple process that is only carried out for the selected pages (or all pages) of a project.

An export can be launched via the PAGE / EXPORT [EXPORT TYPE] menu; an import via the PAGE / IMPORT [IMPORT TYPE] menu.

Export menu

Fig. 8.1 Menu of miscellaneous export options

Import menu

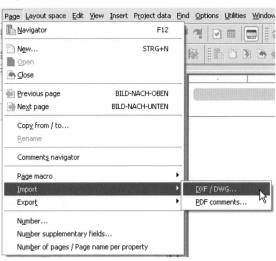

Fig. 8.2 Menu of miscellaneous import options

 NOTE: Exported pages in an image format or DXF/DWG format do not contain any (electrical engineering) logic after the export.

This is why this export function is not suitable for exchanging projects. To exchange projects, you should always do a proper data backup.

NOTE: Pages can only be exported from **one** project at a time. Pages from different projects selected in the **page navigator** cannot be exported together.

The export options are then grayed out and cannot be selected.

8.1 Export and import of DXF / DWG

Files in the DXF or DWG format have a neutral exchange format that EPLAN can use. However, these files are rendered fully graphical following the export or import.

The DXF and/or DWG formats are widely used and popular in various CAD/CAE programs, for example, to represent machine layouts or similar elements. Therefore, they are ideal for exchanging non-program-specific drawings, etc., even if they are purely graphical.

8.1.1 Exporting DXF and DWG files

To export pages in DXF or DWG format, you first select the desired pages in the **page navigator** and then use the mouse to start the function from the PAGE / EXPORT / DXF/DWG menu (the ALT + S keyboard shortcut does not work when the page navigator is open).

EPLAN opens the DXF/DWG export dialog and loads the **Source** field with the selected pages (this field cannot be edited by the user).

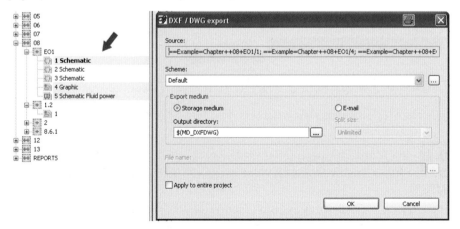

Fig. 8.3 Export pages to DXF / DWG format dialog

The **Export** scheme is more interesting and can be modified by the user. Existing schemes can be selected in the selection field here and then adopted.

Fig. 8.4 Schemes can be adjusted in any manner.

The ⋯ button opens the **Settings: DXF / DWG export and import** dialog, where the current scheme can be edited.

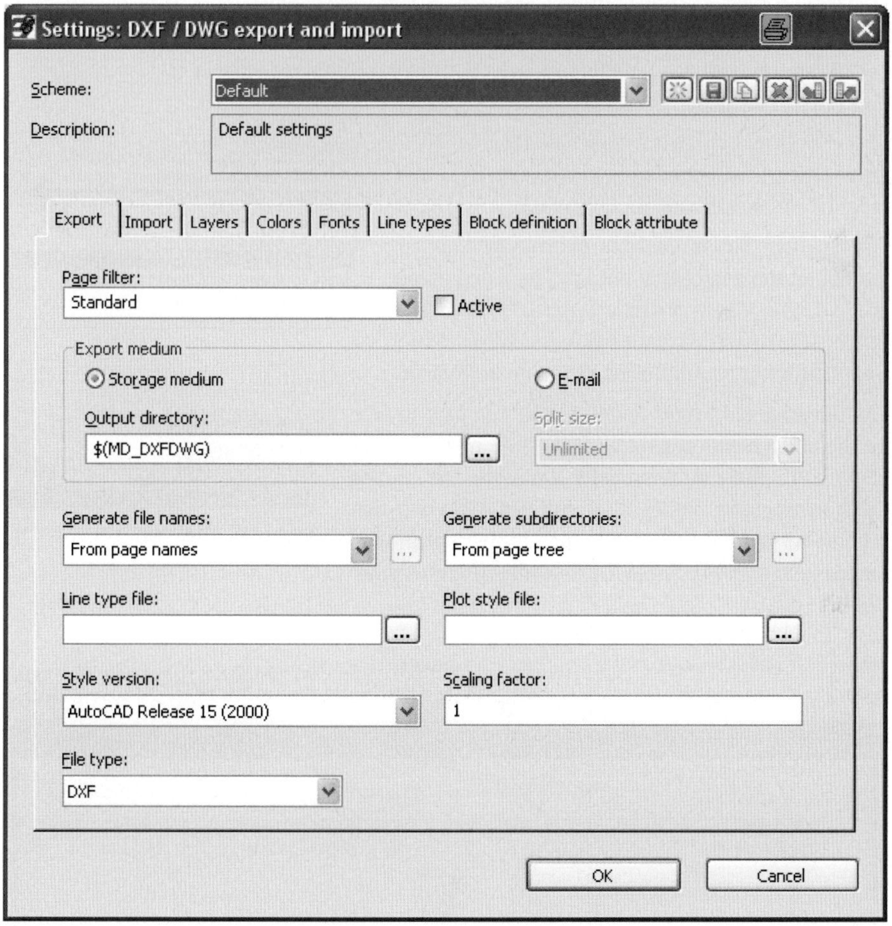

Fig. 8.5 Settings for the export of DXF / DWG files

These settings are not project-specific and are regarded as user settings. You can find them under OPTIONS / SETTINGS / USER / INTERFACES / DXF / DWG EXPORT AND IMPORT.

The **Settings: DXF / DWG export and import** dialog offers the user many options/parameters for influencing or controlling the export and import of DXF/DWG files.

A number of parameters are important for exporting pages in the DXF/DWG format. However, you should create a new scheme at this point via the NEW button and not edit the currently set scheme.

An exception to this is when the scheme settings are not optimal. You can then change these and save the scheme under the same name.

Once a name and optional description for the scheme have been defined, you close the dialog by clicking OK. EPLAN enters the new scheme name into the **Scheme** selection field. The desired export settings can now be adjusted.

The **Output directory** can be set to a special directory for this specific project here, instead of the usual global output directory for DXF/DWG files.

The **Style version** should be adjusted to suit the target system. EPLAN provides a number of different entries for this in a selection field. EPLAN provides a certain number of different target systems.

The **File type** can be set to either DXF or DWG. This setting is also made via a selection field and must be set to suit your requirements.

Fig. 8.6 Selection of formats

The **Scaling factor** controls enlargement or reduction of the page being output. The usual value is 1, i.e., the page is written to the DXF/DWG format on a 1:1 basis without enlargement or reduction. This entry should be left at the default value.

The *Generate file names* option allows the user to define the name of the generated DXF/DWG files. You can allow EPLAN to automatically generate the names using *consecutive numbers*, or you can use your own scheme to generate the file names individually from various page and/or project properties (*From properties* option).

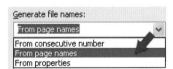

Fig. 8.7 File name options

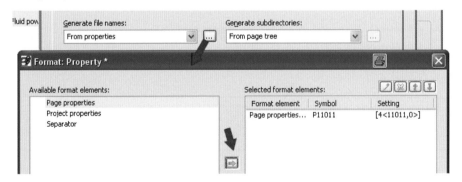

Fig. 8.8 Generate file names from properties (format elements)

Once all options or parameters have been set, you save the scheme via the graphical [SAVE] button. It is then always available and can also be used for other projects, as described previously.

It is also possible to output the DXF/DWG files to be generated in defined and fixed **subdirectories**. This would be practical, for example, when you want to structure the output in such a way that a subdirectory is generated for each higher-level function, with all pages belonging to the higher-level function being output there.

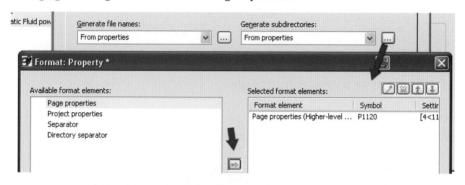

Fig. 8.9 Option of output in subdirectories

The two remaining settings Line type file (*.lin) and Plot style file (*.ctb) are provided by the target system and must exist as files so that they can be included in reports as well.

Fig. 8.10 Options to let the target system define types/styles

You exit the **Settings: DXF / DWG export and import** dialog by clicking the OK button. The CANCEL button discards all changes you may have made to the scheme settings.

EPLAN then applies the newly defined scheme in the **DXF / DWG export** dialog and also adjusts the target (directory) for the output, if this was changed in the scheme.

When you click the OK button, EPLAN generates a DXF or DWG file for each selected page using the defined output parameters.

 NOTE: If this should result in duplicate file names, EPLAN will send a message and then cancel the export without any further messages.

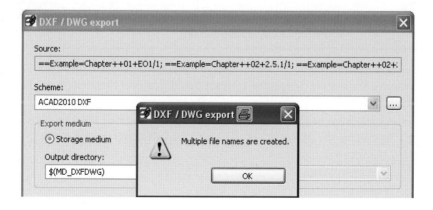

Fig. 8.11 Error message when generating the output

If the output goes through without any problems, the output will be generated. EPLAN then closes all dialogs, and the focus is returned to graphical editing.

Fig. 8.12 Generated DXF files in the stored directory structure

8.1.2 Import of DXF and DWG files

EPLAN cannot only export (generate) DXF/DWG files but can also import CAD drawings in DXF and DWG formats and create project pages from them.

Fig. 8.13 Import DXF / DWG files

Selecting the PAGE / IMPORT DXF/DWG menu, you open the **DXF / DWG file selection** dialog. By default, EPLAN opens the DXF/DWG directory defined at installation for selecting the files.

Of course, you can select any directory.

Fig. 8.14 Dialog for importing DXF or DWG files

The desired file type for the import (DXF or DWG) is set in the lower area of the dialog.

Fig. 8.15 Selection of the file type

You can switch on the PREVIEW option to make initial selection of the file easier. EPLAN then displays a small preview of the relevant file. If you select multiple files, then no preview is shown.

You can now select one or more files in the **DXF/DWG file selection** dialog and apply this selection by clicking the OK button. EPLAN opens the **DXF/DWG import** dialog.

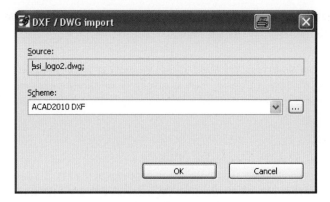

Fig. 8.16 Import dialog

The scheme set in the lower area is the same as that used for the export of DXF/DWG files (because this scheme also contains the import setting).

To adjust the import scheme, or create a new one, you click the [...] button to open the now-familiar **Settings: DXF / DWG export and import** dialog.

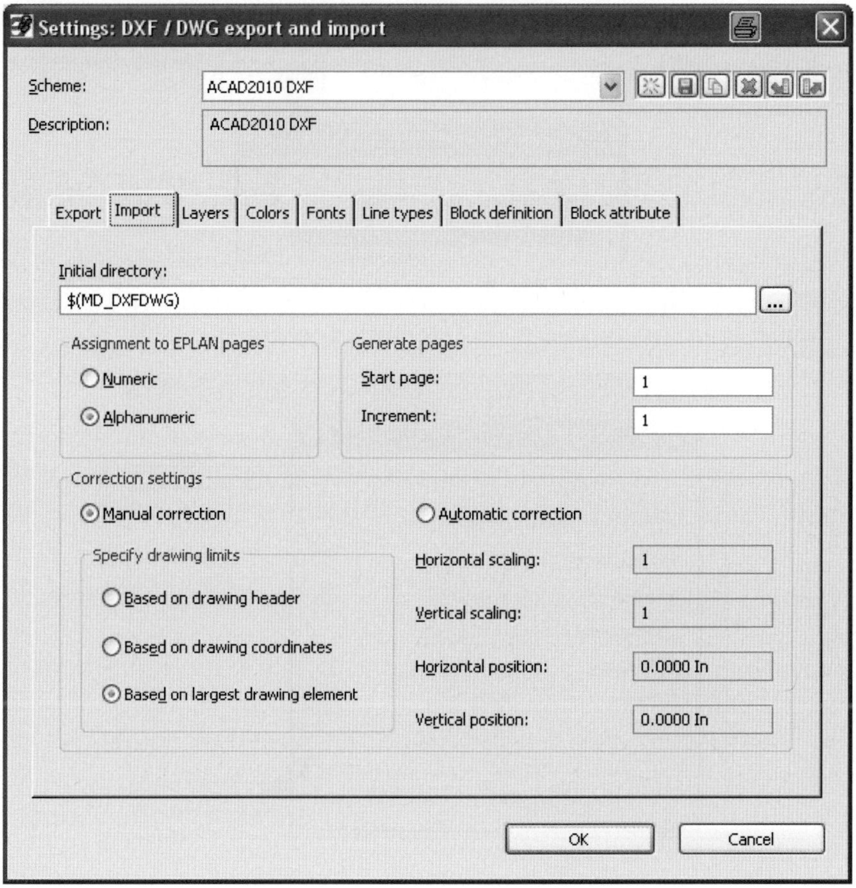

Fig. 8.17 Settings for the import of DXF / DWG files

On the **Import** tab you can now define parameters and options for importing the DXF/DWG files before starting the actual import.

 TIP: It is a good idea to initially accept the settings as they are defined in the EPLAN default scheme. You can specify the **Generate pages** setting in advance, if you wish. However, this is queried again in a later dialog, with more options for placing the DXF/DWG files at specific locations in the project.

If nothing has been changed, the dialog can now be closed via OK. EPLAN returns to the **DXF/DWG import** dialog. When you click OK here, EPLAN starts the import.

This is followed by the **Assign pages** dialog, which asks for important information defining how the DXF/DWG files are to be sorted into the existing page structure of the project.

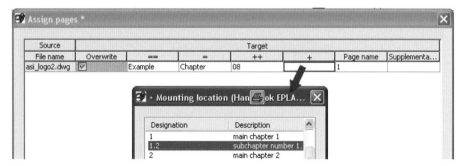

Fig. 8.18 Assign pages prior to the actual generation of the project pages

You enter the desired target designations into the dialog. You can use the NUMBER button to renumber the target pages before importing them if necessary.

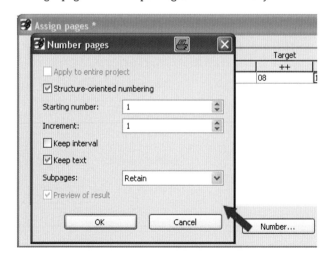

Fig. 8.19 Numbering pages automatically

When you click on OK in the **Assign pages** dialog, the pages are imported (possibly accompanied by an **Import formatting** prompt, which can be confirmed by clicking on OK) and sorted into the page structure of the project as specified.

This completes the import of the DXF/DWG files. The DXF/DWG files are read in and displayed on pages of the graphic type. They can now be edited in the usual manner with the normal graphical editing functions.

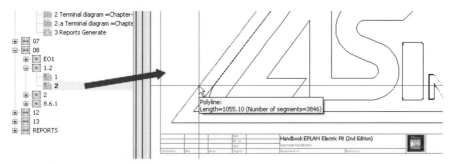

Fig. 8.20 Read-in example of a DXF file

Just to remind you one more time: These pages are graphical pages only without any (electrical engineering) logic. When schematic pages are exported in the DXF/DWG format, the pages are "dismantled" into graphics.

If these DXF/DWG files are imported again (e.g., you import the previously exported schematics), then after importing they are still graphics files and therefore all the imported elements on the pages are also graphics. These imported DXF/DWG files are **not** converted back to logic pages.

8.2 Image files

In addition to the DXF/DWG export and import for various CAD systems, EPLAN can also export project pages as image files.

Schematics and all other pages belonging to the project, such as reports, cover sheets, or the table of contents, are normally always exported as image files when they need to be archived.

The DXF/DWG formats, for example, are not suitable for long-term archiving because these formats may change over time. In contrast to this, image formats will remain unchanged in the future and are recommended for long-term archiving.

8.2.1 Export of image files

You export image files in a similar manner to exporting DXF/DWG files.

Fig. 8.21 Export / Image file menu

You first select the pages to be exported in the **page navigator**. If you wish to export the entire project, then you just need to select the name of the project in the page navigator.

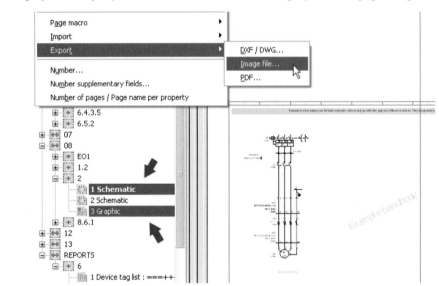

Fig. 8.22 Selected pages in the page navigator

 NOTE: The **Image file export** function also only allows selections from a single project. You cannot select pages for export from different projects at the same time!

You use the PAGE / EXPORT / IMAGE FILE menu to open the **Image file export** dialog.

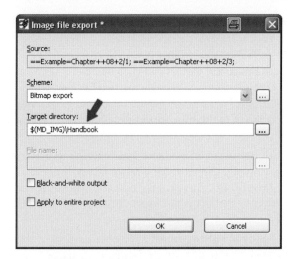

Fig. 8.23 Image file export dialog

You can now select and apply an existing scheme defining the output in this dialog. EPLAN also allows you to define the output settings, such as compression or the color depth of the generated image files, via your own scheme.

To create your own scheme for outputting image files, click the ![] button next to the **Scheme** selection field. EPLAN opens the **Image file export** settings dialog.

Fig. 8.24 Settings for image file export

Click the graphical NEW button to create a new scheme. In the subsequent **New scheme** dialog you define a name for the new scheme and an optional description. After confirming and applying your entries by clicking OK, the **Image file export** dialog is displayed again. The new scheme name is now loaded by default and you can edit it.

Do not forget to save the scheme after making your settings, such as the selection of the file type, the compression, or color depth. Click the SAVE button for this. You can now exit the dialog via the OK button.

EPLAN returns to the **Image file export** dialog. In addition to allowing selection of a different scheme or output directory, this dialog has two further options. The first option affects the output of the pages themselves: Should the output be generated in color or black and white? Selecting **Black and white output** generates image files without color information.

Exceptions to this are embedded images, such as a company logo in a plot frame or cover page form. These are usually output in the original colors. If the black-and-white output option is not used, then EPLAN generates the output pages in color. In this case, "color" means that the output in the image file looks the same as that which you see on the screen.

The second option **Apply to entire project** affects the range of the output pages. At this point you can decide to export all the project pages to image files, even when you have only selected some pages and not the whole project for exporting. You do not need to cancel the export, you can simply activate the **Apply to entire project** option.

 TIP: You can directly enter a new directory into the *Target* field (output directory). EPLAN can also create a [New directory] automatically on export. If EPLAN cannot find the output directory then it creates one.

If all settings have been made, EPLAN can start the export. You can exit the dialog via OK. EPLAN now generates the individual image files and stores them in the set output directory.

Fig. 8.25 Exported schematic pages as images

8.2.2 Insertion of image files (import)

You cannot import image files as pages in EPLAN. The PAGE / IMPORT menu has no function for importing image files. You cannot compare this type of function with the function for importing DXF/DWG files. These are completely different functions.

NOTE: EPLAN handles the import of image files in a different way. Image files are not imported but rather inserted and their size or resolution can be changed later.

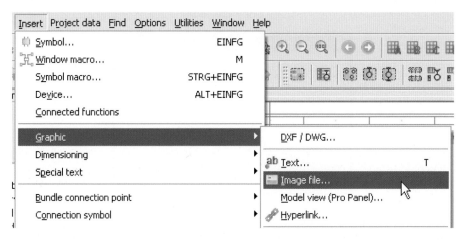

Fig. 8.26 Insert image files menu

When you select the INSERT / GRAPHIC / IMAGE FILE menu, EPLAN opens the **Select image file** dialog. As usual in EPLAN, the corresponding default user directory is displayed (in this case for images). Other directories can, of course, be selected.

Fig. 8.27 Dialog for the selection of image files

EPLAN can insert numerous different image formats. The corresponding image format is preselected in the **Select image file** dialog in the lower area.

Fig. 8.28 Selection of the image format

File types such as *.bmp or *.jpg are the most common, but space-saving formats such as *.png are also possible. There are no limitations on the directory structure or file names used. Special characters or deeply branched directory structures (including spaces) are no problem for EPLAN.

After selecting the desired image file (only one image file at a time can be inserted) you click the OPEN button. The important **Copy image file** dialog is displayed.

Fig. 8.29 Copy image file

TIP: You should take the message in the dialog seriously. If the image file is not copied to the project directory (the **Keep source directory** option), then the possibility exists that this „linked" (referenced) image file will not be saved during a data backup (depending on the settings). It is, therefore, a good idea to always use the **Copy** option in this dialog.

If necessary, you can enable the **Overwrite with prompt** option. If an image file with the same name already exists in the target directory, then a security message is displayed before overwriting. You can still cancel the copy operation at this point.

If you click OK in this dialog, then EPLAN copies the image file into the \IMAGES project directory and returns to graphical editing. You now define the first corner of the image by pressing the ENTER key, or by clicking the left mouse button.

EPLAN displays a window that you can enlarge by pulling with the mouse or via the cursor keys. The size of this window defines the initial size of the image on the page.

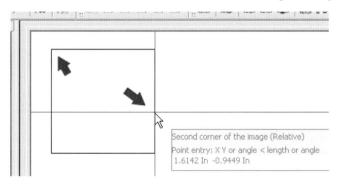

Fig. 8.30 Pull open window

Clicking the left mouse button or pressing ENTER again causes EPLAN to display the **Image properties** dialog. Here you can make settings defining the **size in coordinates** or in **percent** of the original size.

The Keep aspect ratio setting is important. This avoids distortion of the image. This setting should therefore always be activated.

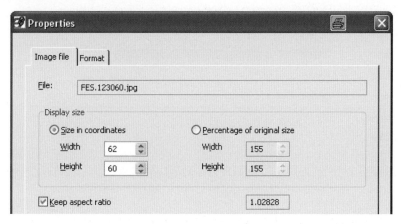

Fig. 8.31 Formatting options dialog

After accepting the settings by clicking OK, the image is now inserted (placed).

To edit images at a later date, for example to change the size or enter something into the **Description** field, you just select the graphic by clicking it once with the left mouse button. EPLAN changes the display and "draws" a box with small handles around the graphic.

Each of these handles can be used with the mouse to make the graphic larger or smaller. You can, of course, directly specify the size of the image in the image properties at any time.

To now open the **Properties** dialog, you first select the graphic as described above and then use the CTRL + D shortcut key to open the properties dialog. This function for calling up the properties also exists in the popup menu via the right mouse button.

8.3 Print

The print command for pages in EPLAN is accessed via the PROJECT / PRINT menu item.

The subsequent **Print** dialog offers a number of settings that will now be briefly described. The page print settings should be adjusted once to suit your particular printer.

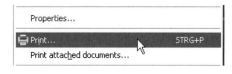

Fig. 8.32 The Print menu

8.3.1 The Print dialog and its options

If you wish to print pages, then you first select them in the **page navigator** and then prepare them via the PROJECT / PRINT menu. EPLAN then displays the **Print** dialog.

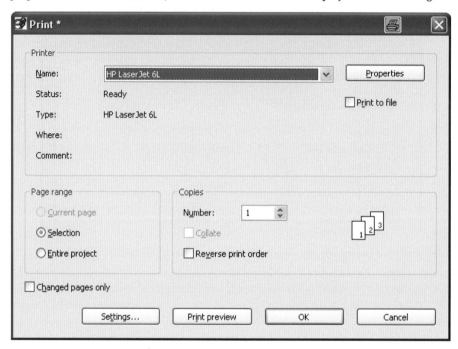

Fig. 8.33 The Print dialog

With a few exceptions, the **Print** dialog is very similar to the normal and familiar Windows print dialog.

Familiar options are the selection of the installed printer and the option allowing printing of multiple copies with different output sorting options.

8.3.1.1 Page range / Changed pages options

At this point, in the **Page range** options, you can select the **Selection** option to print only the currently selected pages in the page navigator or change your mind and select the **Entire project** option. The **Current page** option is automatically set to deactivated as soon as more than one page has been selected in the page navigator.

Fig. 8.34 Print options

The **Changed pages only** setting prints only those pages that have changed since the last print operation.

8.3.1.2 The Settings button

The **Settings** button contains a number of print settings for the pages themselves.

Here you can define the print size, whereby **Print to scale** means that if a page was assigned the A3 plot frame, for example, then it is printed 1:1 in A3 format. Printers smaller than A3 format (i.e., generally A4 printers) will only print the upper left corner of such pages.

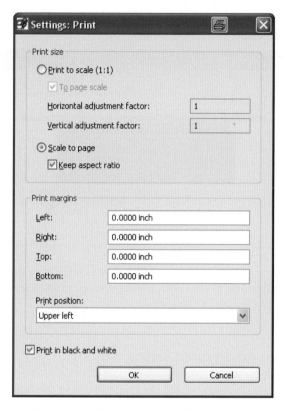

Fig. 8.35 More page settings

The **Scale to page** setting and the **Keep aspect ratio** parameter should be used as default. This causes all larger dimensions to be scaled down to the printer model and a complete page is always printed. Scaling usually is to an A4 page size.

 TIP: The **Print margins** settings must always be adjusted once, to suit the installed printer(s).

Unfortunately, one cannot provide generally applicable suggestions for the print margins because there are simply too many printer models with widely varying technical properties. Here you can only try things out and make a number of tests with the connected printer.

Selecting the **Print position** allows you to position the page on the printed sheet of paper. EPLAN offers a number of different print positions. It is a good idea to use the **Upper center** setting with an upper print margin of 10 mm.

The **adjustment factors** offer an additional setting to compensate for small deviations in printouts to scale. Here, too, it is a good idea to experiment a bit, because these values depend on the printer model as well as on the printer driver in use. It is therefore not possible to specify generally applicable values here.

8.3.1.3 The Print preview button

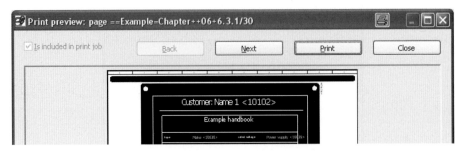

Fig. 8.36 Print preview

The **Print preview** button in the **Print** dialog provides a rough preview before performing the actual printing. However, not all the information on a page is displayed. This is not so important because, at this point, the print preview is mainly used to check the position of the page.

If the print preview looks ok, you can start printing directly from within the print preview via the PRINT button. EPLAN then starts printing on the printer specified as the default printer in the operating system without further prompts.

8.3.2 Important export/print setting

Aside from the simple print/export and the print settings in the **Print** dialog, there are some other important user settings that concern the printing and/or export of the project. As the name implies, these settings are always user-defined.

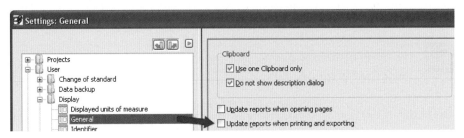

Fig. 8.37 Update reports when printing/exporting setting

To ensure that reports are current, the **Update reports when printing and exporting** setting should be activated. This prevents that the project is printed with outdated reports.

This setting is located in the OPTIONS / SETTINGS / USER / DISPLAY / GENERAL menu.

8.4 Export and import of projects

In addition to exporting and importing image files or DXF/DWG files, EPLAN also allows exporting and importing of entire projects in the XML format.

Fig. 8.38 Read project data in and out

8.4.1 Export of projects

Projects are exported from the **page navigator** via the PROJECT / ORGANIZE / EXPORT menu.

To do this, you must first select the desired project and then start the EXPORT function. After calling the Export function, the **Browse For Folder** dialog is displayed where the directory is defined to which the project to be exported is to be saved.

Fig. 8.39 Define folder for export

 NOTE: Using the page navigator, it is only possible to export one project at a time. If several projects are selected, then the *Export* function cannot be chosen (the entry being grayed out).

After you have selected the export folder and clicked on the OK button, EPLAN immediately starts the export without further prompts. The project is exported and is assigned the file extension *.epj.

8.4.2 Import of projects

The counterpart to exporting is the importing of XML projects. In the graphical editor, you execute this function via the PAGE / ORGANIZE / IMPORT menu.

When you select the IMPORT function, EPLAN opens the **Import project** dialog. Here you select from the directory structure the directory containing the project (*.epj) to be imported.

Fig. 8.40 Import a project (*.epj)

The project is selected and adopted via the OPEN button in the lower area of the dialog. EPLAN opens the **XML project** dialog and inserts the selected project into the **XML file** field.

Fig. 8.41 Import XML project

The **Synchronize parts** parameter enables EPLAN to synchronize project parts with the parts master data during the import. You enter the name of the project into the **Target project** field and the imported project will be shown in the project management under this name.

You can also use the button in this dialog to select a different directory. After you click on OK, the import process is started, and EPLAN generates the project in the defined directory structure.

8.5 Print attached documents

Fig. 8.42 Print attached documents

The PROJECT menu has a PRINT ATTACHED DOCUMENTS function that allows you to print all documents stored in the project, e.g., PDF documents, in a single step rather than individually.

When you select this function, EPLAN opens the **Print attached documents** dialog for the current project. In this dialog, you now have the option of excluding individual documents from printing. To do this, you remove the checkmark in the **Print** column.

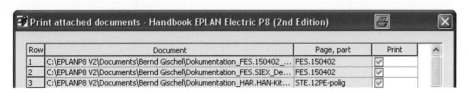

Fig. 8.43 Dialog for selection of attached documents

When you click the OK button, EPLAN prints all documents selected in the **Print attached documents** dialog, one after another, to the connected printer.

 NOTE: External documents (External document page type), linked documents on report pages and linked documents on Internet pages cannot and/or will not be printed.

8.6 Import PDF comments

EPLAN can import comments (notes, etc.) stored in PDF documents (schematics generated from EPLAN) and store these in an extra layer.

This allows (e.g.) a service technician on-site to electronically directly add his / her own comments to the PDF schematics. These can then be imported into EPLAN, viewed, edited and evaluated.

 NOTE: To subsequently import into EPLAN again a PDF with comments, you will have to use the internal PDF output of EPLAN! You can find it in the menu PAGE / EXPORT / PDF. Similarly, it is important to note that no write protection is activated for the PDF output to be generated. You can find this setting in the OPTIONS / SETTINGS / USER / INTERFACES / PDF EXPORT menu, where on the *General* tab you will see the *Read-only* setting.

8.6.1 Import of commented PDF documents

To post comments to a PDF, it is necessary to release the PDF generated by EPLAN for comments. You can do that with Adobe Acrobat Professional (version 7 and higher) or the freeware PDF XChange Viewer (tested with version 2.0 Build 42.4).

The comment function for Adobe Reader (version 7 and higher) can be enabled in Acrobat Professional via the COMMENTS menu followed by the ENABLE FOR COMMENTING IN ADOBE READER menu item. With the freeware PDF XChange Viewer, all you need to do is open the PDF, enter the comments and then save the file.

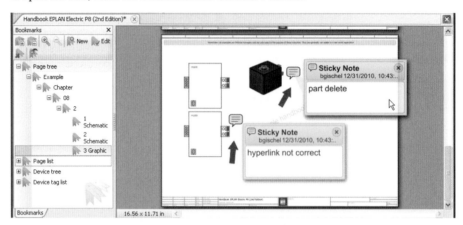

Fig. 8.44 Example of comment entries on a schematic page

After commenting the PDF document in Acrobat Reader or PDF XChange Viewer, it must be saved and closed.

Subsequently, in EPLAN, you can import (read in) this commented PDF document from the corresponding directory via the PAGE / IMPORT / PDF COMMENTS menu.

 NOTE: Prior to the import, the project must be in the same directory to which it was exported!

After successfully importing the PDF document, the **Import PDF comments** dialog containing various import information is displayed.

Fig. 8.45 Successful import

You can see the results of the import in the comments navigator. You can access the comments navigator from the PAGE / COMMENTS NAVIGATOR menu.

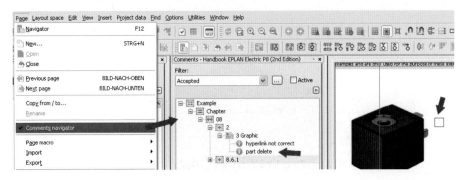

Fig. 8.46 The comments navigator

Here, you can, among other things, use the properties of the respective comment to edit it further - e.g., "Rejected", "Accepted", etc.

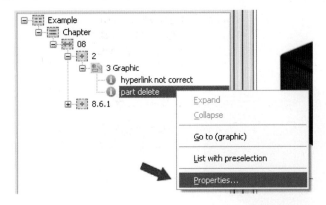

Fig. 8.47 Properties of a comment

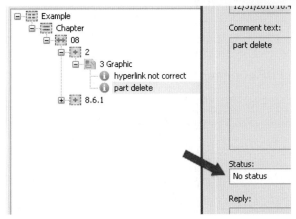

Fig. 8.48 Status of a comment

These properties are, of course, also available directly at the comments in the schematic. You can open them by double-clicking on the yellow box.

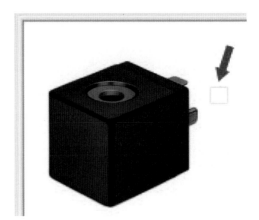

Fig. 8.49 Comment selection

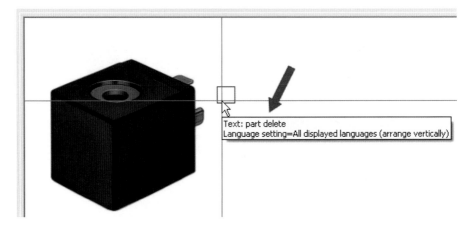

Fig. 8.50 Comment

8.6.2 Delete PDF comments

You cannot only edit PDF comments (status of comments, etc.) but also delete them in the usual way, by selecting them and then calling the normal EPLAN delete function.

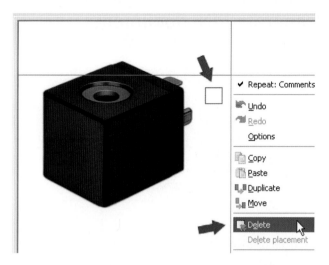

Fig. 8.51 Delete comments

Another much more convenient method is to compress the project and create your own "Remove all PDF comments" scheme (via the PROJECT / ORGANIZE / COMPRESS menu and then adjust an existing scheme or create a new one).

This automatically and finally removes all imported and unnecessary PDF comments from the project in a single step.

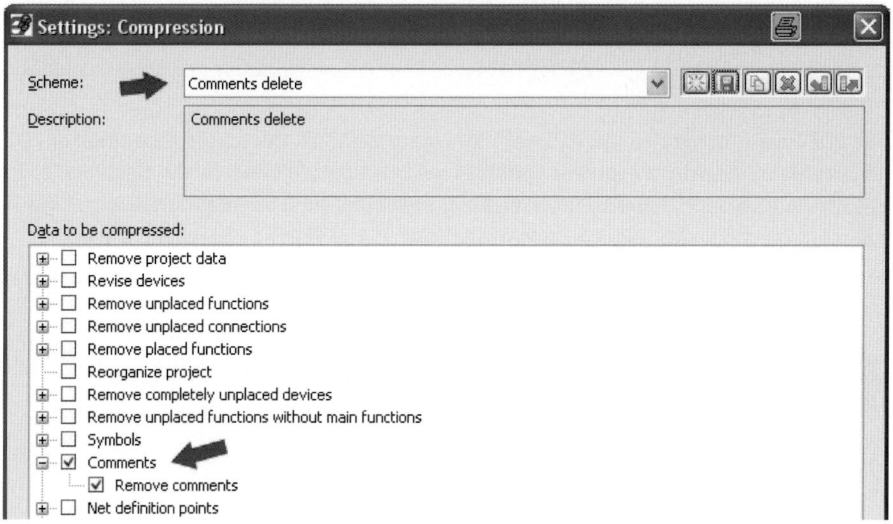

Fig. 8.52 Compression scheme to remove PDF comments

8.7 Generate PDF documents

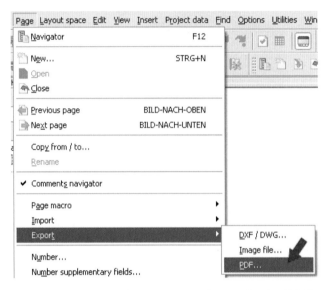

Fig. 8.53 Export as a PDF file menu

To exchange documents without problems, one usually opts for the *PDF format*. The PDF format has become a "de-facto standard" for worldwide document exchange. It therefore makes sense to also use the PDF format for exchanging schematics and other schematic documents.

The PDF format integrated into EPLAN offers more than just the simple generation of PDF documents. A certain amount of "intelligence" can be included in the documents. In EPLAN, "intelligence" means that navigation elements, such as jump functions at cross-referenced components, etc., can also be included in the PDF.

Apart from the intelligent PDF, EPLAN can also generate a PDF in the PDF/A-1b format for longer-term archiving.

 NOTE: If the PDF/A-1b format is used, linked documents and models will **not** be included in the PDF output!

The setting whether the PDF is a standard PDF or one in the PDF/A-1b format is defined in the settings for the PDF output. For this purpose, first select the PAGE / EXPORT / PDF menu and then open the scheme for the PDF output (press More button) in the **PDF export** dialog.

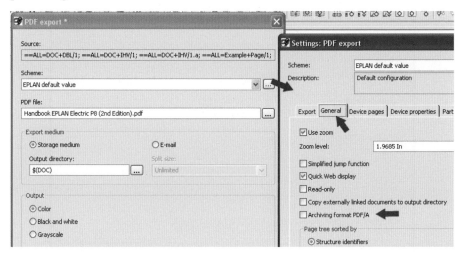

Fig. 8.54 Activate archiving format PDF/A-1b

In the now-open scheme with the settings for the *PDF export*, open the *General* tab and activate the Archiving format PDF/A option.

8.7.1 Export of PDF files

Generating PDF documents is child's play in EPLAN. In the same way as when exporting DXF or image files, you must first select the desired pages or the entire project in the page navigator. The PDF export is also subject to the restriction that you can only export pages from one project at a time. You cannot export pages from different projects that are open in the page navigator at the same time.

After selecting the pages, you use the PAGE / EXPORT / PDF menu to open the **PDF export** dialog.

Fig. 8.55 PDF export dialog and its settings options

The **PDF export** dialog is similar to the DXF/DWG export dialog or the image file export dialog. EPLAN automatically enters the selected pages into the *Source* field. If the project name was selected in the page navigator, then the Source field contains the project name, and the *Apply to entire project* option is grayed out.

Enter the name of the PDF file into the *PDF file* field. By default, EPLAN sets the target directory for the file name automatically to the project directory \[PROJECT NAME]\ DOC in the *Export medium* selection and, here, in the *Output directory* field. The PDF file to be generated also automatically receives the project name.

The target directory and the name of the PDF file can, of course, be freely chosen. To do this, click the button next to the PDF file field. EPLAN opens the **PDF export** dialog. Here you can use the familiar Windows functions to change the directory, create new directories, or change the PDF file name.

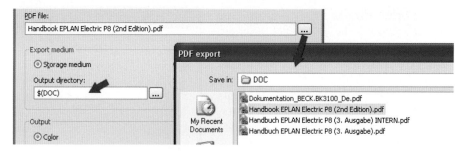

Fig. 8.56 Selection of the directory and file name for export

These modifications are saved with the SAVE button. EPLAN then applies the new settings of the PDF file name to the PDF file field.

In the settings of the **PDF export** dialog, it is also possible to use the *E-mail* option, including its settings (split size), directly as an export medium. If this setting is active, EPLAN will generate the PDF, open the e-mail client and attach the PDF (zipped) to a new mail.

The PDF file can now be generated. But EPLAN also provides other settings defining the format of the output: the settings *Color*, *Black and white*, *Grayscale* as well as *Use print margins*, *Output model* and *Apply to entire project*.

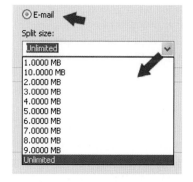

Fig. 8.57 Use E-mail option

You can subsequently activate the **Apply to entire project** option if you have decided to output the entire project and not just the pages you selected. If the entire project has already been selected, this setting will appear grayed out.

The **Color** output option generates a color PDF - as you can see in the screen view. Generally, this looks quite nice but is unsuitable for printing on a black-and-white printer. The color settings of the background scheme *White* are applied. This setting can be found under the SETTINGS / USER / GRAPHICAL EDITING / GENERAL options.

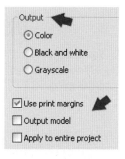

Fig. 8.58 Other settings

When the **Black and white output** option is activated, a black and white PDF document is output. This means that all colored elements are printed in black on a white background. However, embedded image files are output in color.

The last output option, **Grayscale**, generates a grayscale PDF. Depending on what you want to do, you should first experiment with this - and it is the option that I recommend.

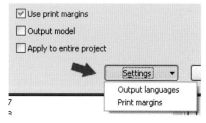

Fig. 8.59 Settings: Print margins

The **Use print margins** setting uses the print margins defined in the workstation print settings. These are located under OPTIONS / SETTINGS / WORKSTATION / GRAPHICAL EDITING / PRINT. They can, however, be changed directly in the **PDF export** dialog at any time. To do this, click the SETTINGS button and make your desired settings in the PRINT MARGINS menu item.

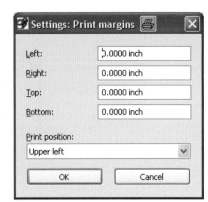

The **Output model** option enables EPLAN to output the models in the project (3D panel view) at the same time.

Fig. 8.60 Print margins dialog

After making your settings, you generate the PDF document by clicking OK. EPLAN generates the PDF document and saves it in the specified target directory. After the successful PDF export, there will be no other messages.

8.7.1.1 Other settings for the PDF export

Aside from the "simple" settings and options for the PDF export described, EPLAN offers an additional great number of other settings on the basis of a PDF export scheme, which can be used to generate information and define how the PDF document is to be generated.

To open the scheme, click on the More button in the **PDF export** dialog.

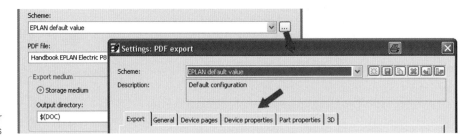

Fig. 8.61 Scheme for other detailed settings

This scheme has several tabs. The *Export* tab contains various settings about how and where the PDF is to be generated, as already described previously.

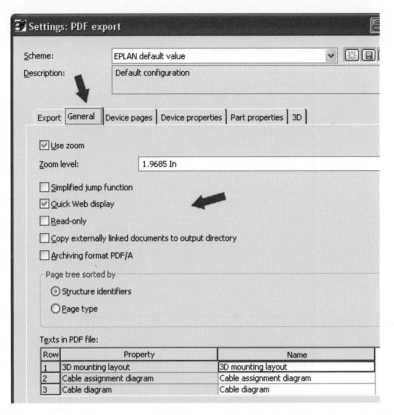

Fig. 8.62 General settings

The *General* tab contains settings and options regarding the basic functions to be given to the PDF. This includes, for example, *simplified jump functions*, *quick web display* or whether the PDF is to be sorted according to structure identifiers or page types.

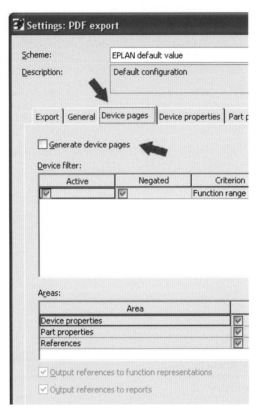

Fig. 8.63 Settings for device pages

The *Device pages* tab defines whether device pages are to be generated at all. If so, you can make other settings to refine the level of detail. For example: Is the output also to include *device properties* or *part properties*? To which *function ranges* do these settings not apply?

8.7 Generate PDF documents

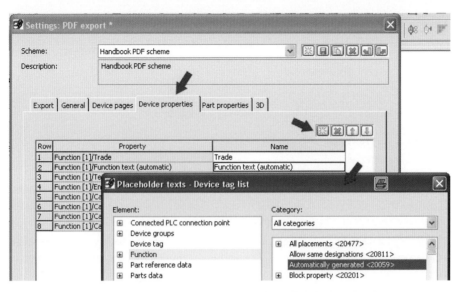

Fig. 8.64 Define device properties

Using the *Device properties* tab, you can output in the PDF very specific properties of the devices. Using the usual EPLAN approach, you can add *placeholders* here or delete them from the list.

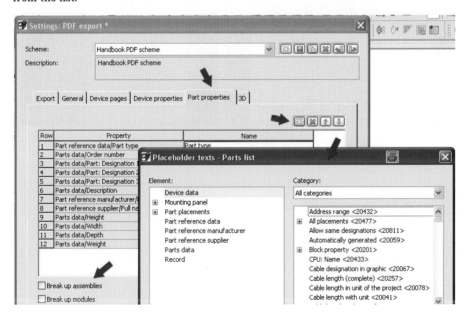

Fig. 8.65 Define part properties for the PDF export

The *Part properties* tab allows you to select the desired part properties to be written into the PDF. This could include, for example, only the *part number* and the first *designation* of a part.

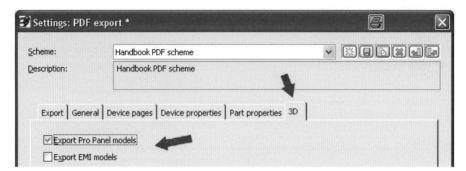

Fig. 8.66 Export 3D components at the same time

Using the *3D* tab, you can also include 3D models in the output of the PDF to be generated. Of course, this applies only if such models exist. This refers to the models that exist in the *layout space* and from which a 2D derivation has been created.

If all settings have been changed accordingly and the scheme has been saved, this scheme will be available to all other projects. The settings contained in the scheme could then be used to let EPLAN generate the PDF.

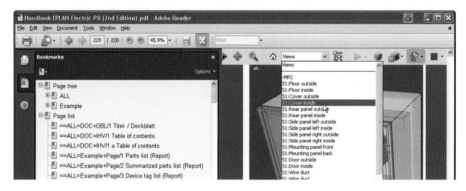

Fig. 8.67 Example: generated PDF, opened with a PDF reader

9 Data backup

An important feature in EPLAN is the data backup. Data backup not only includes zipping, unzipping, backing up, and restoring of projects, but also includes an e-mail function that provides a convenient way of electronically exchanging EPLAN projects between users.

Zipping, backing up, and restoring projects is not all that EPLAN is capable of, however. Individual files, such as a plot frame, the parts database or dictionary (foreign language database), can also be separately backed up and restored.

Fig. 9.1 Menu showing backup methods

■ 9.1 Zipping and unzipping of projects

In EPLAN, zipping a project means that the entire project directory is compressed into a single project file (plus the project information file ProjectInfo.xml), thus saving space.

9.1.1 Zip projects

Zipping is started via the PROJECT / ZIP menu. After a confirmation prompt, EPLAN zips the project immediately. The Zip function itself does not require any special settings.

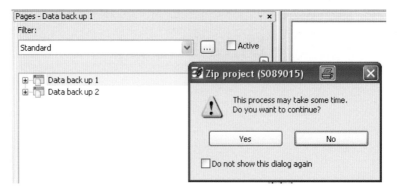

Fig. 9.2 Prompt whether the project is to be zipped

When doing this, **all data** in the project directory is zipped. EPLAN leaves only the project properties information (ProjectInfo.xml) and the zipped project file ([Projectname].zw0) in the project directory.

 NOTE: Only one project at a time can be zipped via the PROJECT menu! Zipped projects are no longer displayed in the page navigator because after being zipped, they are automatically closed and visually removed from the page navigator.

Once EPLAN has completed the ZIP function (of the project), it opens a dialog with a message, and the project "disappears" from the page navigator.

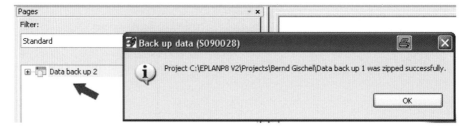

Fig. 9.3 Message that project was zipped successfully

9.1.2 Unzip projects

A zipped project can be unzipped for editing in various ways.

One possibility involves selecting the project via the PROJECT / UNZIP menu and then unzipping it by clicking the OPEN button. Once the project has been unzipped, it will then not be displayed in the page navigator. The unzipped project must additionally be opened first via PROJECT / OPEN.

Another way is to select the zipped project as usual via PROJECT / OPEN and then to open it for editing. Here, too, the project is unzipped, but unlike in the previous option, it is opened in the page navigator immediately after being unzipped.

EPLAN then automatically displays the **Restore** dialog, unzips the project and then opens the page navigator. Depending on the unzipping method, the unzipped project is added to any other, possibly open, projects there.

The project can now be edited.

9.2 Backup and restoring of projects

Compared to zipping a project, the BACK UP and RESTORE options have many more settings for customizing the data backup and restoration processes.

9.2.1 Back up projects

In contrast to the Zip function for projects, the purpose of the Back up function for projects is that they are filed off or can be externally sent to customers.

This is the main difference between these two methods. A zipped project can, of course, also be sent to customers, but it makes more sense to use the Back up function for this.

When working on a project, the backup function can also be directly called from the page navigator via the PROJECT / BACK UP / PROJECT menu.

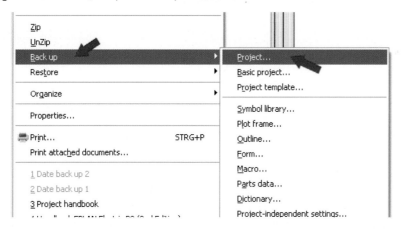

Fig. 9.4 Back up menu and its options

9.2.1.1 Backup options (in the page navigator)

Before starting the BACK UP function, you should first select the project to be backed up in the page navigator. Now you can back it up via the PROJECT / BACK UP / PROJECT menu. EPLAN then displays the **Back up projects** dialog.

The dialog allows you to define further settings affecting the backup scope for the project.

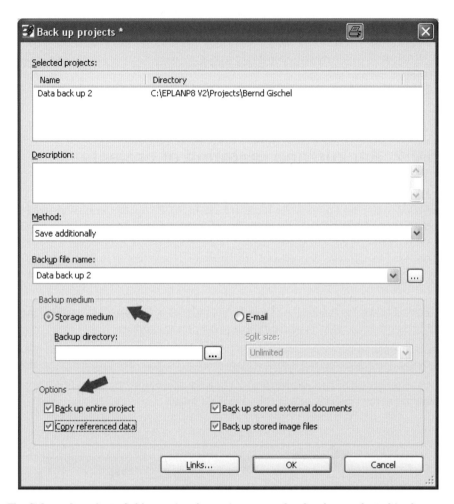

Fig. 9.5 Back up projects dialog

The **Selected projects** field contains the project name that has been selected in the page navigator for backup purposes. EPLAN adopts it automatically and, next to it, displays the additional information of the directory for the project.

 TIP: The Description field is informal. This means that you can optionally enter something here. It has no effect on the actual backup process. The description is displayed in the dialog when restoring (restoring back) a project.

The **Method** selection field defines the manner in which the backup is to occur. This will also affect the choice of whether you wish to edit the project further or not.

EPLAN provides the user with the following options:

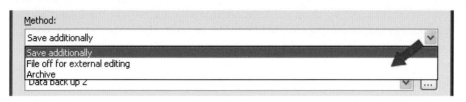

Fig. 9.6 Possible options for the backup method

- **Save additionally** – the project is **additionally** saved on another drive (as *.zw1) or prepared for sending as an e-mail attachment. It still remains in the original project path and is neither deleted nor changed. This is the option that I recommend.
- **File off for external editing** – the project is also additionally saved (as *.zw1) but the original project becomes write-protected so that no further changes are possible. This method is useful when the project (e.g.) is to be filed off to a customer because he / she wishes to make changes and no other changes should be made to the original project until it is restored.
- **Archive** – the project is not additionally backed up but is filed off to a different drive (as *zw1). Apart from the project information (ProjectInfo.xml), the original project is deleted from the original project path.

Apart from the selection of the backup method, EPLAN also provides two options for backup media regarding the project to be backed up.

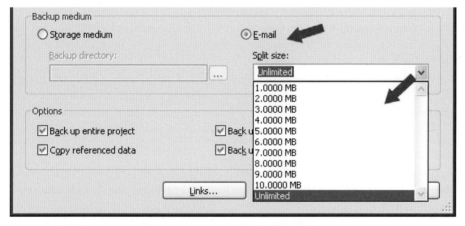

Fig. 9.7 Backup medium selection

- **Storage medium** – usually another drive. You can use any desired directory structure with absolutely no limitations. If this is supported by the operating system, the project can be directly backed up to a CD / DVD burner.
- **E-mail** – the project is backed up and given a *.zw1 file extension. The installed e-mail client is then opened and the project added as an attachment. Here, EPLAN provides the option of dividing the e-mail file into several files (parts) (some variables have been preset via a selection list, but these can be changed in any way; that is, even 6.5 MB is possible). This is useful when the recipient cannot receive large attachments. The recipient **must** then save all the files (parts) in the received e-mails in the same directory of choice in order to restore the project later on.

A project was backed up using the **Save additionally** method and the **E-mail** backup medium. The split size was set to 2 MB.

After you click on the OK button, EPLAN starts the data backup. If the project directory differs from the default project directory, EPLAN opens a dialog with a corresponding message regarding the directory.

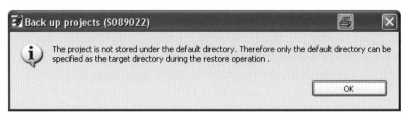

Fig. 9.8 Note indicating that the directory differs from the default

EPLAN then generates several files internally, opens the installed e-mail program and attaches the first part (Part 2 of [Total]) to a new e-mail.

Fig. 9.9 First e-mail generated automatically by EPLAN

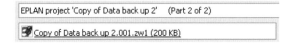

The subject is set automatically to contain the project name, information on which part is attached, and the total number of parts to be sent.

EPLAN does not continue with the data backup until the e-mail is sent. In this way, EPLAN automatically generates all parts and the corresponding e-mails. You only need to enter the recipient and send the mail yourself.

The last part (Part 1 of [Total]) is the most important part. When restoring the project later on, EPLAN first looks for this part in the selected directory, and all the other parts must also be in this directory.

Fig. 9.10 Second e-mail containing the last part of the project backup

This completes the e-mail backup. After the final part has been sent, EPLAN displays a success message.

Fig. 9.11 Success message when all parts have been sent

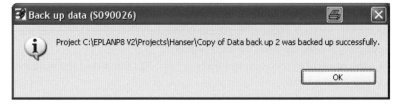

NOTE: If errors exist in any of these parts, then EPLAN cannot restore the project later on. If this occurs, then EPLAN displays a message window stating that the project cannot be restored. In this case, the only solution is to request another data backup.

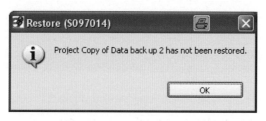

Fig. 9.12 Error message when the project could not be restored

In order not to allow a backed-up project to grow unnecessarily "big" (data volume), EPLAN offers the user four additional options in the **Back up projects** dialog to reduce the data volume.

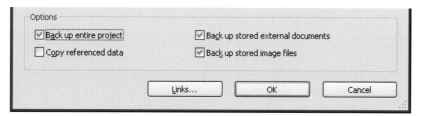

Fig. 9.13 Reduce data volume using these options

- **Back up entire project** – this backs up the entire project, including all data in the project directory and all subdirectories.
- **Copy referenced data** – if this option is set, then EPLAN backs up everything, including referenced data. This increases the amount of data in the project! For this, however, the options for stored external documents and stored image files must also be activated.

 NOTE: Referenced data is data that is not stored in the project, but is available only as reference. In short: EPLAN only knows the link (file path) where, for example, the graphic is stored on the server.

- **Back up stored external documents** – this option allows the user to include in or exclude from the data backup external documents stored in the project, such as PDF or Excel documents. However, only the external documents in the \[project name]\ DOC directory are included in the backup.
- **Back up stored image files** – similar to the previous option (external documents), you can define whether image files are included in or excluded from the backup process. This also only applies to image files in the [project name]\Images directory.

External documents or image files that are **not** in the project directory will **not** be automatically included in a backup. The previously mentioned **Back up external documents** and **Back up image files** options have **no effect** at this point.

If these files should also be backed up, then they must either be stored in the project (copied into the project directory on insertion during project editing) or they must be additionally and automatically stored in the project directory by EPLAN via the LINKS button.

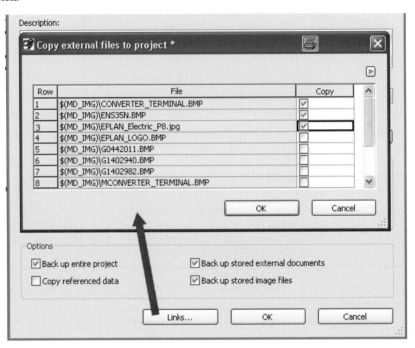

Fig. 9.14 Copy to project externally linked files (back up projects)

For external documents and image files that are linked into the project via hyperlinks, there is another way of collecting them and copying them together into the project directory before performing a data backup.

To do this, you use the EDIT / OTHER / LINKS menu to open the **Copy external files to project** dialog. This dialog is identical to the one that you can call up during the backup of projects.

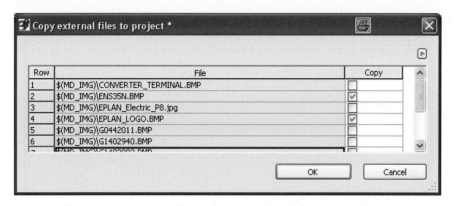

Fig. 9.15 Copy to project externally linked files (graphical editor)

EPLAN lists all linked external documents in this dialog. You must then select the **Copy** function for all desired documents. When you confirm the dialog by clicking OK, all selected files are copied to the \[project name]\DOC or \IMAGES project directory.

9.2.2 Restore projects

Backed-up projects can be restored.

To do so, first call the PROJECT / RESTORE / PROJECT menu and then select in graphical editing one or several projects you wish to restore (restore back).

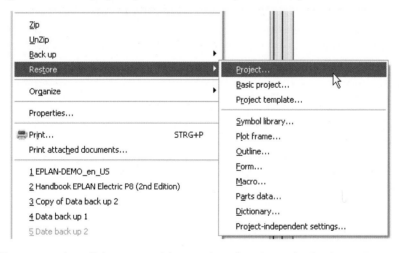

Fig. 9.16 Restore projects menu

The **Restore project** dialog opens with a number of setting and selection options.

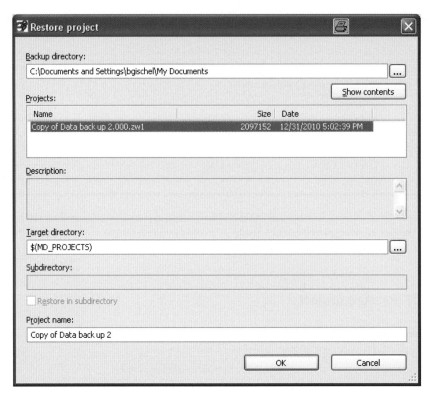

Fig. 9.17 Restore project dialog

- **Backup directory** - in the **Restore project** dialog, you can select the backup directory via the [...] button. You can use any type of backup directory, which is not bound to any specific structure.

TIP: However, it is a good idea to define a permanent directory for the backup files and/or the e-mail files. It is also a good idea to integrate this directory into the normal EPLAN directory used for the installation.

- **Projects** - the project to be restored must be selected here.
- **Target directory** - before restoring a project, you should first check the **target directory** field and/or set this to the correct project path. To change the target directory, click on the [...] button to select a different directory.
- **Project name** - you can assign a new one in the field of the same name before you restore the project, or you can leave the current project name as is.

Via OK or a double-click on the project (in the Projects field) with the left mouse button, you can start to restore the project using the above set parameters. EPLAN then begins the restore process. No further query dialogs are displayed.

If a project of the same name already exists in the target directory, then EPLAN displays a warning message asking if the target project should be overwritten. If you click the YES button, EPLAN restores the project and overwrites the old project.

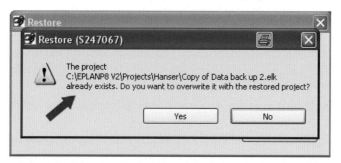

Fig. 9.18 Possible message if a project already exists

EPLAN can also overwrite opened projects if you wish. For this purpose, EPLAN closes the project following a prompt and then restores it.

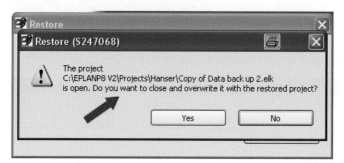

Fig. 9.19 Another message if a project already exists

Once the project has been fully restored, EPLAN opens it in the **page navigator**.

If you click the NO button, EPLAN cancels the restore process and displays a message stating that the project was not restored.

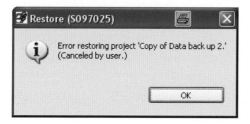

Fig. 9.20 Note displayed when canceled by user

By clicking on OK, you are automatically returned to graphical editing.

9.3 Other important settings

Related to the backing up and restoring of data are the default settings, e.g., for defining backup directories or for cleaning projects of old data that are no longer required.

It is also possible to automate the process of backing up data and cleaning projects.

9.3.1 Default settings for project backup (global user setting)

You can define global default settings for projects in the settings. They are then applied to the specific project.

You can adjust this and other default settings in EPLAN via OPTIONS / SETTINGS / USER / DATA BACKUP / DEFAULT SETTINGS / PROJECTS, where you can set the target directory, split size of e-mail attachments and other options, such as back up entire project, save external documents and image files and the default setting for the backup description field.

These settings are then applied when backing up a project.

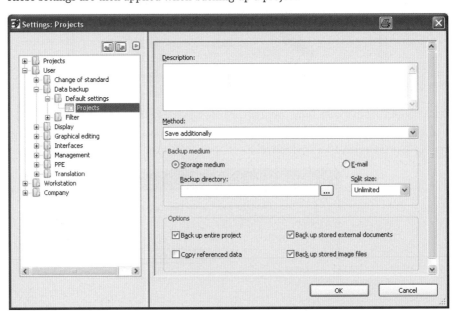

Fig. 9.21 Default settings to back up projects

TIP: If you subsequently change these settings yourself when backing up a project, then EPLAN remembers these settings precisely for this project. The global user settings then no longer apply. The updated data backup settings for a project always have priority over the data backup settings in the user settings and will always be used in the future for this project. You should note this.

The backup directory and the split size of e-mail attachments can, of course, be changed later, also for projects that have already been backed up.

The new settings are then applied again for the next backup process.

9.3.2 Compress project (remove unnecessary data)

While editing a project, it can become "filled" with a great deal of data.

Unplaced devices may be inserted or forms (master data) may be created for test purposes only. Since EPLAN stores all data in the project automatically, even when used only once, a project may end up containing (e.g.) ten different terminal diagram forms or dozens of parts even though only one of these is actually used for reports at the end of project editing or the parts are no longer needed.

Data that has become unnecessary in the course of project editing can be removed in EPLAN via the Compress project function. You reach this function via the PROJECT / ORGANIZE / COMPRESS menu.

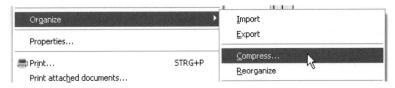

Fig. 9.22 Project / Organize / Compress menu

EPLAN then starts the **Compress project** dialog.

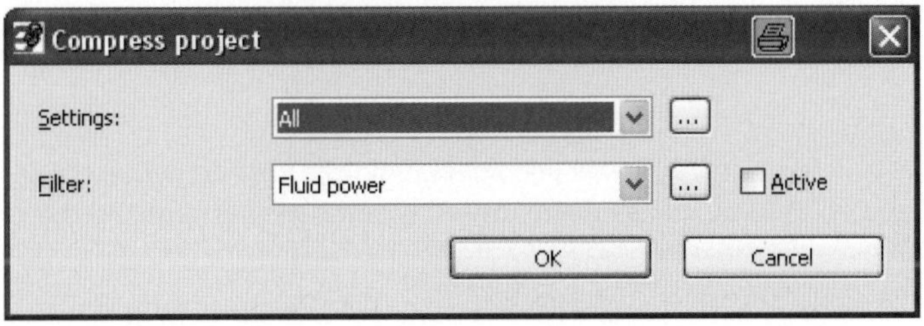

Fig. 9.23 Compress project dialog

The dialog offers several ways of affecting the data volume of the project.

You select the desired compression scheme to compress the data in the **Settings** selection field. EPLAN provides a number of default settings (finished schemes, which can be selected from a selection list), but here too, you can also define your own personal settings by creating and saving your own schemes.

To do this, you click the [...] button next to the **Settings** selection field. Then, the **Settings: Compression** dialog opens.

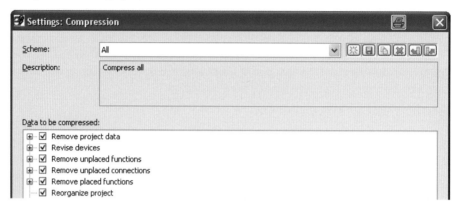

Fig. 9.24 Settings options for a compression scheme

Here you can use the familiar graphical buttons to create a new scheme or copy and change an existing scheme.

In the lower **Data to be compressed** field, EPLAN offers a number of actions that can be performed during the compression.

One example is the item **Remove project data / Unused forms**. This setting removes all unused forms from the project and the project then contains only forms that are actually used.

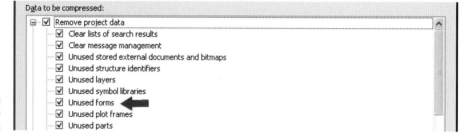

Fig. 9.25 Setting to remove unnecessary forms

 NOTE: If, for example, forms documentation was created before the compression, all forms (including the unused ones) remain stored in the project, because they are needed for the forms documentation. Here, the reports should be removed, especially the forms documentation, prior to the compression.

The other, second, method is to use filters for the compression process. As usual, filters can be set and are either created or edited as schemes.

In the **Filter** for compression you can (e.g.) exclude specific areas from the compression process, such as particular functions in the schematic or entire structure identifiers.

EPLAN thus provides various options for specifying which data is to be removed (using filters) and which areas or data the compression process should include.

9.3.3 Automated processing of a project

So far, we have looked only at individual steps. So, the Organize / Compress function is used to free the project of unnecessary ballast and the project is then backed up using the **Project / Back up / Projects** function.

The **Automated processing** function in EPLAN allows such actions to be combined together and to automate them. In our example, these are the compress project and data backup actions.

The Automated processing function can be found under the UTILITIES / AUTOMATED PROCESSING menu. This contains two menu items: SETTINGS and RUN.

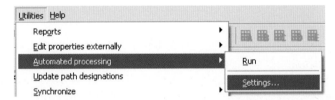

Fig. 9.26 The Automated processing menu item and its options

Under the SETTINGS menu item, we find the **Settings: Automated processing** dialog.

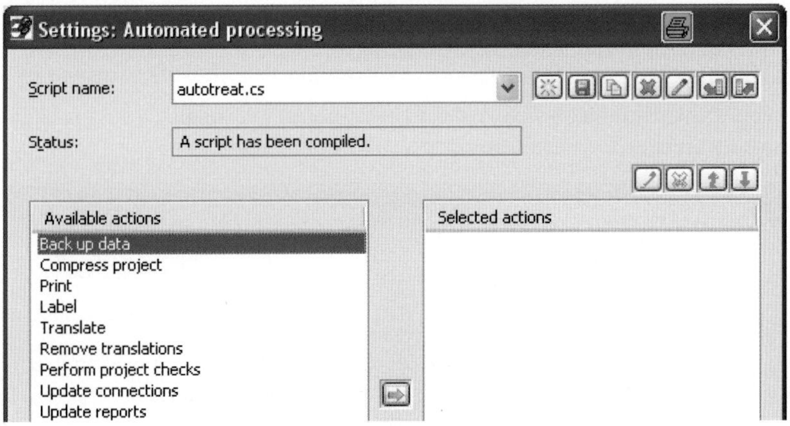

Fig. 9.27 The possible settings of automated processing

The **Automated processing** function is a script function. This means that all selected actions are "translated" into a script language and can be externally adjusted using an editor if necessary.

The dialog consists of the **Script name** and **Status** fields (not editable, these only provide information on whether the script was already created), the **Available actions** field (at the left), and the **Selected actions** field (at the right).

You use the [→] button to add selected available actions to the **Selected actions** field. If further settings are available for this action, EPLAN opens the corresponding settings dialog for further settings for this action after you move it.

For example, the **Back up data** action opens the dialog of the same name and you can make special settings that apply to this script.

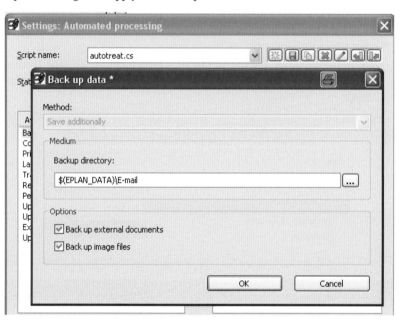

Fig. 9.28 Further dialog for data backup

If you now combine these settings, i.e., compress project followed by a data backup, then you can now process it automatically.

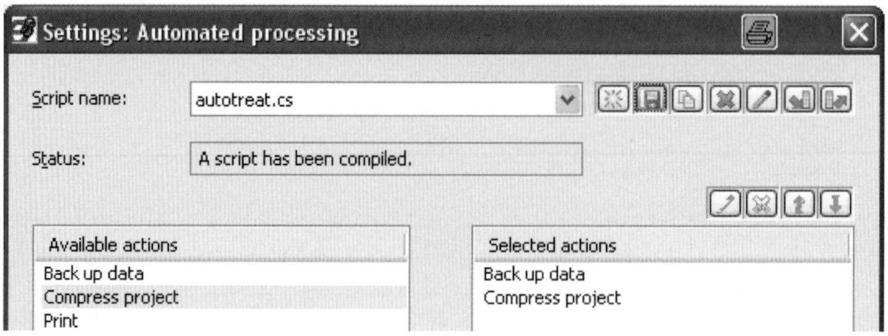

Fig. 9.29 Finished script with selected actions

This means that there are no more individual steps, only this one script is called up and EPLAN fully automatically performs all the different actions, which were previously defined, in a single step.

You can later edit the available actions via the top toolbar on the right. To do so, select the corresponding action in the **Selected actions** field and open it with the Edit button.

 NOTE: Created scripts are always stored on a project-specific basis. This means that a script **[script name].cs** can always be found in the corresponding project directory.

In general, the last edited script is stored in the settings and can be started via the UTILITIES / AUTOMATED PROCESSING / RUN menu item.

After starting the script, EPLAN executes the stored actions and displays the current progress of the script in a dialog.

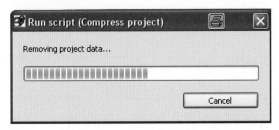

Fig. 9.30 Status of the script

Once the script, including the actions contained therein, is finished, EPLAN opens the **System messages** dialog containing the descriptions of the actions that were performed.

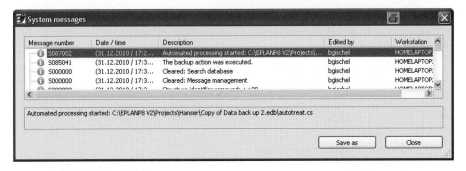

Fig. 9.31 Generated system messages on the process of the script

This dialog allows the user to check what actions were performed and what the results of these were. After clicking the CLOSE button, the **System messages** dialog closes and you can continue using EPLAN.

9.4 Backup and restoring of master data

As well as projects, you can also back up and restore master data in EPLAN.

This is a great advantage because it provides a unified, simple method of exchanging data between EPLAN users. EPLAN uses file extensions (among other features) to distinguish between the different types of data backups.

- **zw1** – file extension for projects
- **zw2** – file extension for symbol libraries
- **zw3** – file extension for plot frames
- **zw4** – file extension for forms
- **zw5** – file extension for macros
- **zw6** – file extension for parts data
- **zw7** – file extension for dictionaries
- **zw8** – file extension for user and workstation data

9.4.1 Back up master data

Backing up of master data generally always follows the same process.

Master data is saved via the PROJECT / BACK UP menu item, and subsequently the desired type of master data is selected that is to be saved.

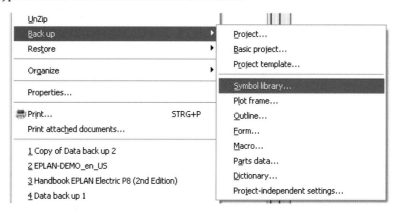

Fig. 9.32 Project - Back up - [Master data] menu

If master data is backed up via the menu item, then EPLAN opens the corresponding **Back up [master data]** dialog, whereby **[master data]** represents the various types of master data such as symbol libraries, plot frames, or the parts database.

The title bar of the respective window also indicates which type of data backup dialog is currently being used.

Fig. 9.33 Title bar

The **Back up parts data** dialog here is used as a representative example, since all the dialogs have the same structure and are used in the same way.

A data backup dialog for master data has a window with an entry indicating which master data is to be backed up. The example illustrates backing up of parts data.

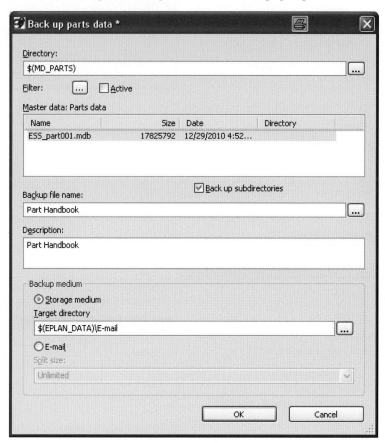

Fig. 9.34 Dialog as an example of how to back up master data

- **Directory** - A selection field defining the directory containing the master data to be backed up.
- **Filter** - you can filter certain files using the [...] button. Here you can define the files to be backed up and those that are not to be backed up. For these filter settings to become effective, you must activate the *Active* check box.
- **Backup file name** - an important entry since a backup cannot be performed without a name for the backup file. EPLAN uses this name as the file name for the backup file and appends the appropriate file extension.
- **Description field** - this can (but does not have to) contain a description of the current data backup. There are no limits to the contents of this field. This description is later displayed in the *Description* field when restoring from the backup.

The fields / options such as **Storage medium** or **E-mail** have already been sufficiently described in the section on data backups for projects. This will not be further discussed here.

When all settings are correct, you start the data backup process by clicking the OK button. EPLAN begins to back up the data and displays a final message after a successful backup.

Fig. 9.35 Success message

This completes the backing up of master data.

9.4.2 Restore master data

You use the PROJECTS / RESTORE [MASTER DATA] menu to restore (restore back) master data.

As described in the previous chapter, examples of master data are plot frames, forms, parts data or symbol libraries.

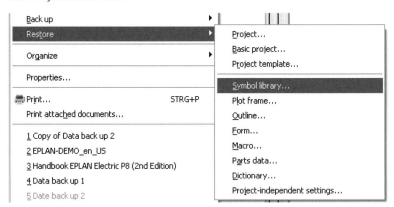

Fig. 9.36 Project - Restore menu

If you wish to restore this data in EPLAN, you use the PROJECT / RESTORE [MASTER DATA] menu to open the dialog of the same name.

Fig. 9.37 Example of a Restore master data dialog

Again, EPLAN has standard dialogs for various types of master data, so that the dialog shown here - to restore master data - is identical to that of all other types of master data.

You set the **Backup directory** to the directory containing the data you wish to restore. You can use the button to select the appropriate storage location if this differs from the default storage folder.

Once the correct folder is selected, the **Master data: [Master data type]** field lists the available backups (files).

NOTE: You can only select one file here. Multiple selection is not possible.

The **Description** field displays the information entered when the master data was backed up. This field is purely informative.

To restore the master data to a different specific directory, you must set up a suitable folder that differs from the default directory in the **Target directory** field. The default directory is the original source directory of the master data, i.e., where it was located when the backup was performed.

Once all entries match your requirements, you start to restore master data by clicking the OK button.

EPLAN restores the selected master data to the specified directory and then displays a message with information about the restore process.

Fig. 9.38 Success message

9.4.3 Default settings for the backup of master data (user)

To avoid always having to manually fill the individual filters with values (which master data is to be backed up and which not), there are user settings available under OPTIONS / SETTINGS / USER / BACKUP / FILTERS / [MASTER DATA TYPE] allowing you to exclude from (**Do not back up**) or include in (**Back up**) backups all areas of the master data fields in backups.

These settings then apply globally to every backup process. This allows (e.g.) specific master data to be excluded from a backup based on their file names.

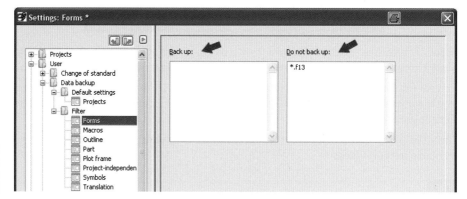

Fig. 9.39 Default settings to back up master data (user-specific)

9.5 Send project by e-mail directly

In addition to sending an e-mail as part of a normal backup process, EPLAN also allows a project to be directly sent via e-mail.

You can access this function directly via the PROJECT / SEND BY E-MAIL menu item.

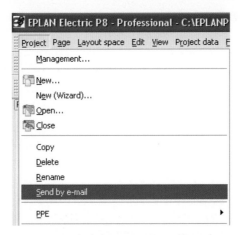

Fig. 9.40 Send project by e-mail directly

To do this, you first select the desired project in the **page navigator** and then call up the **Send by e-mail** function.

EPLAN starts a prompt.

Fig. 9.41 First note regarding e-mail sending

Then, EPLAN handles the default settings for sending by e-mail (e.g., the stated size of split files) and begins to generate the backup files.

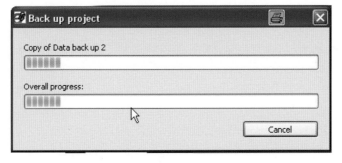

Fig. 9.42 Progress dialog for Back up project by e-mail

Depending on the size of the files, EPLAN now generates again one or several e-mails with the split project files and then waits for the user to send these e-mails.

Fig. 9.43 Newly generated e-mail with the split files of the project

Once all e-mails have been sent, EPLAN shows again a success message, which means that at this point the complete project has been sent.

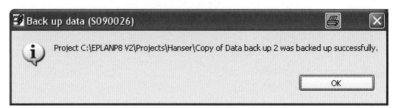

Fig. 9.44 Success message when sending e-mail messages

 NOTE: You can send only one project at a time by e-mail from the **page navigator**. Multiple selection allowing several projects to be sent by e-mail at once is not possible in the **page navigator**.

10 Master data editors

First off: EPLAN supplies a large number of finished system master data for forms, plot frames and also symbols.

These are normally sufficient for most applications. However, if special conditions for appearance exist or if particular project or device data is to appear in the forms and this is not possible with the standard EPLAN reports (forms) supplied, then some new system master data such as reports (forms) must be created from scratch.

 NOTE: In EPLAN, forms are for subsequent reports, also under normal circumstances. This is why we can speak of reports or forms. But not all reports are forms. A report can also be an Excel table created with the *Labeling* function!

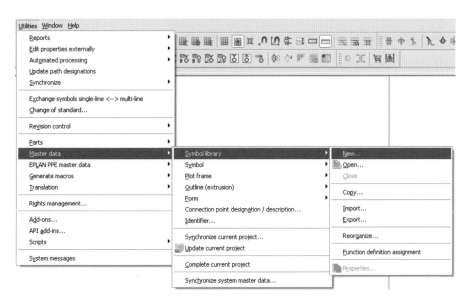

Fig. 10.1 The master data menu with the possible editors

NOTE: EPLAN always opens master data for editing on a temporary basis in the current project. But which project that is does not matter in this context.

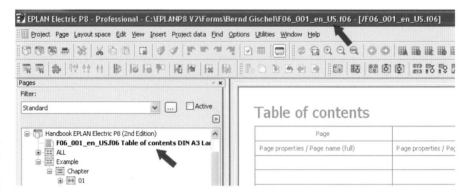

Fig. 10.2 Example representation for master data opened for editing

Generally, you always edit system master data. Project master data is never modified by the master data editors; rather the form or plot frame in the system master data directories is modified and/or newly created and then saved in the system directories of the master data.

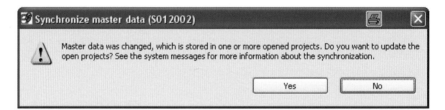

Fig. 10.3 Prompt when stored master data have been modified

If modified master data is also to be used in the current project, following the modification of master data the system master data must be synchronized with the project master data, i.e., the automatic prompt whether data is to be synchronized must be answered with YES. Otherwise, the modifications to the master data will not take effect in the current project, because they will not have been updated yet at that time.

You access the various editors via the menu UTILITIES / MASTER DATA / [TYPE OF MASTER DATA]. [TYPE OF MASTER DATA] can be: symbol libraries in general, symbols, plot frames or forms.

The option to create outlines is not discussed further at this point, because this is a subject related to mechanics and of less use to the actual (electrical) enclosure project planning.

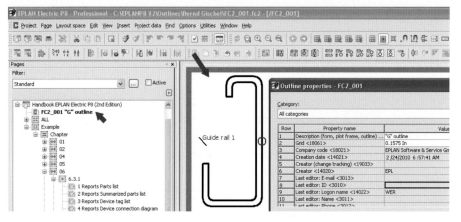

Fig. 10.4 Example of a finished outline (created by EPLAN)

 NOTE: Before you edit master data, note the following: Original master data supplied with EPLAN should never be modified. It is possible to modify these but, if EPLAN master data is a suitable basis for your own master data, then it is highly recommended that you copy them and assign a new name. You can then modify this copy as you wish.

If you follow this recommendation, then, when performing updates etc., the new, possibly modified original EPLAN master data in your system data directories can simply be replaced with the EPLAN master data of the same name. This means that you can enjoy improvements or extended functions immediately, and you will not have to integrate them yourself with your own master data.

 NOTE: It should generally also be noted that it is easier and faster to use existing master data as basis for editing. You can, of course, create this from scratch if you wish. Naturally, this will take more time and requires a more thorough understanding of the master data editor.

This chapter provides a brief explanation of the basic function steps required for editing and understanding.

Not all the details and possibilities of the form editor, plot frame editor, symbol editor or symbol library editor can be explained here. These editors are simply too complex for this, because there is an (almost) infinite number of options to implement various requirements in connection with the master data.

10.1 Preparatory measures

Although it is convenient to edit the system master data of any project, this also carries certain „dangers".

The more you follow this procedure, the greater the „danger" of losing your perspective at a later stage: What forms/plot frames have I already created, what do they look like, and which ones can I now use as templates for a new form/plot frame?

The „final story" is that you must eventually click through many forms/plot frames, and the small preview does not always provide you with all the information on the form/plot frame that you need. You must then load the form, only to realize it is not the correct one that you used last week in another project.

> **TIP:** If a multiple screen solution is used, you can, ideally, enable the separate graphical preview. The window size of the preview can be modified, and you can view the forms/plot frames in a convenient size. The graphical preview is activated in the VIEW / GRAPHICAL PREVIEW menu.

10.2 Overview of forms

To avoid any such „master data chaos" to some degree, you have various options at your disposal to display forms/plot frames in a structured manner.

10.2.1 First option - manual overview

A form overview would be a type of „form project", which is created and maintained manually. But it is really nothing else than a normal schematic project. Depending on your personal wishes and organization, all forms and plot frames are stored as extra pages.

All forms/plot frames are gathered in a project and created as separate pages. As well, the corresponding form or plot frame is assigned to such a page.

The selected page structure covers all identifiers, which allows deep structuring of the pages by customer and page types (in this case report types). However, this is a matter of personal taste and users can decide for themselves.

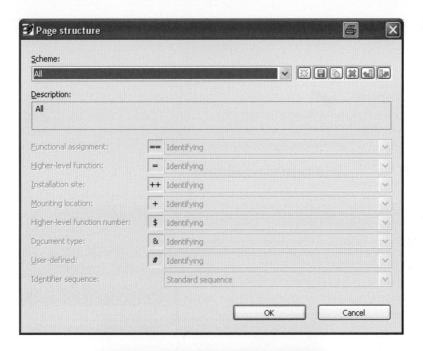

Fig. 10.5 A possible page structure

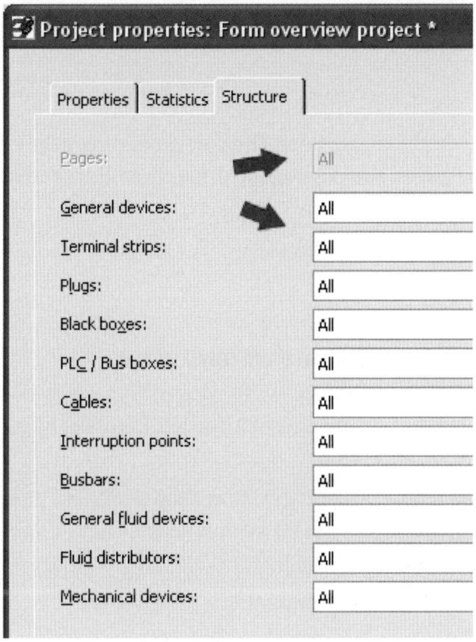

Fig. 10.6 Structuring of the remaining objects

Of course, you cannot directly edit the form or plot frame on these pages. We do not want to do this anyway. When the project is always maintained with the master data, you have a good overview of which forms/plot frames have been created over time, and, if needed, you can select a particular form or plot frame for subsequent editing.

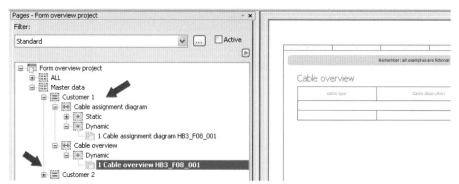

Fig. 10.7 Possible layout of the page structure with the various types of identifiers

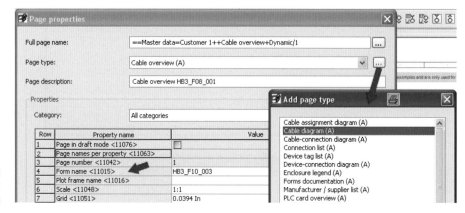

Fig. 10.8 Page properties of a page in the project

The pages are created via the usual steps, the corresponding page types (e.g., here a cable diagram) are set for the pages, the page descriptions are filled, and the form is selected via the selection in the *<11015 Form name>* property.

This stores it in the page and displays it.

10.2.2 Second option - automatic overview

A second option to document forms and plot frames involves using EPLAN's own media to create an overview.

To do so, you also create a new project and store all master data (forms, plot frames) in the project from the corresponding customer system master data directory.

The easiest way to do this is via Synchronize master data (UTILITIES / MASTER DATA / SYNCHRONIZE CURRENT PROJECT menu).

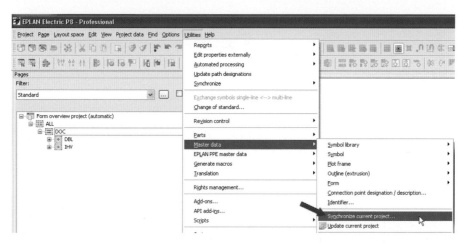

Fig. 10.9 Synchronize master data menu for a project

In the following **Synchronize master data** dialog, you can then copy to the project all (non-)existing system master data.

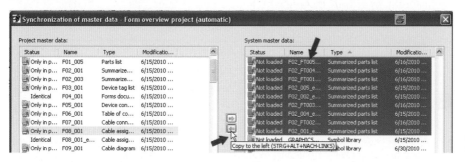

Fig. 10.10 Synchronize master data dialog

If all master data like forms and plot frames have been copied to the project, you can then generate the reports *Forms documentation* and *Plot frame documentation*.

EPLAN then generates for all forms and plot frames in the project the corresponding overview in the project automatically. You can also use this as an overview of the existing master data in the system directories.

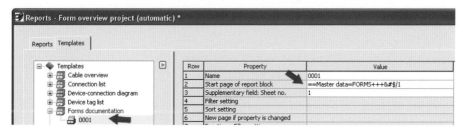

Fig. 10.11 Templates for the forms and plot frame documentation reports

The final result could look something like this:

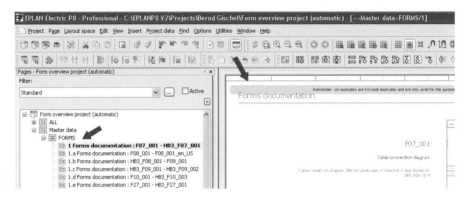

Fig. 10.12 Display in the page navigator

Unfortunately, there is no other sorting possible for forms documentation or plot frame documentation. But it is sufficient for a quick overview.

The great advantage of an automatic generation of forms documentation or plot frame documentation is clearly found in the fact that the corresponding pages are generated automatically by EPLAN literally with no more effort than „pushing a button". It really could not be any faster.

■ 10.3 Forms

Forms are used to graphically insert evaluated project data into the project documentation. Examples of forms are terminal diagrams, cable diagrams, device tag lists or a table of contents.

The supplied master data does not always exactly match the existing project data.

This may be rooted in the page references being too long, the function texts not fitting the intended rows and columns, or the display of PE terminals having to be „spruced up" with a small graphical symbol. Consequently, reports will then have to be adjusted or created from scratch.

10.3 Forms

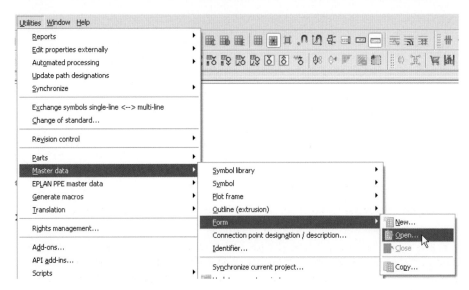

Fig. 10.13 Form editor menu

EPLAN provides a number of different ways of adjusting a form to suit the new layout. Forms can be created from scratch, edited (i.e., existing forms are opened and modified), or copied, and the copied form is then adjusted.

The form structure can be reached via the UTILITIES / MASTER DATA / FORM menu.

 NOTE: The form editor does not have a Save function. If changes are made and the form editor is exited via the CLOSE function in the UTILITIES / MASTER DATA / FORM menu, then the form is automatically stored with these changes in the system master data.

The project master data, however, still contain the previous "old" form, which you could, if necessary, copy back to the system master data directory.

This means that, as long as no (manual) synchronization of the master data (system master data with the project master data) is performed (e.g., via the UTILITIES / MASTER DATA / SYNCHRONIZE CURRENT PROJECT menu), this method can be used to bring the edited form in the system master data back to the state of the project master data.

After starting the function, EPLAN opens the SYNCHRONIZE MASTER DATA [PROJECT NAME] dialog. The form is now selected in the project master data area and the ➡ button is used to copy this into the system master data area.

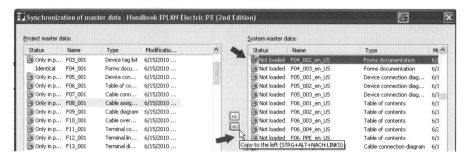

Fig. 10.14 Synchronize master data dialog

When existing master data is found, EPLAN opens the COMPARE MASTER DATA dialog. A decision must be made here as to whether the system master data should be replaced with the project master data.

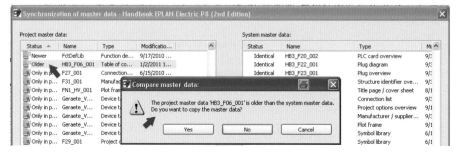

Fig. 10.15 Prompt asking whether the system master data should be overwritten

If the dialog is confirmed via the YES button, the system master data is overwritten with the project master data and EPLAN displays a notification message that the system master data was updated.

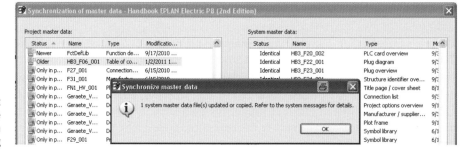

Fig. 10.16 Result message in the Synchronize master data dialog

 TIP: This method is really only an „emergency plan" for restoring master data that were unintentionally changed. As already mentioned, EPLAN allows automatic synchronization of master data when projects are opened.

If master data is incorrectly reset to an old state, unwanted master data synchronization may take place in existing projects (if the automatic master data synchronization has been activated)!

You should therefore be extremely careful when using this feature!

10.3.1 Create new form (from copy)

The easiest method is to create a form based on an existing one. It makes no difference what type of page is currently open in the project. You use the UTILITIES / MASTER DATA / FORM / COPY menu for this.

EPLAN opens the **Selection / Copy form** dialog.

Fig. 10.17 Copy form dialog

In this dialog, you select the form to be copied. You confirm the selection by clicking the OPEN button in the lower area.

EPLAN then opens the **Create form** dialog. The new form name is entered into the **File name** field of this dialog.

Fig. 10.18 Entry in the File name field

The form can now be saved via the SAVE button. EPLAN closes the dialog, opens a temporary page with a **cable diagram page type** and displays the copied form on this page for editing.

Fig. 10.19 Temporary page with the form in the page navigator

The form can now be edited. In the form editor, all graphical functions such as **Insert line** or **Insert circle** (via the INSERT / GRAPHIC menu) are available, as well as functions for inserting normal text.

The INSERT menu contains a number of other functions relating only to the form editor. The graphical functions have already been explained or are well-known functions in the graphical editor.

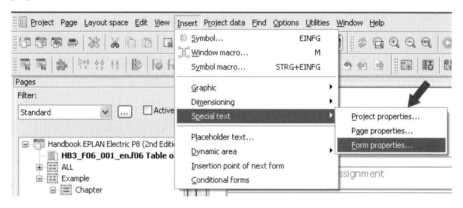

Fig. 10.20 Special texts

Special texts such as project, page, or form properties can be placed in forms. The actual form texts (data for reports) are, however, called "placeholder text" in EPLAN.

In contrast to the project and page-specific texts or form properties, these placeholder texts are only filled with project data after reports are graphically generated.

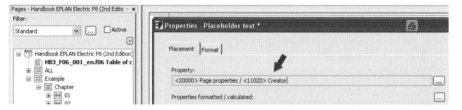

Fig. 10.21 Display placeholder text dialog

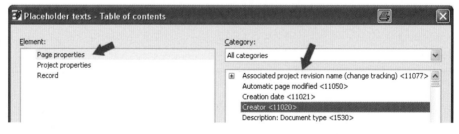

Fig. 10.22 Select placeholder text dialog

After starting via the INSERT / PLACEHOLDER TEXT menu, in the **Placeholder text** dialog on the **Placement** tab, use the ▭ button in the subsequent **Placeholder texts – [Report type]** dialog to select a placeholder text.

The report shown here is an example - other placeholder texts apply to all other forms.

In our example, the element in the left area is preselected (left selection area in the dialog), as is the desired property that is required in the right area of the dialog.

You can now confirm this selection by clicking OK.

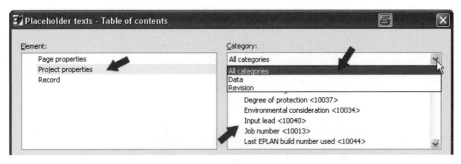

Fig. 10.23 Preselection (left) and selection (right) of the property

EPLAN closes the dialog and transfers the selected placeholder text into the dialog of the same name.

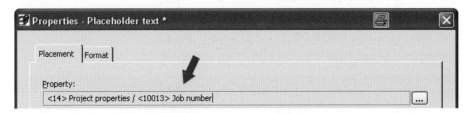

Fig. 10.24 Applied placeholder text

Now confirm the **Placeholder text** dialog by clicking OK; the placeholder text now hangs on the cursor and can be placed at any (sensible) position in the form.

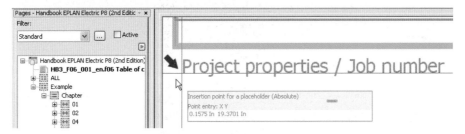

Fig. 10.25 Choice of placement

As with any other (free) text, placeholder text can be freely formatted.

This allows the form to be appropriately „constructed". It is also possible to subsequently modify placeholder text and select a different one. To do this, select the placeholder text and display its **properties** by double-clicking or via the popup menu.

You then proceed as described above. You use the ⋯ button to call up the subsequent dialogs and select, apply and place other properties in the usual manner.

Upon completion, the form can be checked. Using the function in the UTILITIES / CHECK FORM menu, EPLAN can check the form for errors. If the form is OK, the following message is displayed: The form is OK.

Fig. 10.26 Check form   Fig. 10.27 Success message

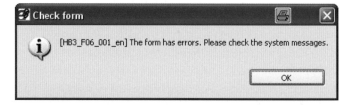

Fig. 10.28 Error in the form

If the form has errors, EPLAN indicates this and writes the errors to the system messages. These should be corrected to avoid incorrect system master data.

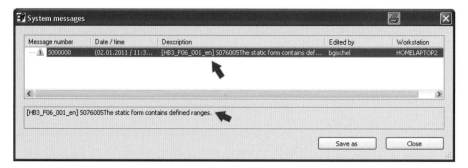

Fig. 10.29 System messages on incorrect form

When handling forms, EPLAN distinguishes between dynamic forms, where only certain areas exist (e.g., a data area, which may contain an X amount of data), and static forms, where the amount of data is permanently "wired" into the form.

Dynamic forms expand their graphical display area depending on the amount of project data. The maximum amount of data that the dynamic form should display is defined in the form properties. In our example (the procedure is the same for all forms) you can access the form properties via the key combination CTRL + M + D or in the page navigator via the PROPERTIES function in the right-click popup menu.

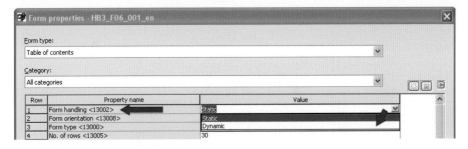

Fig. 10.30 Form properties dialog

EPLAN opens the **Form properties** dialog. All the relevant form data is defined here.

How should the form be handled, dynamically or statically, which entries in the automatic page description property should be written when project data is evaluated, or what is the maximum amount of data (number of rows) that the form should display before a page break or new report page is created?

A sample dynamic form then appears as follows. The rows are dynamically extended depending on the volume of data. If a terminal strip has five terminals, then the form will contain five rows with data. A terminal strip with twelve terminals will show twelve (data) rows in a report.

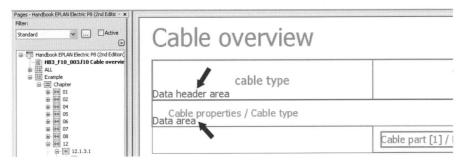

Fig. 10.31 Example of a dynamic form

In principle, static forms have the same properties as dynamic forms.

In static forms, the data areas are fixed, i.e., permanently defined, and the maximum number of data rows is entered in the form properties – in the same way that the graphic for the form was created with the maximum number of data records.

The following image shows an example of a static form. The difference can be clearly seen. Whereas a dynamic form has a data row that is dynamically expanded, a static form already has all data rows as a finished graphic. The data is written to these graphical rows when generating a report.

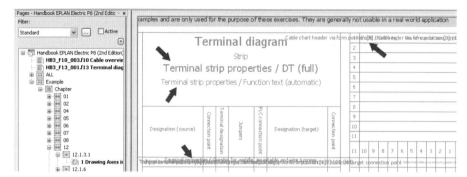

Fig. 10.32 Example of a static form

If the form is ok, then you can exit the form via the UTILITIES / MASTER DATA / FORM / CLOSE menu. At that time, it is then "saved" definitively.

 NOTE: After the very first editing, a copied form is automatically stored in the system master data and (when used) also stored in the project master data. At this point, synchronization of the master data (system and project) does not need to be performed.

When the form is edited a second time, then it is only updated in the system master data. Master data synchronization must now be performed and confirmed in order to update/synchronize the project master data with the system master data!

10.3.2 Edit existing form

An existing form is selected for editing via the UTILITIES / MASTER DATA / FORM / OPEN menu in the **Open form** dialog.

EPLAN then opens a temporary page once more, which contains the form that was selected for editing.

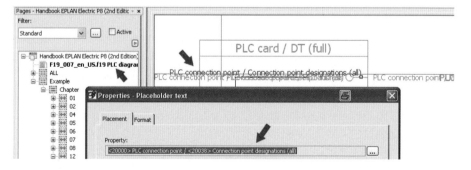

Fig. 10.33 Example: Form is open for editing

TO DO: There will be no explanation on the editing of the form at this point, because the general procedure for editing an open form is the same as the procedure described in the last chapter.

10.3.3 Create new form

You create a new form via the UTILITIES / MASTER DATA / FORM / NEW menu.

EPLAN then opens the **Create form** dialog. The new **Form name** must be entered in the **File name** field.

Fig. 10.34 Define form name

The **Create form** dialog can now be exited via the SAVE button. EPLAN opens, in the now familiar manner, a temporary page in the background and the **Form properties** dialog is displayed in the foreground.

You can already begin making the first entries here. However, this is not absolutely necessary at this time and you can enter this data later.

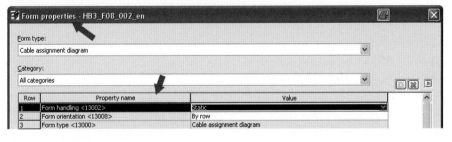

Fig. 10.35 Form properties dialog

As described in the previous sections, form editing is done using the various different functions such as inserting graphical elements or texts.

To create a new form takes time and effort, because you start from a blank form. It is therefore recommended that you copy a similar form and then modify this copy.

 TO DO: The procedure for creating and editing a new form is identical to the procedure for editing existing or copied forms and is therefore not explained further.

10.4 Plot frame

Plot frames are a sort of frame around the actual project pages. They contain, for example, structure identifiers, column and/or row information and other data like page description, etc.

Editing plot frames in the plot frame editor is very similar to editing forms in the form editor. The editor is accessed via the UTILITIES / MASTER DATA / PLOT FRAME menu.

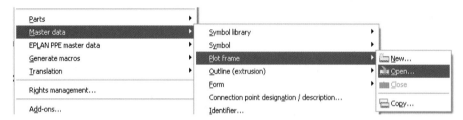

Fig. 10.36 Plot frame editor menu

As with the form editor, this menu has the items New, Open, Close, and Copy. This menu functions in a similar way to the menu used when editing forms. We will therefore only deal with the features that are special to the plot frame editor.

 NOTE: The plot frame editor does not have a Save function either. To restore plot frames that were accidentally modified, you should therefore use the Undo function in the plot frame editor or consider using the process of synchronizing existing project master data with the system master data.

10.4.1 Create new plot frame (from copy)

To copy and then edit a plot frame, in the UTILITIES / MASTER DATA / PLOT FRAME menu, select the COPY function.

EPLAN opens the **Copy plot frame** dialog.

Fig. 10.37 Copy plot frame dialog

In this dialog, you select a suitable plot frame and confirm your selection by clicking the OPEN button in the lower area.

EPLAN closes the **Copy plot frame** dialog and opens the **Create plot frame** dialog.

Fig. 10.38 Create plot frame dialog

You enter the new name of the plot frame into the **File name** field. You then close the dialog via the SAVE button, and EPLAN temporarily opens in the current project a page with the copied plot frame.

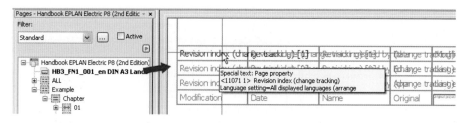

Fig. 10.39 Open plot frame in the page navigator

The plot frame can now be edited. You use the same functions as already described in the form editing section. Graphical elements and texts can also be placed in the plot frame editor.

In contrast to forms, it is possible (and, in part, for project planning necessary) to specify column text (paths) and row text for the plot frame. You can also embed a special text watermark in the plot frame. This special text is then automatically filled with data from revision management.

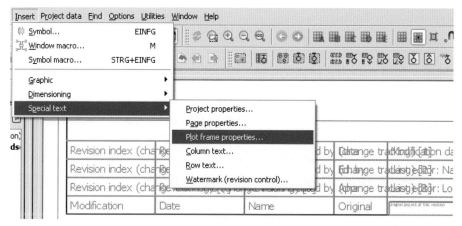

Fig. 10.40 Insert / Special text menu

Column and row texts are treated as normal text. This also includes the formatting (font size, color, etc.).

Column and row texts are placeholder texts, defined by entries in the **Plot frame properties** dialog (via the CTRL + M + D shortcut key or in the **page navigator** by selecting the temporary page, opening the popup menu, and selecting the PROPERTIES item), and which are later automatically filled in the project.

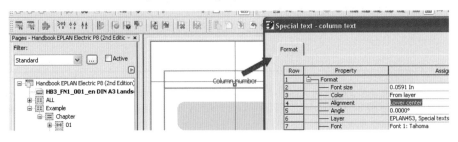

Fig. 10.41 Formatting options for column and row texts

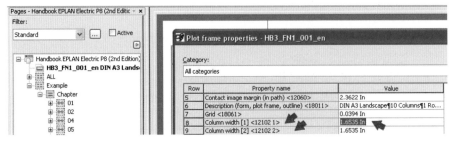

Fig. 10.42 Plot frame properties

Once all entries have been made or the plot frame has been graphically edited to suit, it can be closed ("saved") by selecting in the UTILITIES / MASTER DATA / PLOT FRAME menu the CLOSEfunction.

EPLAN closes and "saves" the plot frame in the system master data definitively. If the plot frame is needed for the project, it will be stored in the project from the system master data automatically. This functional process is identical to working with forms; the same is true of the synchronization in case of subsequent modifications to the plot frame.

10.4.2 Edit existing plot frame

In addition to copying plot frames, existing plot frames can also be edited. To do this, select the UTILITIES / MASTER DATA / PLOT FRAME menu and then select the OPEN function.

EPLAN opens the **Open plot frame** dialog.

Fig. 10.43 Open plot frame dialog

Here you select the plot frame you wish to edit and open it via the OPEN button in the lower area of the dialog. EPLAN closes the dialog and opens the plot frame again as a temporary page in the page navigator of the current project.

It can now be edited using the familiar functions (graphics, text, etc.) and the plot frame properties can also be edited or extended.

Close the plot frame using the UTILITIES / MASTER DATA / PLOT FRAME / CLOSE menu. If the plot frame has been used in the current project, you will be prompted whether you wish to synchronize with the project master data.

10.4.3 Create new plot frame

Instead of copying or editing existing plot frames, you can create a completely new plot frame via the NEW function in the UTILITIES / MASTER DATA / PLOT FRAME menu.

EPLAN then opens the **Create plot frame** dialog.

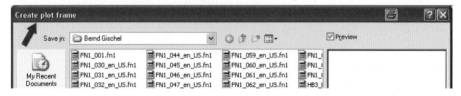

Fig. 10.44 Create plot frame dialog

You enter the new name of the plot frame into the **File name** field of the **Create plot frame** dialog.

Fig. 10.45 New plot frame name in the File name field

Clicking the SAVE button creates the plot frame and EPLAN then opens the **plot frame properties** of the plot frame and also displays an empty temporary page in the page navigator of the current project.

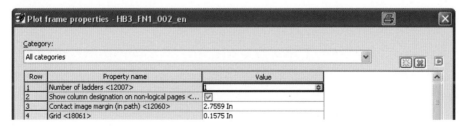

Fig. 10.46 Plot frame properties of the new plot frame

Now, you can edit the plot frame using the familiar functions.

TO DO: It is recommended, instead of creating a new plot frame, that you use here as well a copied plot frame as a template for a new plot frame and then modify it accordingly.

This is done in the same way as editing an existing plot frame and will not be discussed further here.

11 Old EPLAN data (EPLAN 5...)

Generally speaking, importing data from older versions of EPLAN 5, EPLAN 21, PPE and FLUID is a good way of continuing to work with these projects in EPLAN Electric P8. The conversion is relatively trouble-free.

TIP: It is a better solution to start new projects completely using EPLAN Electric P8 and to not base them on data from projects created with older versions.

It is also not absolutely necessary to convert all projects from old versions into EPLAN Electric P8. Since EPLAN will continue to support the earlier versions for a long time to come, there is no reason to panic. Older or currently running projects can still be edited or finished with the earlier versions.

This chapter will only provide a very brief introduction to the import functions. For additional information, I suggest that the reader consult the first edition of the manual, which contains a thorough chapter on importing older data, or use the online help for EPLAN Electric P8, or read the support documentation in the EPLAN support area on importing data from old versions to EPLAN Electric P8.

In general, we must recall that there were numerous options for creating schematics in the old versions of EPLAN. It is therefore unavoidable that deviations will occur here and there in a converted (imported) project.

NOTE: If projects are imported into EPLAN Electric P8 from an old version, then the old project remains intact. After importing, the original project exists (that of the old version) and the same („new") project converted into the EPLAN Electric P8 data model. None of the old data is deleted or changed, so the original project can be edited further in the previous version without problems.

If the project is edited further in the previous version, then the project no longer agrees with the project that was previously converted to an EPLAN Electric P8 project. Therefore it must be re-imported into EPLAN Electric P8. You should note this.

11.1 Import options

EPLAN Electric P8 offers an import assistant for the various old versions of EPLAN, as well as for the different data from older versions. This can encompass entire projects or individual data sets, such as a form, a macro, parts data or cable data.

You will find this import assistant in the PROJECT menu in the EPLAN 5 - / FLUIDPLAN DATA IMPORT (for EPLAN 5 and Fluid projects, as well as EPLAN 5 / Fluid master data), EPLAN21 DATA IMPORT (for EPLAN 21 projects and/or EPLAN 21 master data) and EPLAN PPE DATA IMPORT (for PPE older projects and PPE older data) menu items.

In general, it is possible to import both entire projects and individual data with EPLAN Electric P8. This is why additional functions and calls (for EPLAN 5 / fluidPLAN, EPLAN 21 and EPLAN PPE) were integrated into the individual data import option menu items. These enable, for example, an individual plot frame, or exported parts data from the older system, to be imported into EPLAN Electric P8.

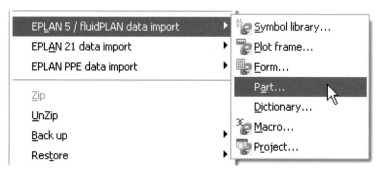

Fig. 11.1 Menus for (old) data imports

 NOTE: The sequence of menu items in the menu is important during data importation. This sequence was not arbitrarily designed; it represents the standard sequence of a data import.

For example, an EPLAN 5 data import should begin with the symbol library, then the plot frame, followed by the forms, and then the parts data and the foreign language file (this is called the dictionary in EPLAN Electric P8), and any macros. At the end, the actual EPLAN 5 project can be imported.

In addition to these individual steps, EPLAN Electric P8 offers a „one-click assistant" for data importation. When importing old data, you would start with the PROJECT / EPLAN 5 / FLUIDPLAN DATA IMPORT / PROJECT menu item, provide some information in the subsequent dialogs, and then EPLAN Electric P8 can automatically take care of the rest of the import procedure.

12 Extensions

In addition to the standard functions delivered with EPLAN Electric P8, there are a series of further function extensions that are covered by additional modules, also called add-ons. These add-ons make it easier to work with projects and their data, or help in other ways to accelerate and optimize engineering.

This chapter will explain some of these add-ons. In addition to those that are discussed here, there are still a series of further add-ons that can be requested directly from EPLAN.

■ 12.1 EPLAN Pro Panel

The most interesting new feature in EPLAN Electric P8 Version 2.0 – and this is not just my personal opinion – is without question the **EPLAN Pro Panel** module. The module offers mounting layouts in 3D, thus providing significant advantages to the EPLAN user.

12.1.1 EPLAN Pro Panel – what is it?

EPLAN Pro Panel is a very useful alternative to conventional 2D mounting panel layouts. The 3D approach is based on an existing schematic, but of course it can also be started with the selected parts without there being a schematic. This feature is provided by the common data platform upon which all the EPLAN products – such as EPLAN Electric P8, EPLAN Fluid, and now EPLAN Pro Panel – are based.

EPLAN Pro Panel supports the user in projection and mounting layout, e.g. with a collision check so that devices are not planned in the wrong position in the layout. Of course, EPLAN Pro Panel doesn't relieve users of (mental) work, but it does help to avoid wrong decisions (for example, misplacements), which saves time in the end.

For reasons of space, this chapter unfortunately cannot cover, describe and explain all of the facets of EPLAN Pro Panel. It is too expansive and there are simply too many choices with regard to how you can approach projects with EPLAN Pro Panel. But we will explain the basics of the first steps in a project with its devices, to 3D layout of a mounting panel, to the final step of creating the familiar 2D model view.

12.1.2 General

In addition to the various uses offered by EPLAN Pro Panel, there are a few conditions of use to consider. These affect the hardware as well as which data will be necessary to have fast and successful experiences with EPLAN Pro Panel.

12.1.2.1 What are the hardware requirements?

In terms of hardware, we don't need powerful "CAD machines"; instead, the hardware that runs EPLAN Electric P8, for example, is sufficient.

The graphics card will have to pay some tribute to the 3D feature. The graphics card must at least be able to handle OpenGL Version 2.1. In principle, you can also do without, but without OpenGL, something will be displayed incorrectly sooner or later, and the joy of construction with EPLAN Pro Panel will be lost.

Of course, we need not mention that using multi-screen solutions with EPLAN Pro Panel is better than a small one-screen solution.

12.1.2.2 What are the data requirements?

Basically, there are no major demands here with regard to data and its contents.

You do not need an elaborate 3D model; simple values for width, height and depth are enough to get started.

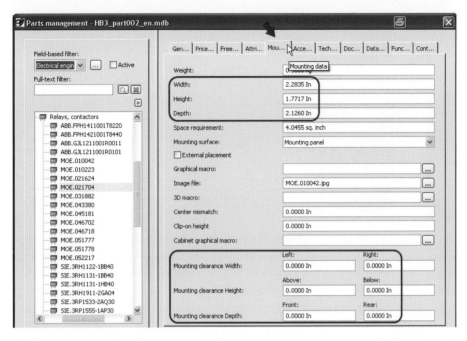

Fig. 12.1 Required technical data for a part

Ideally, of course, data for mounting clearances, center mismatch and clip-on height should also be filled out correctly.

If instead of a single „faceless" body you would like to have a detailed body with the smallest screw details in the layout, you can add a macro for this (*Graphical macro* or *3D macro* field). An image file is always advantageous, and not just for EPLAN Pro Panel.

These files can of course then also be used in different reports (e.g. a parts list with the technical data and an image of the device), and their use is not limited to EPLAN Pro Panel.

12.1.2.3 What is possible with EPLAN Pro Panel?

In addition to the visually appealing layout of a 3D display (better understood by customers), useful "by-products" are also possible in this display.

For example, a 2D derivation of a 3D layout.

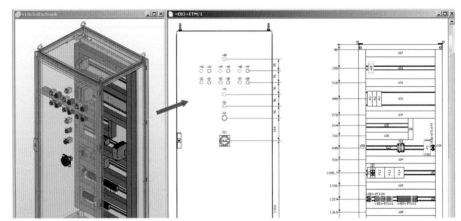

Fig. 12.2 3D representation and the 2D view derived from it

The clear advantage here is obvious. The enclosure is projected in 3D, and the 2D representation, with its dimensions, distribution, front views and side views, and maybe additional follow-up documentation such as an enclosure legend, can be created automatically and transferred to the workshop for construction.

Accordingly, a layout is not done twice; instead, the mounting layout is only created, changed and expanded in 3D, and the 2D derivation and all of the other automatic follow-up documents, such as reports, are simply updated.

12.1.2.4 General information on operation

In this chapter and the following one, we will deal with a few basic settings, notes on operation, and the meaning of some buttons, menu items, etc.

12.1.3 Toolbars in EPLAN Pro Panel

In addition to the menu functions, a large number of functions in EPLAN Pro Panel is also available as toolbars. As usual in EPLAN Electric P8, toolbars can be called up and activated by right-clicking in the toolbar area.

Fig. 12.3 Toolbars

EPLAN's standard scope of delivery includes two toolbars (**3D viewpoint** and **Pro Panel options**), and you can set up a third one (**Insert Pro Panel...**) yourself; they contain the insertion of devices such as enclosures, mounting rails, or areas locked for placing. This saves you the alternative route via the INSERT / [DEVICE FOR PRO PANEL] menu, in which [Device for Pro Panel] means mounting panels, mounting rails, etc.

12.1.3.1 3D viewpoint toolbar

The first six buttons of this toolbar offer different *2D view directions*, e.g. on the enclosure, the terminal box, or other parts.

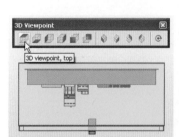
Fig. 12.4 View from above

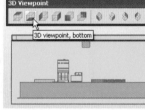
Fig. 12.5 View from below

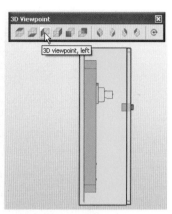
Fig. 12.6 View from left

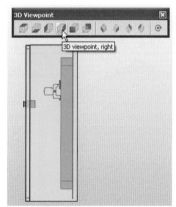
Fig. 12.7 View from right

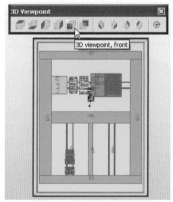
Fig. 12.8 View from front

Fig. 12.9 View from rear

The next four buttons activate the *3D view* and its four different isometric viewpoints.

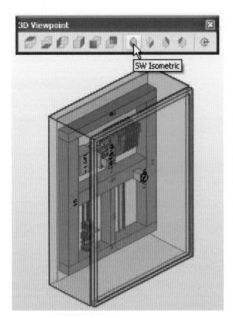

Fig. 12.10 South-West isometric view (SW)

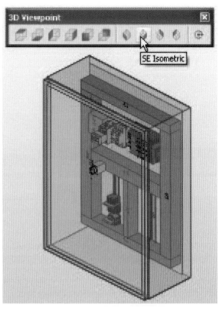

Fig. 12.11 South-East isometric view (SE)

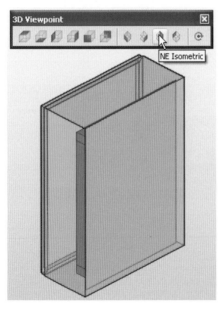

Fig. 12.12 North-East isometric view (NE)

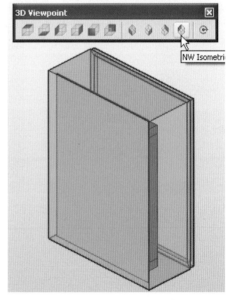

Fig. 12.13 North-West isometric view (NW)

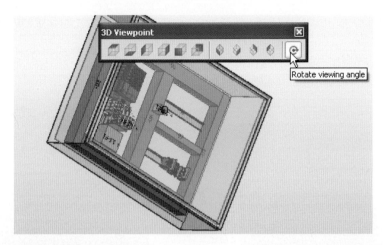

Fig. 12.14 Manually select rotation angle

The last button, ROTATE VIEWING ANGLE, allows you to hold down the left mouse button to freely and steplessly rotate an object to get the exact angle that you want or need.

12.1.3.2 Pro Panel options toolbar

This toolbar enables various display and control functions, among others.

When the COLLISION CHECK button is switched on, EPLAN calculates, before placement, whether items will collide, showing this behavior in color and not allowing the device to be placed.

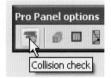

Fig. 12.15 Collision check

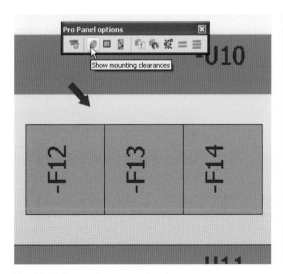

Fig. 12.16 Do not show mounting clearance

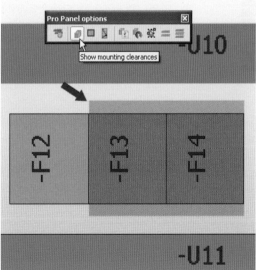

Fig. 12.17 Mounting clearance is shown.

The DISPLAY MOUNTING CLEARANCES button enables a quick visual check of whether there is still some more space around the placed device, or if the mounting clearances are already as minimal as possible.

The SIMPLIFIED REPRESENTATION button activates a function that displays all devices in a simple "chunk" without additional fancy graphics.

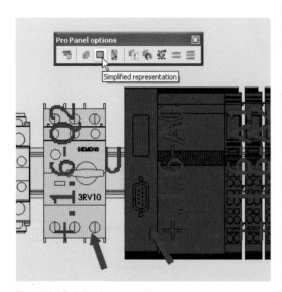

Fig. 12.18 Detailed representation

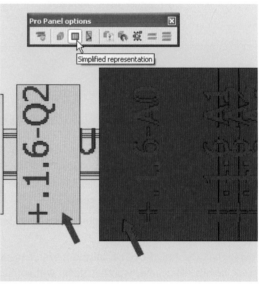

Fig. 12.19 Simplified representation

In the left image, all devices (even if present and drawn in the macro) are displayed with lots of details, such as screws, setting ranges, switches, etc. If this is not necessary or if the display (the graphical rendering) becomes too slow, then you can switch on the simplified representation (right image).

This also applies, for example, to large, extensive terminal strips where the focus is less on the individual terminals than it is on identifying the terminal strip as a whole in the mounting layout.

The CONNECTION PREVIEW button contains a preview of the devices that are connected to each other.

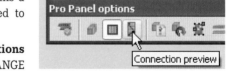

Fig. 12.20 Connection preview

All of the other buttons in the **Pro Panel options** toolbar, such as PLACEMENT OPTIONS, CHANGE ROTATION ANGLE, CHANGE HANDLE, ADOPT LENGTH, and PLACE CENTERED, are addressed in greater depth in the following chapters. At this point, this is just a brief overview of the functions.

PLACEMENT OPTIONS button. If a device is placed and is hanging on the cursor, this button can call up the **Placement options** dialog.

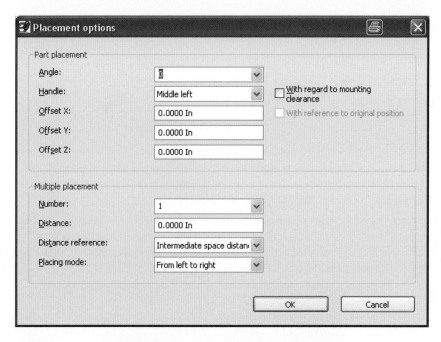

Fig. 12.21 Placement options dialog

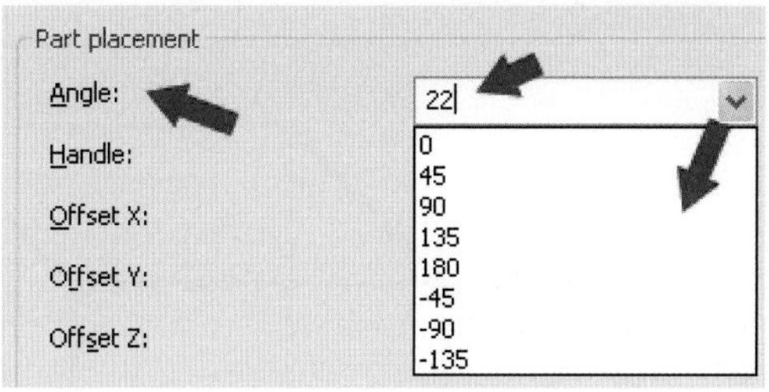

Fig. 12.22 Adjustable rotation angle

There are several options in this dialog that affect placement. In the *Rotation angle* selection, you can apply a value from the selection list, but you can also enter any degree value; for example, you could place a device rotated by 50 degrees.

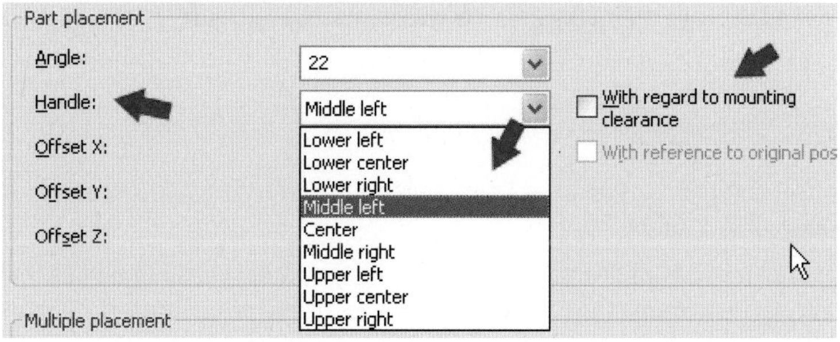

Fig. 12.23 Selecting a handle

It is also possible to preselect the *handle*. You can also apply a value here from the selection list. This selection list, however, cannot be extended. The handle is closely associated with the *With regard to mounting clearance* setting. If this setting is activated, the mounting clearance is calculated at the handle and the device is accordingly placed "further" (meaning with mounting clearance).

Fig. 12.24 Offset settings

Offset setting X, Y, Z – You can enter values directly here to achieve a precise placement at specific angles. This means that EPLAN Pro Panel does not allow another placement, such as for example the placement of a device 50 mm from the edge. A manual deviation is therefore not possible - "trembling mouse" during placement.

The adjacent setting, *With reference to original position*, can only be actively switched on during actions such as copying, duplicating, etc. This allows you to enter an offset value directly, and EPLAN Pro Panel automatically calculates the additional placement (e.g. for a copy of a device) from the original position.

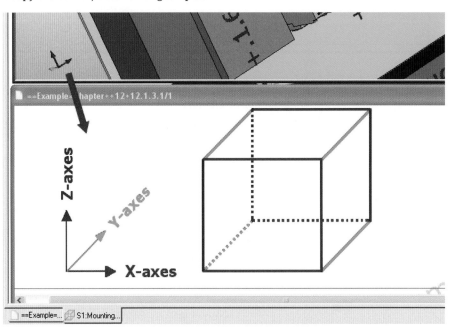

Fig. 12.25 Axes

For the offset settings, it is good to know that we are working in 3D space in the three drawing axes of X, Y, and Z.

The three drawing axes are represented in EPLAN Pro Panel by colored lines that always run at right angles to each other. These axes provide useful orientation in 3D space during placement of objects, for example.

NOTE: The Y axis is the green line, the X axis is red, and the Z axis is blue in EPLAN Pro Panel's 3D display.

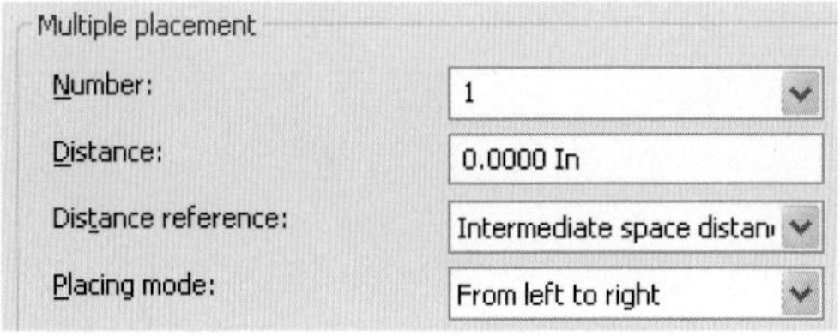

Fig. 12.26 Multiple placement

The setting for multiple placement is for when you want to place several objects together at once. Then all of the objects selected in the 3D mounting layout navigator hang on the cursor, and after placing the first device on the mounting surface (e.g. a mounting panel), all of the following devices are automatically placed with the settings as defined in the multiple placement area.

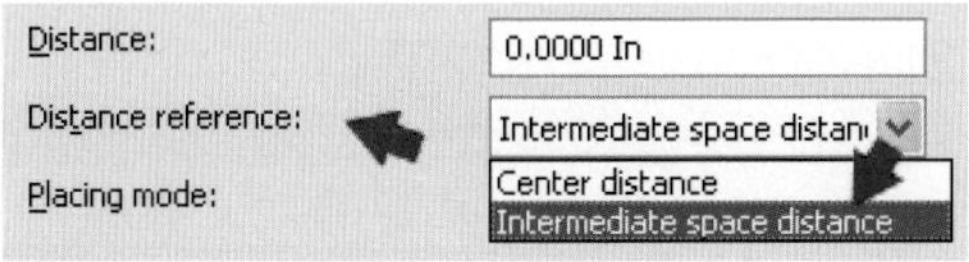

Fig. 12.27 Setting for distance reference and distance

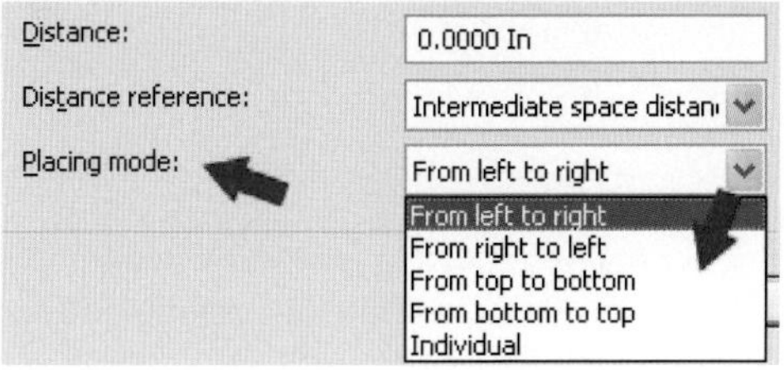

Fig. 12.28 Setting for placing mode

One example would be the placement of control elements in a door. EPLAN Pro Panel automatically calculates the desired distances from the first control element, without having to manually place every single control element yourself.

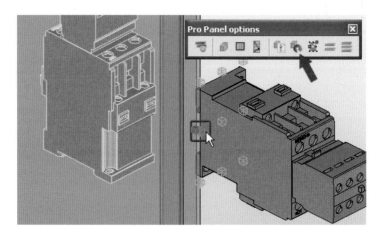

Fig. 12.29 Changing the rotation angle before placement

Another button, CHANGE ROTATION ANGLE, offers the option of easily changing the rotation angle before placing a device. However, you can only rotate in fixed 90-degree increments.

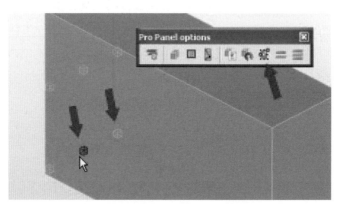

Fig. 12.30 Changing the handle

With the CHANGE HANDLE button (or just pressing the A key), you can quickly change an object's handle. The image clearly shows that the contactor has nine handles: external and internal contacts that are arranged in the center, left and right. This allows the object to be handled in a number of ways and to be placed with the desired point.

Fig. 12.31 Adopting the length

To place an additional mounting rail or cable duct later, for example, that should have the precise length of an existing mounting rail or cable duct, you can use the ADOPT LENGTH button.

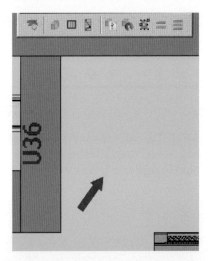

Fig. 12.32 Free space

Following procedure: An additional cable duct is to be placed on a mounting panel in free space with the same length as an adjacent cable duct. The cable duct is selected and applied via the INSERT / WIRE DUCT menu. The cable duct now hangs on the cursor.

Now this cable duct should adopt exactly the same length as the duct displayed at left (U36). To do this (the new duct is still hanging on the cursor), you click on the ADOPT LENGTH button. The cursor now changes its apperance.

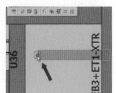

Fig. 12.33 Cable duct hanging on cursor.

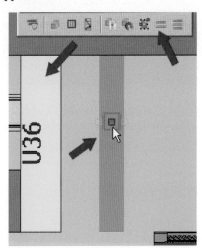

Fig. 12.34 Adopted data

Then the cable duct (U36) follows, the length and alignment of which are to be adopted, and EPLAN Pro Panel immediately adopts this data. The new cable duct, with its new changed alignment and length, now hangs on the cursor for placement. The duct can now be placed.

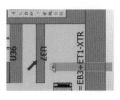

Fig. 12.35 Placed cable duct

After placement, the next duct can be placed, or the action can be ended.

Fig. 12.36 The Place centered button

The last button, PLACE CENTERED, located in the **Pro Panel options** toolbar, enables you to place a mounting rail or cable duct exactly centered between two other mounting rails or cable ducts. Of course, this requires that at least two mounting rails or two cable ducts are already placed.

Fig. 12.37 Placing a mounting rail

A mounting rail is selected and applied from the INSERT / MOUNTING RAIL menu from the parts management. The mounting rail is now hanging on the cursor.

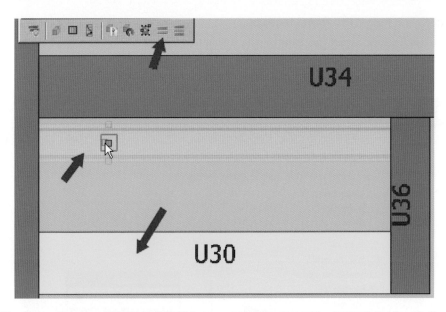

Fig. 12.38 Adopting existing length

Now you click the ADOPT LENGTH button and then the cable duct whose length the mounting rail should adopt (U30). EPLAN Pro Panel adopts the length of the cable duct for the mounting rail to be placed.

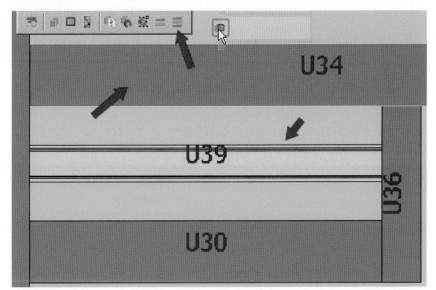

Fig. 12.39 Mounting rail placed centered

Now click the PLACE CENTERED button, and the cursor changes its appearance, and then the "counterpiece", the upper cable duct, follows (U34). EPLAN Pro Panel then places the mounting rail automatically in the center between the two cable ducts. Now the next mounting rail can be placed (it is still hanging on the cursor), or the action can be ended.

12.1.3.3 Insert Pro Panel toolbar...

The last toolbar, **Insert Pro Panel...**, is a customized toolbar that is not part of the standard EPLAN Pro Panel delivery. EPLAN does, however, provide a similar menu bar (EPLAN Pro Panel), but this only has the option of placing a locked area via a button.

TIP: The **Insert Pro Panel...** toolbar is provided in the sample data for this manual and can be imported into EPLAN if necessary.

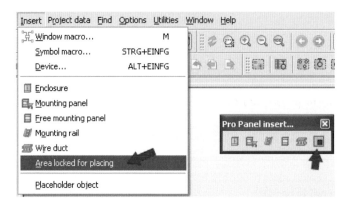

Fig. 12.40 Insert Pro Panel toolbar ...

This toolbar runs the menu functions from the INSERT / ENCLOSURE or INSERT / MOUNTING PANEL menu, etc., after clicking a separate button, thereby making working with EPLAN Pro Panel a bit faster, since you don't have to go through the INSERT menu.

Fig. 12.41 Mounting rail button

There is little to say about the function itself. If you click the MOUNTING RAIL button, **Parts management** opens in the correct area and only offers mounting rails for selection.

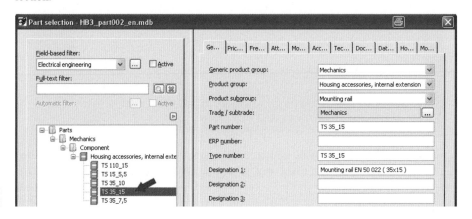

Fig. 12.42 Selecting a mounting rail

This can be adopted directly, and after clicking the OK button in parts management, EPLAN closes parts management and the mounting rail can be placed.

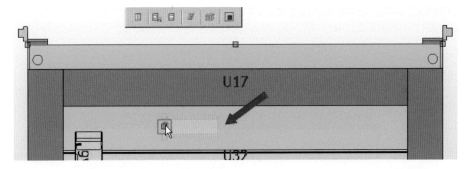

Fig. 12.43 Placing a new mounting rail

12.1.4 Menus in EPLAN Pro Panel

In addition to the toolbars, EPLAN Pro Panel offers new menus and menu expansions with additional functions.

 NOTE: Part of the menus and menu expansions overlap with the description of the functions of the individual toolbar buttons. Therefore, we will only mention these here; for a more thorough treatment, please see the „Toolbars" chapter.

12.1.4.1 Layout space menu

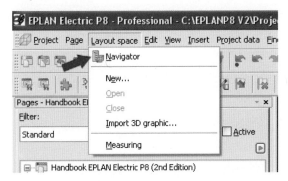

Fig. 12.44 Layout space menu

The LAYOUT SPACE menu incudes general functions. The LAYOUT SPACE / NAVIGATOR menu item opens the layout space navigator. The NEW, OPEN, and CLOSE menu items create a new layout space or open and close an existing layout space. To use the OPEN and CLOSE menu items, the layout space navigator must be open; otherwise, these menu items are grayed out.

A layout space is essential to be able to display 3D models. If a layout space does not exist yet, then one must be created. This is done by clicking the NEW item in the LAYOUT SPACE menu.

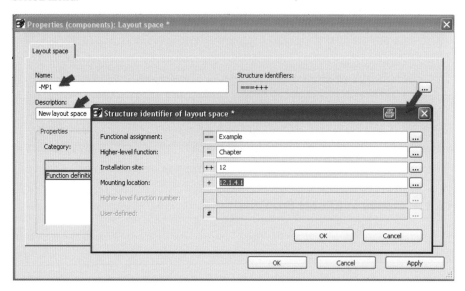

Fig. 12.45 Creating a new layout space

EPLAN opens the **Properties (components): Layout space** dialog. A *name* is assigned to the layout space, and structure identifiers, as well as additional properties such as a *layout space description*, can also be assigned.

After you click the OK button, the new layout space is created in the layout space navigator and is available immediately for 3D objects.

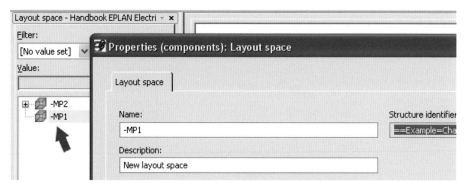

Fig. 12.46 Available layout space

The LAYOUT SPACE / IMPORT 3D GRAPHIC menu item enables you to import CAD data into EPLAN.

12.1 EPLAN Pro Panel

TIP: The 3D CAD data to be imported must be saved in the neutral STEP format (file extension *.stp, *.step or *.ste). But: Not all of this CAD data is suited for use in EPLAN Pro Panel because some manufacturers process this data in a very detailed and complex way that can push a standard computer to its limits. Please consider this before importing!

After selecting the LAYOUT SPACE / IMPORT 3D GRAPHIC menu item, the **Open** dialog opens.

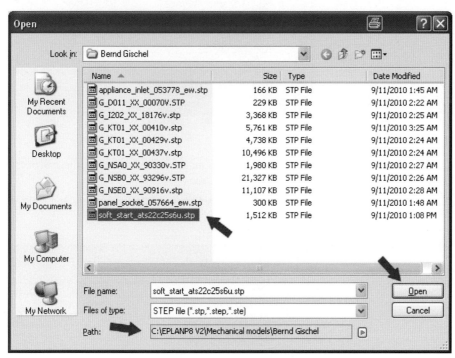

Fig. 12.47 Importing a 3D file

In this dialog, you select a file to import and confirm your selection by clicking the OPEN button.

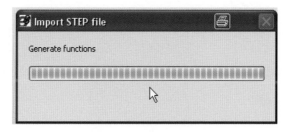

Fig. 12.48 The external STEP file is imported.

EPLAN then creates the 3D representation without further queries and displays it as a 3D object in the layout space.

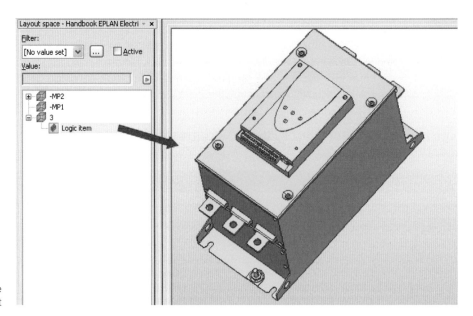

Fig. 12.49 Imported file as a 3D object

You can perform measurements, for example on an equipped mounting panel, with the MEASURING function in the LAYOUT SPACE menu.

When the Measuring function is started, the **Measuring result** dialog opens and you can click the first measuring point with the cursor.

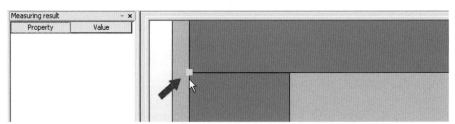

Fig. 12.50 Setting the first measuring point

Then you click the second measuring point. EPLAN Pro Panel determines the corresponding values and displays them in the **Measuring result** dialog.

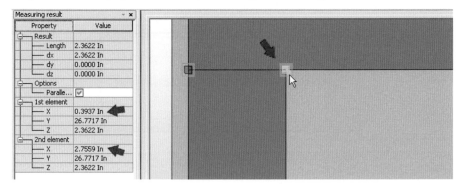

Fig. 12.51 The second measuring point was set and the length was determined.

12.1.4.2 Expansion of the Edit menu

EPLAN Pro Panel has a new menu item, DEVICE LOGIC, in the EDIT menu. This menu item has additional submenu items.

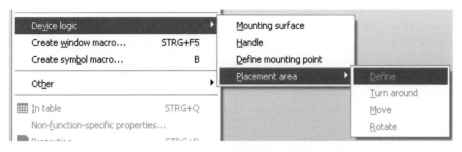

Fig. 12.52 Device logic menu

The *Device logic* menu item has the following main task. Because devices that are used for mounting layouts must have certain properties – such as the ability to be placed in the layout space or on other objects, where the handle points are, etc. – you can determine this device logic with these menu items.

For a correct placement, it is at least necessary to define a placement area for this object. You can do this in the EDIT / DEVICE LOGIC / PLACEMENT AREA menu items and the functions connected to them.

Overall, this device logic area belongs to the range of graphical editing functions.

12.1.4.3 Expansions in the View / Insert / Options menu

A few menu items for working with objects in the layout space have been added to the VIEW menu.

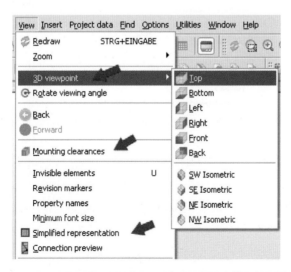

Fig. 12.53 Expansions in the View menu

These menu items, such as 3D VIEWPOINTS or SIMPLIFIED REPRESENTATION, have the same function as the existing **3D viewpoint** and **Pro Panel options** toolbars, which were described in greater detail in previous sections. Therefore, we will not repeat the function descriptions or their functionalities.

Like the VIEW menu, the INSERT menu was also expanded to include menu items for working with Pro Panel.

The corresponding menu items are also identical on the **Pro Panel** toolbar, or my personal toolbar, **Insert Pro Panel...**. These toolbars were already explained in detail, so we will avoid repeating ourselves.

Fig. 12.54 Expansions in the Insert menu

The same applies to the OPTIONS menu as to the two previous menus, VIEW and INSERT.

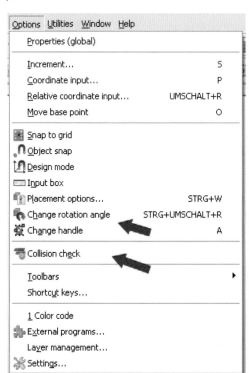

Fig. 12.55 Additional Pro Panel menu items

Functional expansions, such as CHANGE ROTATION ANGLE or CHANGE HANDLE, have already been explained in detail.

12.1.4.4 The Insert / Generate model view (2D) menu

Using the INSERT / GRAPHIC / MODEL VIEW (PRO PANEL) menu item, you can create a 2D view of a 3D mounting layout and its components (such as a mounting panel or a side panel).

12.1 EPLAN Pro Panel

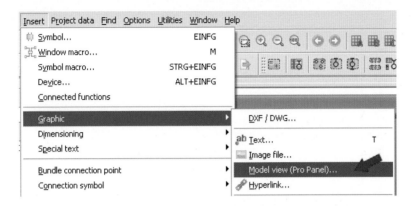

Fig. 12.56 Generating a model view (2D)

To generate a model view (a 2D representation) from a 3D representation, you have to switch first to normal editing. You open both the graphical editor and the page where the model view is to be placed.

 NOTE: The page type where a 2D derivation is to be generated must be a *model view*-type page!

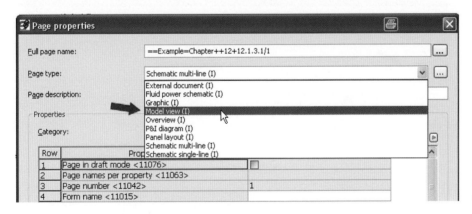

Fig. 12.57 New model view-type page

Now you click the INSERT / GRAPHIC / MODEL VIEW (PRO PANEL) menu item.

The cursor changes and you select the first corner of the model view. After placing the first corner, you select and click the second corner.

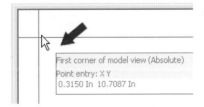

Fig. 12.58 Setting the first corner

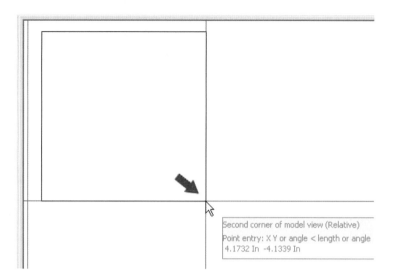

Fig. 12.59 Setting the second corner

EPLAN now opens the **Model view** dialog. In this dialog, a *view name* and an optional *description* are defined for the model view.

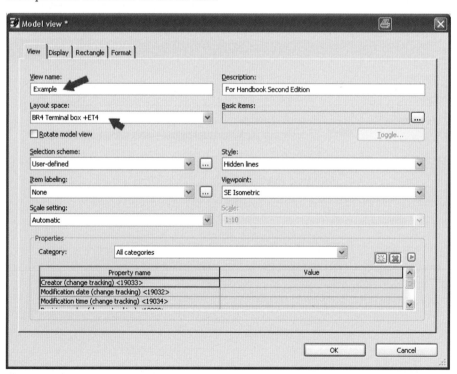

Fig. 12.60 Model view dialog

Then the *layout space* and the *basic items* must be selected. The layout space affects (e.g.) a terminal box or an enclosure. Basic items can be the *mounting panel*, a *flange*, or *side panels*. You can select all of these from the selection lists.

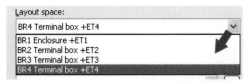

Fig. 12.61 Selecting the layout space

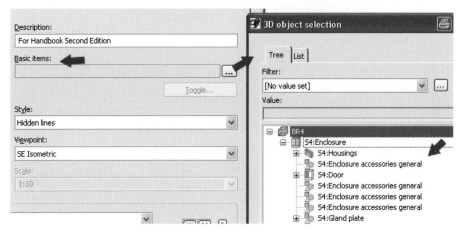

Fig. 12.62 Selecting objects

If the selections above have been made, the following entries must still be made: definitions that generate a *viewpoint* (view) and whether an *item labeling* should take place, and if so according to what filter schemes.

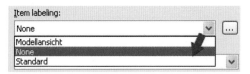

Fig. 12.63 Defining the item labeling

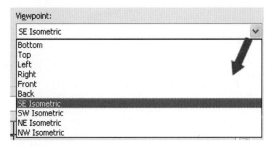

Fig. 12.64 Defining the viewpoint (of the object)

Once these two settings are completed, you can leave the Model view dialog by clicking OK. EPLAN now generates the 2D model view with the settings you just defined and displays it with a progress dialog.

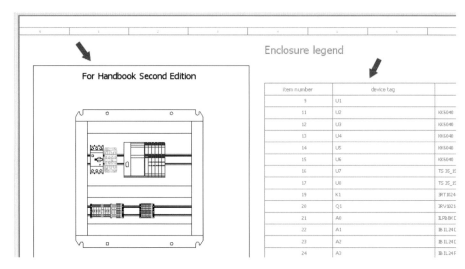

Fig. 12.65 Finished model view with an inserted enclosure legend

To quickly open the 3D model from the 2D model view, select the 2D model view and select the *Open 3D view* entry from the right-click popup menu.

Fig. 12.66 Opening the 3D view

EPLAN then opens the 3D view.

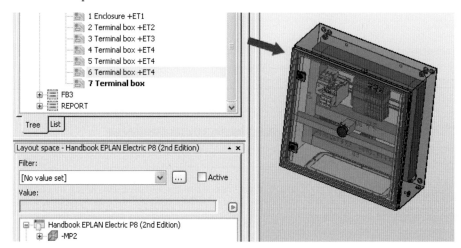

Fig. 12.67 Opened 3D view (manual rotation angle)

12.1.4.5 Expansion of the Utilities / Reports menu

The UTILITIES / REPORTS menu leads to the updating of the model views that exist in a project.

Fig. 12.68 Generating the model view

The MODEL VIEW function in the UTILITIES / REPORTS menu is used to update one or more model views in a project. EPLAN opens the **Model views** dialog.

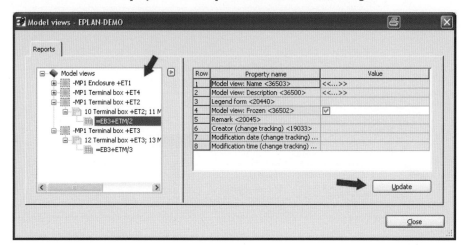

Fig. 12.69 The Model views dialog with the existing views in the project

If one or more views are going to be updated now, these reports (model views) are selected in the left area of the dialog and updated with the UPDATE button. This puts all of the selected model views back to the same status as their 'templates,' the 3D views.

12.1.5 Navigators in Pro Panel

In addition to the helpful toolbars and menus and the menu expansions, EPLAN Pro Panel also has two new navigators.

12.1.5.1 Layout space navigator

As we've said several times, a 3D view is not possible without a layout space. This means that the creation of layout spaces is mandatory. As described previously, this is done via the LAYOUT SPACE / NEW menu.

If a layout space was created, it will be located in the **layout space navigator**. The layout space navigator can be opened via the LAYOUT SPACE / NAVIGATOR menu.

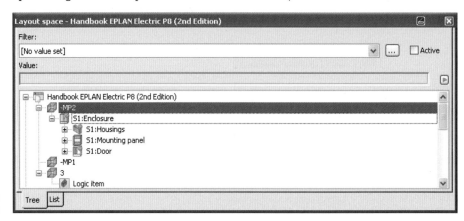

Fig. 12.70 Layout space navigator

The layout space navigator contains all of the objects that belong to a 3D layout. Depending on your needs, the VIEW can be set or changed in the layout space navigator. To do this, right-click in the layout space navigator and select the **View** menu item.

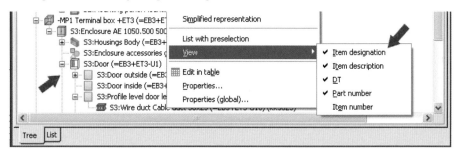

Fig. 12.71 The View entry in the popup menu

In addition to the different views, the popup menu contains a series of additional functions for editing 3D models.

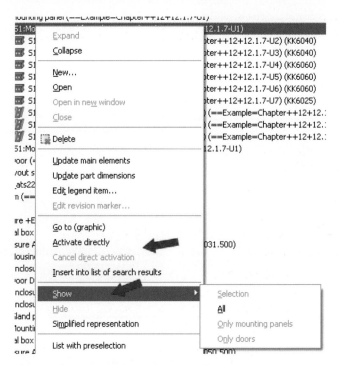

Fig. 12.72 Popup menu

In addition to those entries known from the other navigators such as NEW or GO TO (GRAPHIC), there are a few special menu entries for the layout space navigator, . The ACTIVATE DIRECTLY entry activates precisely the 3D object that you have selected in the layout space navigator.

The *Activate directly* function activates the selected mounting surface and all of the parts contained in it, and shows the mounting surface from the front in a 2D view. All of the objects that belong to this layout space are hidden.

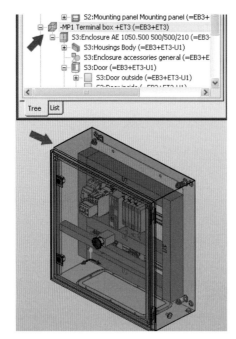
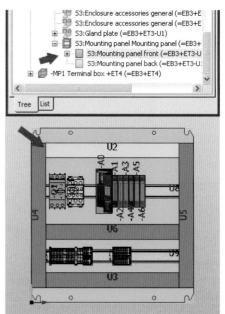

Fig. 12.73 Standard view

Fig. 12.74 Mounting panel directly activated

To undo the direct activation, select the **Cancel direct activation** menu item in the popup menu. This sets the view back to standard.

In addition to the functions named above, there are also the SHOW and HIDE menu entries. The Hide menu entry enables you to hide a layout space's complete contents or selected objects. The Show menu entry enables you to show a few selected specific objects, or to show everything again.

12.1.5.2 3D mounting layout navigator

In addition to the layout space navigator, which displays among other things the end results of equipment and also contains all of the mechanical parts, such as enclosures, bases and orifices, there is also the 3D mounting layout navigator.

12.1 EPLAN Pro Panel

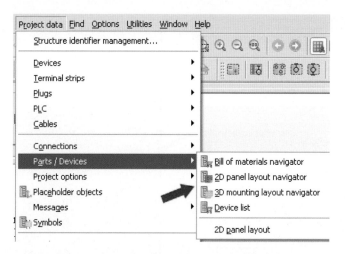

Fig. 12.75 Calling up the 3D mounting layout navigator

The 3D mounting layout navigator has the device view of the project. This is also where all of the devices are listed and offered for placement in the layout space.

The navigator is started via the PROJECT DATA / PARTS / DEVICES / 3D MOUNTING LAYOUT NAVIGATOR menu.

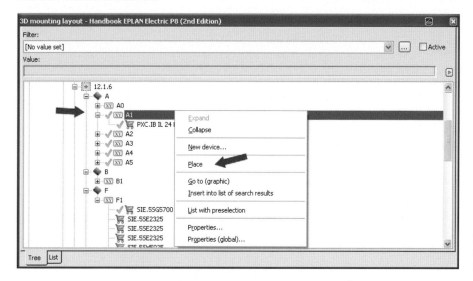

Fig. 12.76 The 3D mounting layout navigator

The desired devices from the 3D mounting layout navigator can now be placed in the layout space.

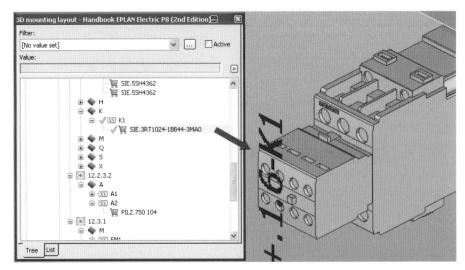

Fig. 12.77 An object to be placed hangs on the cursor.

To do this, one or more devices are selected and the PLACE menu entry in the popup menu is selected. The selected devices are now hanging on the cursor and can be placed in the layout space.

 NOTE: The 3D mounting layout navigator only displays those devices that have a part entry. Regardless of whether this part has additional technical data, such as width, height, macro data, or the like!

12.1.6 Settings for EPLAN Pro Panel

There are a few global and user-specific settings for EPLAN Pro Panel.

The **Project-specific settings** are set in the OPTIONS / SETTINGS / PROJECTS / [PROJECT NAME] / MANAGEMENT / 3D IMPORT menu.

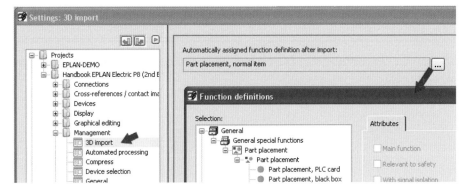

Fig. 12.78 Project-specific setting 1

This is where the automatic function definition required for a 3D import is defined.

In addition to this, there is also a second project-specific setting. This setting can define the background colors of the model views.

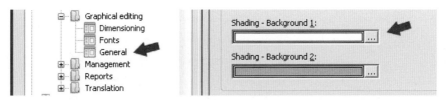

Fig. 12.79 Project-specific setting 2

This second project-specific setting can be found under OPTIONS / SETTINGS / PROJECTS / [PROJECT NAME] / GRAPHICAL EDITING / GENERAL.

The **User-specific settings** for the 3D area are found in OPTIONS / SETTINGS / USER / GRAPHICAL EDITING / 3D.

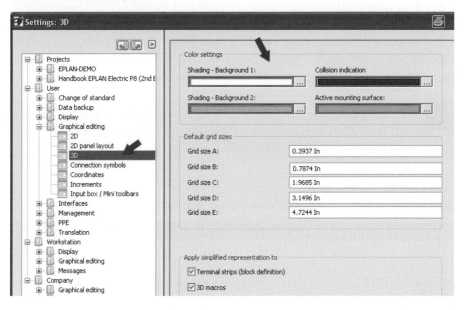

Fig. 12.80 User-specific setting 1

You can select color settings for backgrounds or other grid sizes here. This is also where the default setting is located for how terminal strips (individual terminals) or 3D macros in simplified representation should be handled.

In addition to these settings, it is also possible to show an input box while working in the 3D area. This setting is located under OPTIONS / SETTINGS / USER / GRAPHICAL EDITING / INPUT BOX / MINI TOOLBARS.

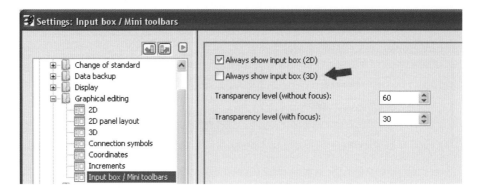

Fig. 12.81 User-specific setting 2

12.1.7 A practical example with EPLAN Pro Panel

Now that you have had a brief overview of the functions and how they work, this chapter provides a short and simple example of how to use EPLAN Pro Panel.

We will proceed step by step so that you can follow this example.

 TIP: Prerequisite for a 3D layout: devices with associated parts in the project. These parts have the minimal requirements in terms of technical data, such as width, height, and depth.

Step 1: We will begin by creating a layout space in the LAYOUT SPACE / NEW menu. The layout space receives a *designation,* a *description*, and a structure identifier.

Fig. 12.82 A new layout space was created.

Step 2: The layout space is opened (LAYOUT SPACE / OPEN menu), the desired layout space is selected in the layout space navigator, and then a terminal box AE 1058.500 is selected from parts management, loaded, and inserted into the layout space via the INSERT / ENCLOSURE menu.

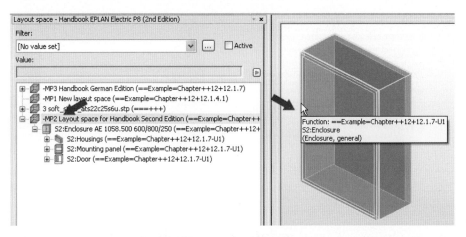

Fig. 12.83 Inserted terminal box

Step 3: Now a few cable ducts are placed on the mounting panel of the terminal box. To do this, you directly activate the mounting panel in the layout space navigator so that the front view is visible in the 3D view. To do this, you select *mounting panel front* in the layout space navigator and select the ACTIVATE DIRECTLY entry in the popup menu.

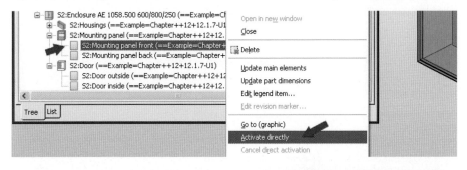

Fig. 12.84 Selecting the mounting panel front.

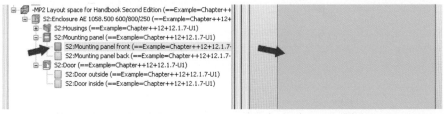

Fig. 12.85 The mounting panel front is now activated.

Step 4: Now you select a cable duct from parts management via the INSERT / WIRE DUCT menu and then place it later on the mounting panel. Before placement, you call up the PLACEMENT OPTIONS. The upper cable duct should be placed 50 millimeters from the upper edge and 10 millimeters from the left edge of the mounting panel. Then you should extend the cable duct's length to the right edge, where it is also placed 10 millimeters from the edge.

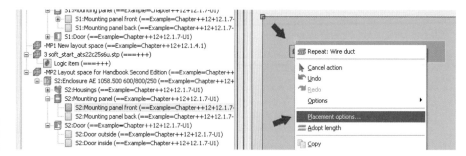

Fig. 12.86 Placing the cable duct via Placement options

If you have clicked the PLACEMENT OPTIONS menu item, then the corresponding X and Y offsets, as well as the handle that serves as the basis for calculating the offsets, are set as the first placement point.

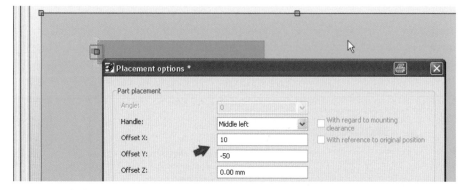

Fig. 12.87 Setting placement options for the cable duct

After you leave the dialog by clicking **OK**, the handle is set to the upper left corner of the mounting panel. To do this, you have to click this corner with the mouse.

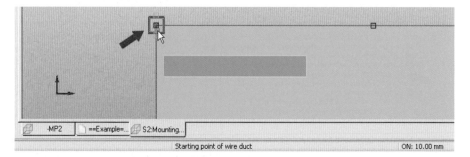

Fig. 12.88 The first handle is set

Now you call up the **Placement options** dialog again (from the popup menu) and enter the corresponding offsets for the second placement.

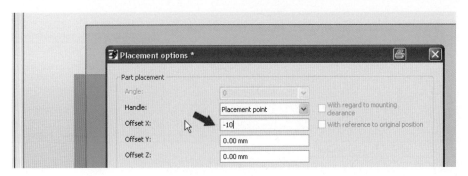

Fig. 12.89 The second handle is set

You can now exit the **Placement options** dialog again, and you left-click on the upper right corner of the mounting panel. EPLAN Pro Panel now places the cable duct as desired.

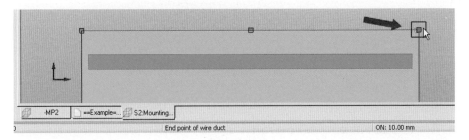

Fig. 12.90 The second handle was placed

Step 5: The lower cable duct is placed with the same steps. The offset settings for the first handle are: offset X = 10 mm; offset Y = 50 mm; handle = lower left. The offset settings for the second handle are offset X = –10 mm; offset Y = 0 mm.

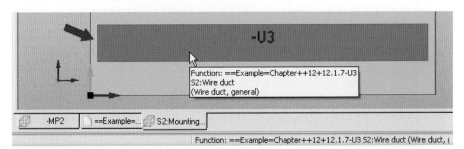

Fig. 12.91 Placement of second duct complete

Step 6: Now the left and right side of the cable duct is connected to the upper and lower cable ducts. To do this, you again select a duct from parts management and load it via the **Insert / Wire duct** menu. The duct hangs on the cursor again. Now you can switch the handle to one of the outside handles with the A key, and then you should left-click one end of the existing duct.

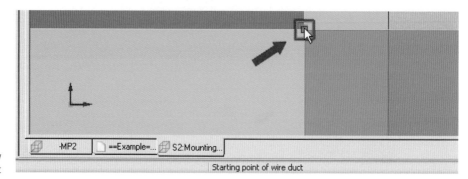

Fig. 12.92 Placing a new duct

Now you drag the duct down with the mouse, thereby placing the duct on the external handles of the existing duct (click with the left mouse button). This also places this duct.

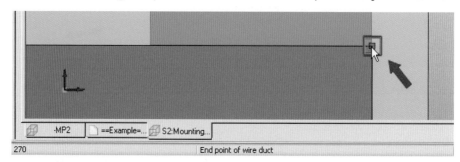

Fig. 12.93 The new duct is placed

Step 7: Now a cable duct is placed in the center between the cable ducts. This is relatively easy because EPLAN finds the centered handles on the existing cable ducts on its own.

You select a duct from parts management via the **Insert / Wire duct** menu and then place the duct on the center connecting point of the left duct.

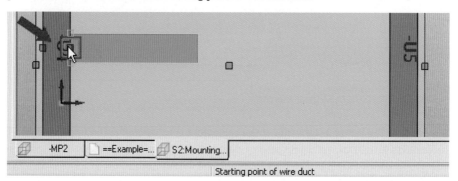

Fig. 12.94 Let the new duct snap into place on the handle

Now you drag this duct with the mouse to the right, onto the right duct. It will again snap into place on the center handle.

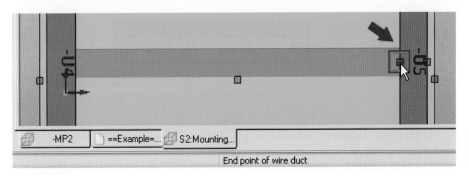

Fig. 12.95 Connecting the new duct with the right duct

This also finishes the placement of the center duct, which sits precisely in the center between the upper and lower duct.

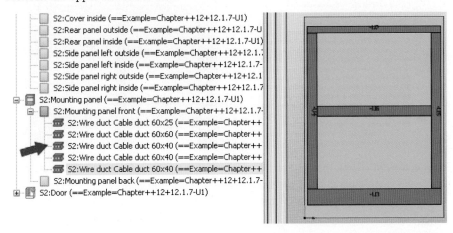

Fig. 12.96 Overall view of mounting panel

Step 8: Now another duct is placed in the lower area, precisely in the center between the left and right ducts. To do this, take a cable duct from parts management and place it on the center of the -U6 cable duct.

The cable duct hangs on the cursor after being selected. Now click the **Adopt length** button and the -U4 cable duct.

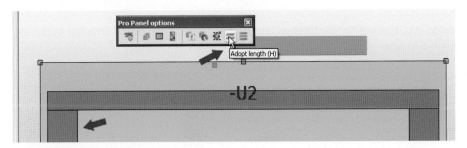

Fig. 12.97 Adopting the length

Then click the **Place centered** button and the -U5 cable duct. EPLAN Pro Panel places the new duct exactly in the center.

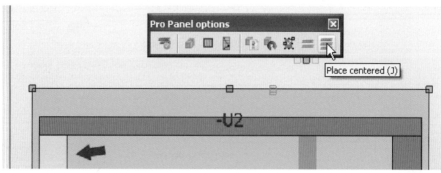

Fig. 12.98 Place centered

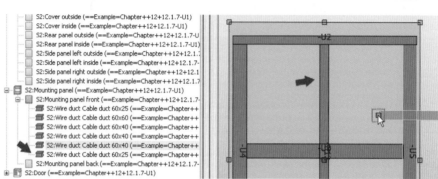

Fig. 12.99 Finished placement of duct

Step 9: Now of course this duct is too long, and if the collision check were turned on, it would not be possible to place the duct because it "cuts" the center duct. This is why you trim, or shorten, the upper part of the duct now.

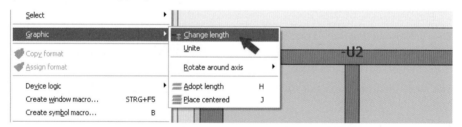

Fig. 12.100 Change length

To do this, you call up the **Change length** menu item in the **Edit / Graphic** menu and click the center duct.

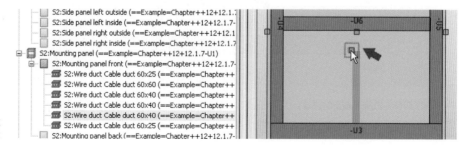

Fig. 12.101 Changing the length of the duct

After clicking the duct, you can easily change the length by dragging it with the left mouse button. The handle is also dragged to the -U6 duct and snapped into place there. This shortens the duct, and all of the desired cable ducts have now been placed.

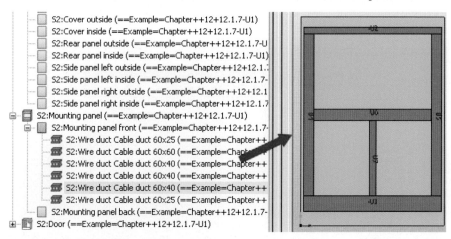

Fig. 12.102 Finished mounting panel equipped with cable ducts

Step 10: Now you place a few *mounting rails*. Mounting rails are placed using the same procedure that we used for the cable ducts. This means that you select them from parts management in the INSERT / MOUNTING RAILS menu and place them on the mounting panel. To do this, you use the handles on the ducts, or the ADOPT LENGTH and PLACE CENTERED functions.

Once the mounting panel is equipped with mounting rails, it looks like the following image.

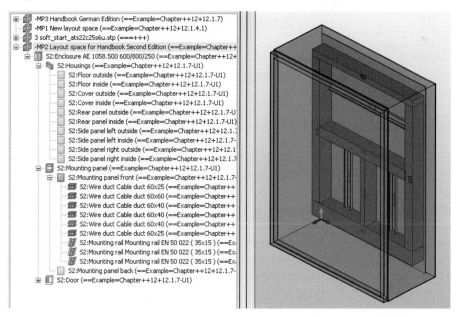

Fig. 12.103 Mounting panel equipped with ducts and rails

Step 11: After you have placed the cable ducts and mounting rails on the mounting panel, all of the devices can be added.

To do this, you open the PROJECT DATA / PARTS / DEVICES menu of the 3D MOUNTING LAYOUT NAVIGATOR. Then the mounting panel will be set to the *3D viewpoint front* view.

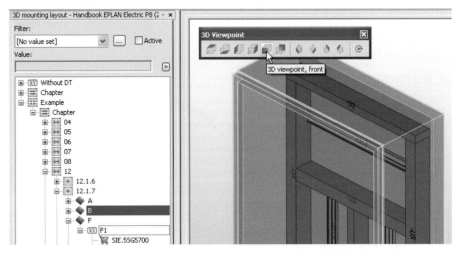

Fig. 12.104 Opened 3D mounting layout navigator

Step 12: Now the devices that exist in the project and that are listed in the 3D mounting layout navigator can be placed on the mounting panel.

To do this, you select the devices in the navigator and right-click to select the PLACE menu item from the popup menu.

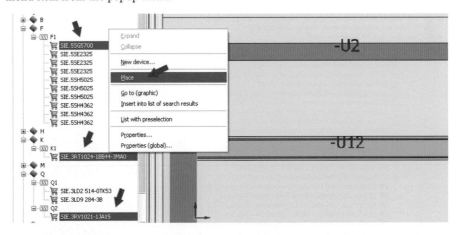

Fig. 12.105 Devices to be placed are selected in the navigator

After clicking **Place**, the devices hang on the cursor and can be snapped onto the mounting rails. Before placement, you can of course call up the placement options once more and enter the desired placement options, either by opening the popup menu by right-clicking, or via the corresponding button.

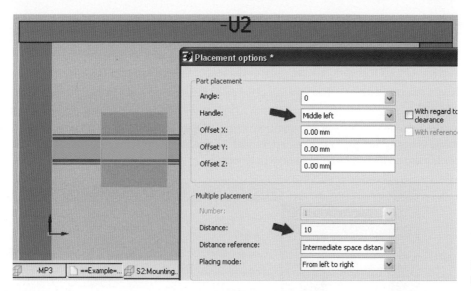

Fig. 12.106 Defining placement options

After clicking **OK** and placing the first handle on the center of the mounting rail (EPLAN snaps it in on its own), all of the devices with set placement options are placed. The same thing happens to the other devices that will also be put on this mounting rail.

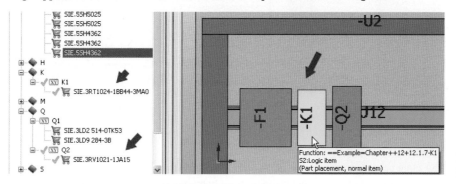

Fig. 12.107 The first placed devices

To facilitate visual checks, EPLAN assigns a small green check mark to the device in the navigator to show that the device is placed.

Step 13: Now the other devices are placed. The terminal strips are placed upright in the lower area on the two shorter mounting rails. To do this, you select the terminals in the navigator and call up the Place function via the popup menu.

Fig. 12.108 Query as to how the terminals should be placed

Before placement, EPLAN asks in the **Place parts of a terminal strip** dialog how the terminals should be placed.

If you answer this dialog with YES, then you can place the first terminal on the mounting rail and EPLAN then automatically places the rest of the marked terminals.

If you respond with NO, then you can place all of the terminals one by one on the mounting rail.

In the example, we answer the question with YES and then call up the **Placement options** dialog before placing the first terminal. You define the placing mode here, and the terminals should be placed from bottom to top.

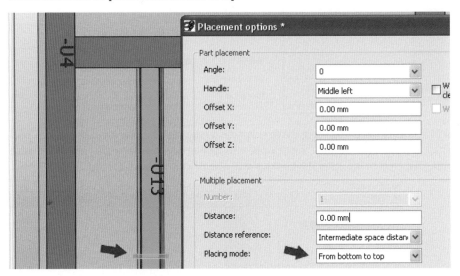

Fig. 12.109 Defining the placing mode for terminals

Then you can leave the dialog by clicking **OK**, and the first terminal can be placed. EPLAN Pro Panel then automatically arranges all of the other selected terminals to this first terminal.

The same process applies to the next terminals. These are selected and then placed on the mounting rail. Then all of the terminals have been placed.

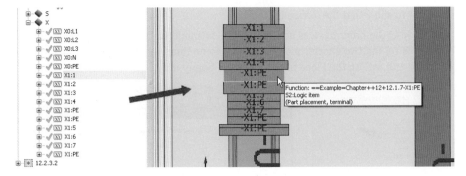

Fig. 12.110 Placed terminals

Step 13: Now you should add the signal lamp into the door. To do this, you directly activate the door's exterior in the *layout space navigator* and set the view to 3D viewpoint front.

You select the signal lamp in the 3D mounting layout navigator, call up the *Place* function, and then right-click the placement options. You enter a distance of 300 millimeters from the upper edge of the door.

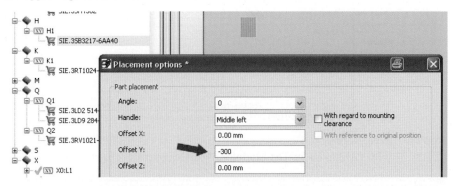

Fig. 12.111 Before placement of the signal lamp

Then confirm the dialog with **OK**; now you can grab the middle of the upper edge with the snap functions and click the left mouse button. EPLAN now places the signal lamp in the center at a distance of 300 millimeters from the edge.

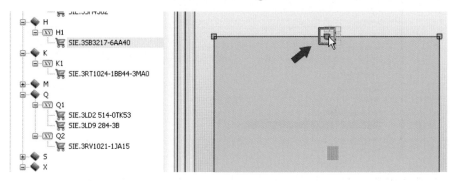

Fig. 12.112 Grabbing the center

This is how the result looks in the 3D view.

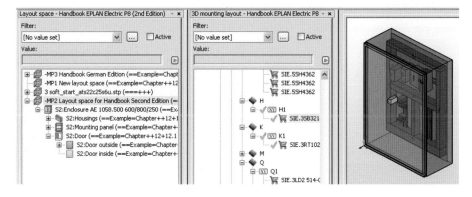

Fig. 12.113 Final result: equipped enclosure

Step 14: To complete the documentation, you now generate a 2D model view from the 3D view. You can close the layout space and 3D mounting layout navigators, switching to the graphical editor and creating a *model view* page.

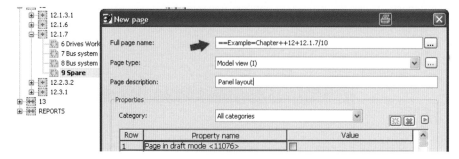

Fig. 12.114 Generating a new model view-type page

Step 15: This page opens and a *model view* is inserted via the INSERT / GRAPHIC / MODEL VIEW (PRO PANEL) menu. The **Model view** dialog opens after placement.

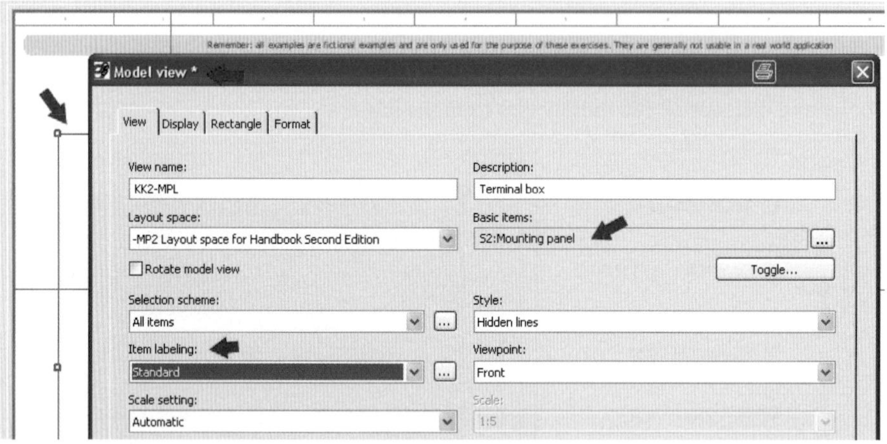

Fig. 12.115 Settings for generating the model view

The *view name*, the *description*, the layout space to be displayed, and the items of the layout space to be displayed (here, the *mounting panel*) are set in this dialog. Also, the scheme for *item labeling* is set to default.

After you click on OK, the dialog is closed and EPLAN generates the 2D model view.

12.1 EPLAN Pro Panel **467**

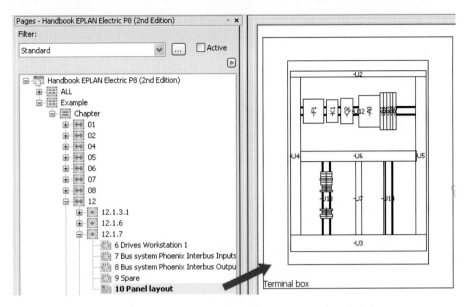

Fig. 12.116 Partial view of 2D model view

Step 16: Now an enclosure legend is placed on the page. To place an enclosure legend, you open the properties of the model view (double-click the model view or use the keyboard shortcut CTRL+D) and choose a form for the *Legend form <20440>* property name in the properties.

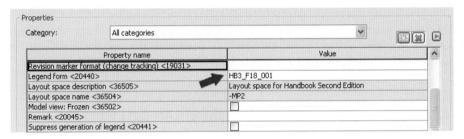

Fig. 12.117 Defining the legend form for this model view

Then the *enclosure legend* report can be placed manually on this page using standard EPLAN Electric P8 procedure.

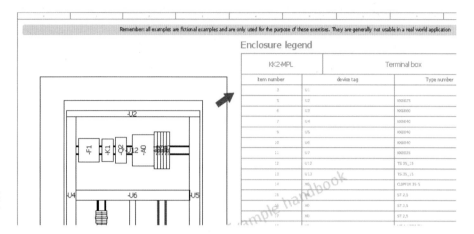

Fig. 12.118 Mounting panel with enclosure legend

12.2 EPLAN Data Portal

Previous chapters made frequent mention of the necessity of correct function templates (the individual function definitions of devices), device selection, and finished macros.

Function templates, among other things, are a major focus when working with EPLAN Electric P8. Users can, of course, fill their own parts with this additional information themselves, or they can use the EPLAN Data Portal solution.

12.2.1 What are the advantages of the EPLAN Data Portal?

The EPLAN Data Portal provides online access to parts data that can be directly used during current and subsequent project planning.

This parts data is usually provided by the manufacturer, is subjected to an incoming inspection by EPLAN (whereby EPLAN as a software manufacturer cannot, of course, check the correctness of the data provided by the manufacturer), and is then provided online in the EPLAN Data Portal.

This allows the user to directly use parts (devices, macros, etc.) from the EPLAN Data Portal, without having to add them to, or create them in, his or her own parts management.

12.2.1.1 Preconditions for use

A valid Software Maintenance Contract must exist in order to use the EPLAN Data Portal. The EPLAN Data Portal can then be used as desired by providing data from the contract.

EPLAN distinguishes specific user groups on the basis of their Software Maintenance Contract levels, but this does not play a major role in the use of the EPLAN Data Portal. As always, the more comprehensive the Software Maintenance Contract, the greater the number of options are for using the EPLAN Data Portal. The following chapters will be limited to standard use.

12.2.2 Before the first start

An account must be created in order to use the EPLAN Data Portal. To do this, you start the OPTIONS / SETTINGS / USER / MANAGEMENT / DATA PORTAL menu item.

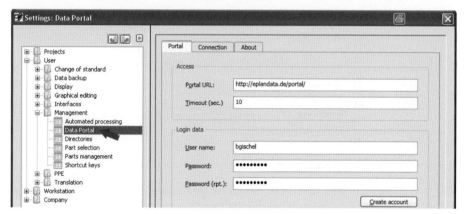

Fig. 12.119 Data Portal settings

The settings for the Data Portal are subdivided into several tabs. Access and login data are defined in the *Portal* tab (default setting already entered by EPLAN - this does not need to be changed). This creates an account, and the user can use the Data Portal after successfully completing and confirming the prompt.

The settings on the *Connection* tab normally do not need to be changed.

12.2.2.1 Data Portal Navigator

The navigator is the main focus of the EPLAN Data Portal. This is started via the UTILITIES / DATA PORTAL NAVIGATOR menu.

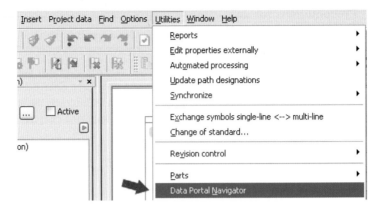

Fig. 12.120 The navigator menu item

Like elsewhere in EPLAN, the navigator can be docked and undocked anywhere. This means that it can be freely placed on the desktop or docked to any desired position.

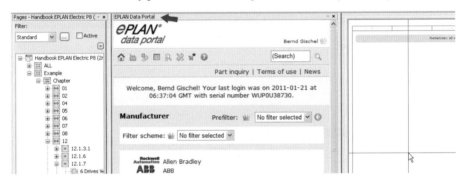

Fig. 12.121 The Data Portal Navigator

The navigation strip, with buttons for functions such as display manufacturers, search, settings, and online help, is located under the navigator dialog logo.

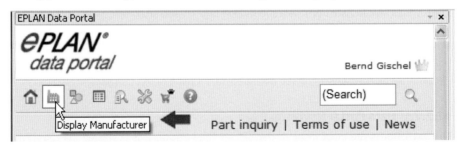

Fig. 12.122 The navigator toolbar

The navigation bar buttons have the following meanings:

Somewhat aside is the ⊙ button, which returns you back to the previously displayed page in the navigator when clicked.

Fig. 12.123 The Back button

Below the navigation bar is the actual "data area" of the navigator. Usually, or when first started, this contains an overview of the manufacturers present in the EPLAN Data Portal. Every manufacturer can be clicked here and the Portal opens the corresponding pages with the current manufacturer data in the navigator. Navigation can then continue through this data.

Fig. 12.124 Listing of manufacturers that are represented in the Data Portal

12.2.3 How the Portal works

How does the EPLAN Data Portal work? This is simple. Using the Portal, missing parts data or macros can be inserted into schematic pages and / or the parts management while working on projects. This, of course, assumes that the missing data is actually available in the Portal.

12.2.3.1 Insert parts, macros and data from the Portal

A brief example will show how a PNOZ device can be found in the Portal and then be integrated into an open schematic page.

Requirement: The Data Portal navigator is started. The search function is then started via the button.

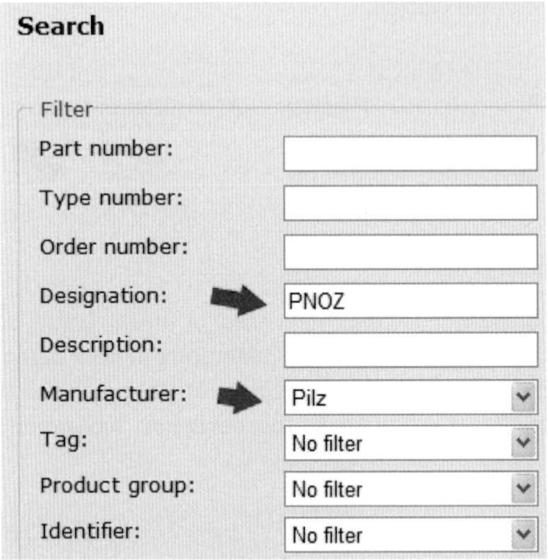

Fig. 12.125 Search function

EPLAN opens the **Search** dialog. Here you can enter the value PNOZ in the *Designation* field and enter Pilz in the *Manufacturer* field. All other filters are not used or are set to the value *No filter*. Pressing ENTER, or clicking the FIND button at the very bottom of the *Search* dialog, starts the search.

If EPLAN finds parts matching the search term, then the Portal lists these in an overview list. In addition to the header data such as manufacturer, subgroup, languages, and characteristics, the list also contains the parts found by the search.

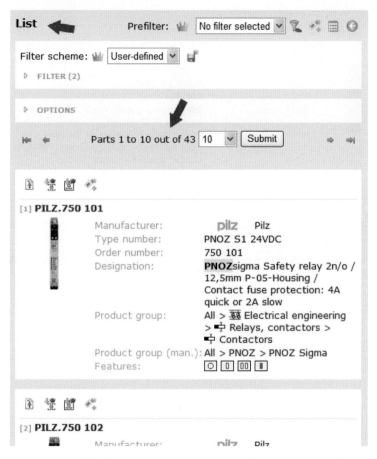

Fig. 12.126 Parts found

In addition to the actual devices and their part numbers, the list also contains other information such as: type and order number, designation, the product group to which the device belongs, the assigned identifier, in what languages the device is available in the database, and what other characteristics the device possesses. Characteristics are additional information.

Simply stated, the more superior a part in the Portal is, the more characteristics it has in the Portal. Characteristics can be: the device has device data (function definitions / templates) or (e.g.) macros are available for the device.

 NOTE: The search results always depend on how the search and the various search criteria were set and also what data is actually available in the Portal.

Fig. 12.127 Selected PNOZ device

In our example this is the part PILZ.750 104. This found device can now be directly adopted in the project planning.

To do this, the Portal list view has several buttons above the part that have the following meanings.

 Back to beginning of page

This button is self-explanatory. It sends you back to the beginning of the pages.

 Insert macro <xyz> in graphical editor (on the schematic page)

If you click this button, EPLAN hangs the macro on the cursor, allowing it to be placed on the schematic page.

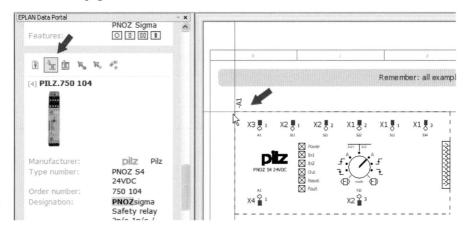

Fig. 12.128 Place selected macro directly in the graphical editor

 Insert part <xyz> as device in graphical editor (on the schematic page)

If you activate this button, the **Import of parts** query dialog appears first.

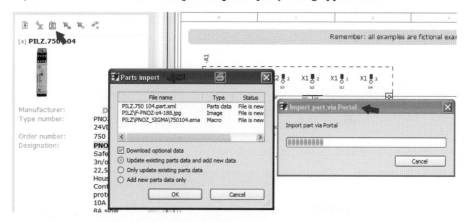

Fig. 12.129 Insert device in graphical editor

Depending on your needs, you must decide here how the data is stored in the existing parts management, or whether for example the *optional data* should be imported as well.

After you click OK in the **Import of parts** dialog, EPLAN imports the data into the parts management and then hangs the macro on the cursor again. Now the macro can be placed on the schematic page.

 Import part <xyz>

When this button is clicked, only the selected parts data is imported into the parts management. EPLAN restarts the **Import of parts** dialog, and depending on the settings, the corresponding parts data is imported into the parts management, and the rest of the data is saved in the directory structure (images, for example).

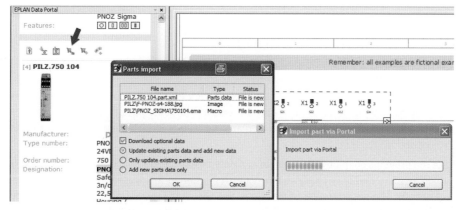

Fig. 12.130 Pure import of parts

After the import, EPLAN closes these dialogs and the graphical editor is active again.

 Put part <xyz> in the shopping cart

Clicking this button adds the selected device to the shopping cart. The shopping cart itself is called up via the 🛒 button.

Fig. 12.131 Displaying the shopping cart

All devices in the shopping cart can be imported here completely (collected).

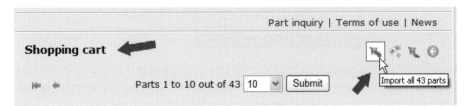

Fig. 12.132 Importing the entire shopping cart

All of the parts can also be removed from the shopping cart.

Fig. 12.133 Completely emptying the entire shopping cart

You can go back to the previous page by clicking on this button.

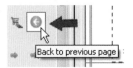

Fig. 12.134 Back button

■ 12.3 Project options

In EPLAN Electric P8, macros with value sets are very flexible, but they are limited in one aspect. Macros with value sets can only be used on the same page in EPLAN. Cross-page macros with value sets have not been available up to now. EPLAN closes this gap with the Project options module.

Like the previous chapters, this chapter can only briefly sketch the possibilities offered by project options, and only provides a short introduction to the procedure for creating project options. The large number of functions, and the wide range of possibilities offered by the Project options module, make it impossible to describe all the details of project options in the limited space available in this book.

This chapter is intended therefore as food for thought, and to illustrate what is possible when using project options. The possible uses are (almost) limitless.

12.3.1 What are project options?

As the name describes, project options are various options for (partial areas) of a project. This can mean that (e.g.) the PLC in a project is implemented with a Siemens PLC for one customer and a Schneider PLC for a different customer.

This cannot be implemented using *value set macros* but is possible using the *Project options* module. To do this, the desired options are displayed or hidden.

Project options cannot be created across projects. They are only available for the project where they have been created. However, template projects with project options can be created.

12.3.2 Terminology in the Project options module

The term project options is always used here. Project options are a type of "generic term". To help to understand project options, the following section contains a brief explanation of the terms used.

- **Project options group** – this function allows the grouping of project options. They can contain several project options but only one project option at a time can be switched on. All other project options in the project options group are then switched off. Project options groups can only be created in the project options navigator.
- **Project options** – these are partial areas of a project that can be switched on and off as desired. They can consist of one or more extracts, pages, or page areas, or of unplaced objects. Project options can also only be created in the navigator.
- **Section** – a section is always assigned to a project option. A section can consist of a partial circuit, a complete page, or several pages. Unplaced objects are also possible in sections. In addition to the navigator, sections can also be created from the PROJECT DATA / PROJECT OPTIONS / CREATE SECTION menu.

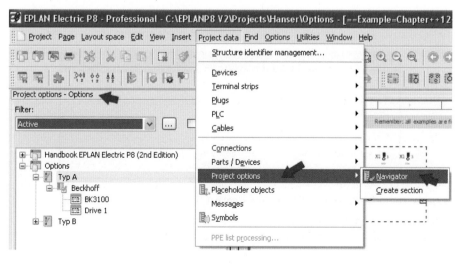

Fig. 12.135 Starting up the project options navigator and the navigator itself

- **Project options navigator** – the project options navigator is used for managing the project options groups, for creating and managing the project options, and switching the options on and off. This is called up via the PROJECT DATA / PROJECT OPTIONS / NAVIGATOR menu. The typical, familiar tree and list views are available within the navigator. Custom filters can also be set for the view in the navigator.

12.3.3 Creating options and sections

- A bit of preparation is required for you to be able to simply switch between different options in a project. Project options groups and their affiliated project options must be created, and finally there is the most important point: creating the sections.
- The following chapter provides short steps that explain the most important points for creating these options and their associated parts.

12.3.3.1 Project options group

A **project options group** is created in the project options navigator. To do this, the navigator is started via the PROJECT DATA / PROJECT OPTIONS / NAVIGATOR menu.

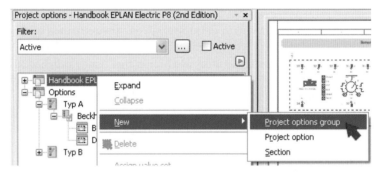

Fig. 12.136 Creating project options groups

After this, the popup menu in the navigator dialog is called and the NEW / PROJECT OPTIONS GROUP entry is selected. EPLAN opens the **Project options group** dialog.

Fig. 12.137 New project options group

A name and description must be entered here. Clicking the OK button saves the group and displays it in the navigator.

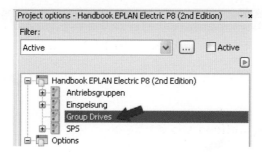

Fig. 12.138 View in the navigator

Examples of project options groups could be: *power supply*, *PLC*, *automatic operation*, *manual operation*, and so on. The actual project options for power supply or PLC can be created beneath these project options groups.

12.3.3.2 Project options

Project options can also only be created in the project options navigator. Project options always belong to a project options group. This means that one of these has to be selected, or that a project options group must be selected when creating a project option. To do this, the navigator is opened once more and the popup menu command NEW / PROJECT OPTION is started.

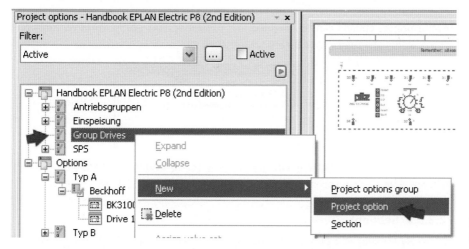

Fig. 12.139 Create new project option

EPLAN then opens the **Project option** dialog. The project options group can be changed here, if desired. The name of the project option is then defined and a description can also be entered.

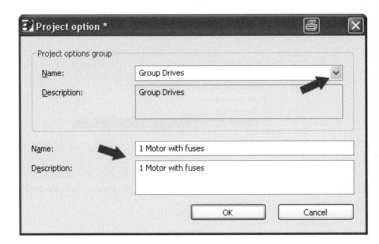

Fig. 12.140 Create new project option

When the OK button is clicked, the project option is saved and sorted under the selected project options group in the navigator.

Fig. 12.141 Newly created project option

Several project options can be assigned to a project options group. However, only one project option at a time can be active, i.e. switched on.

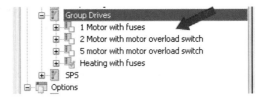

Fig. 12.142 Multiple project options

You can see if a project option is switched on from the icon in front of the project option name. A small green check mark is set here.

12.3.3.3 Sections

The actual sections can now be assigned within project options. As already described, sections can be partial sections, one or more pages, or unplaced functions.

To generate a section, the corresponding project options group and one of the project options below this group are selected in the navigator. The section is then generated via the NEW / CREATE SECTION popup menu entry.

12.3 Project options

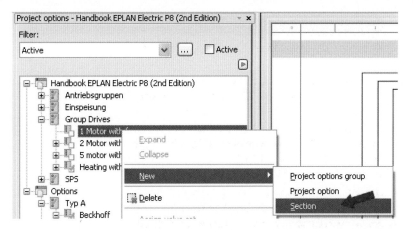

Fig. 12.143 Creating a new section

EPLAN starts the function. The objects belonging to this section must now be defined. This is done using windows. When this is finished, a name and description for the section are entered into the subsequent **Section** dialog.

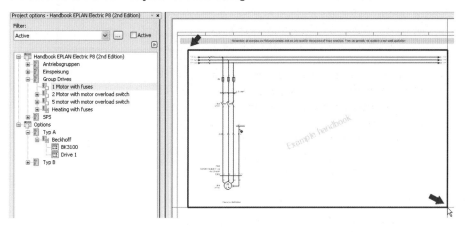

Fig. 12.144 Creating a section via a window

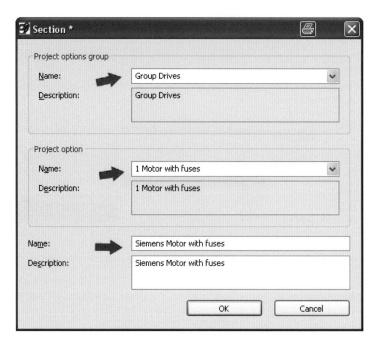

Fig. 12.145 Designating a new section

In the **Section** dialog, it is also possible to assign the section once to a different project option or project options group. After clicking the OK button, the section is saved and sorted under the associated *project options group* and *project option* in the navigator.

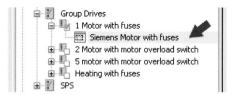

Fig. 12.146 Section in the navigator

The next section can now be created. The first section can be switched to be transparent as a construction aid when doing this. To do this, the section is selected and the MAKE TRANSPARENT (ON / OFF) popup menu item is clicked (switched on). After this, the previous section must be switched off by clicking the green check mark in front of the project option name.

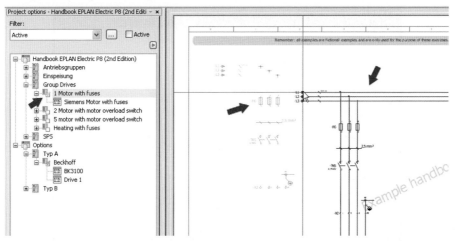

Fig. 12.147 Editing new circuit parts for a new section

EPLAN now visually switches the previous section to transparent. You can see this in the yellow check mark in front of the section name. The "new" section can now be created. To do this, the desired elements are placed and edited and you can use the placements of the (previous) transparent option (the section) to help with the placement of the new devices.

As an editing aid, you can also (e.g.) copy the previous section and place it over the transparent section. The devices can now be replaced, new devices added, or unnecessary devices and other elements deleted. Completely different wiring can also be created, depending on what partial circuit is desired or required for this option.

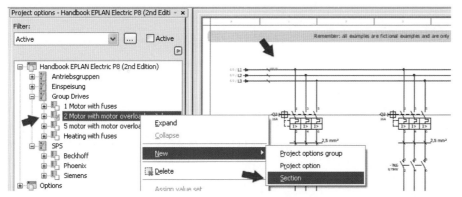

Fig. 12.148 Defining a new section

Once all changes for this section have been made, the section can be defined and saved. To do this, the devices are once again selected using the window method and the new section is generated via the PROJECT DATA / PROJECT OPTIONS / CREATE SECTION menu or via the navigator's popup menu.

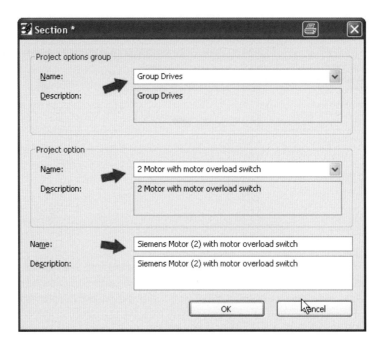

Fig. 12.149 A new section is designated.

This is then provided with a name and description as before. The section is then saved. Both sections are now available as options in the project and can be switched on and off as desired.

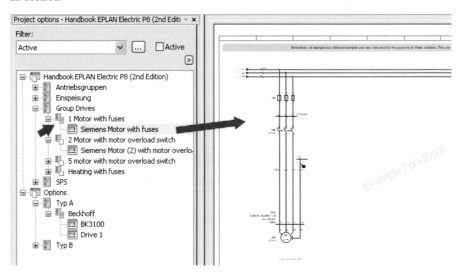

Fig. 12.150 Section with a motor

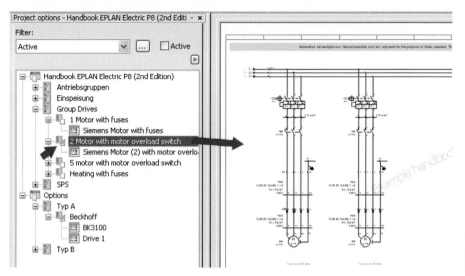

Fig. 12.151 The same page with another section

12.3.4 Generate options overview report

If several options are being used in a project, it may be helpful to produce an overview of the options that are used and not used.

EPLAN offers an easy option for automatically creating reports in a broad range of forms. All that is required is a form of the *project options overview* type and a report or report template.

The image offers an example of how this kind of overview of the options used may appear in a report. Of course, you can work here with filters and sortings so that for example your report only includes the active options in the form.

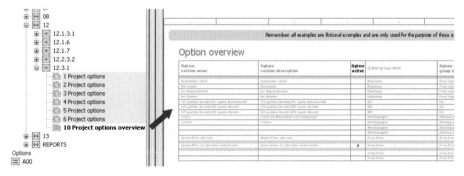

Fig. 12.152 Generated automatic options overview

13 FAQs

Aside from the extensive online help that EPLAN provides, there will always be one or the other question that is not covered in it or whose answer does not precisely fit the situation at hand.

This chapter contains a number of frequently asked questions and their answers regarding EPLAN Electric P8, spanning different categories.

> **TIP**: I should point out that some questions may have several answers. EPLAN is known for not taking a „My way, or the highway" stance; instead, there is always room for various possible solutions.

13.1 General

Question: What is the fastest way to rotate grouped graphical elements?

Answer: By selecting the grouped elements and then defining the center of rotation (EDIT / GRAPHIC / ROTATE menu), you can then use the OPTIONS / RELATIVE COORDINATE INPUT menu (or the SHIFT + R shortcut key) and the subsequent **Relative coordinate input** dialog to enter directly the desired angle around which the grouped element is to be rotated. Confirm the input with OK; then, press the OK button again to allow EPLAN to rotate the element by the angle set.

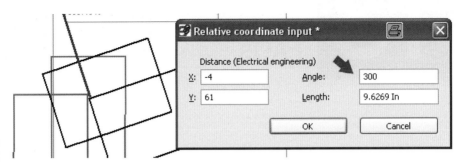

Fig. 13.13.1 Set angle in the dialog

Question: How do I enlarge or reduce graphical objects?

Answer: Select the object to be enlarged/reduced. In the EDIT / GRAPHIC menu, select the SCALE function. Then, specify the origin for scaling and confirm with OK.

NOTE: Only integer values or values with a decimal point are valid here.

Question: What is the meaning of the PROPERTIES (GLOBAL) function in the OPTIONS menu?

Answer: To edit properties globally, one must activate OPTIONS / PROPERTIES (GLOBAL) in the menu. This way, it is possible to edit a contactor and its contacts distributed in the schematic simultaneously (i.e., globally). For example, you could apply a complete change to the DT.

Fig. 13.13.2 Properties (global) option

If the *Properties (global)* option has been activated, it is not possible to change the *displayed DT* in the symbol properties. In that case, the field will appear grayed out. Only the *Full DT* field can be edited.

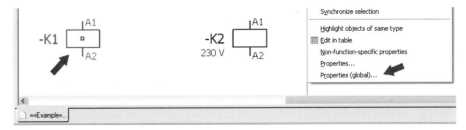

Fig. 13.13.3 Popup menu

The *Properties (global)* function can also be found in the popup menu of the right mouse button. To use this function, you must select, for example, a device. Then, you can call *Properties (global)* from the popup menu. Actually, you can select any device. It may be the coil, or also one of the coil's contacts.

Question: What is the difference between a *basic project* and a *template project*?

Answer: Projects as project templates always contain all project-specific settings and the schemes of the project structure. Project templates have the file extension *.ept. If a new project is created on the basis of a project template, all project-specific settings will be imported and applied. The page structure in such a newly created project can be adjusted subsequently if no pages have been added to the project yet.

The following applies to basic projects. Basic projects contain both pages and project-specific settings. They have the file extension *.ebp. If a project is created on the basis of a basic project, all project-specific settings and the pages already contained in the basic project will be imported and applied. However, the page structure of the new project is defined by the pages of the basic project and, unlike project templates, cannot be changed.

The page numbers, though, can generally be changed for both types of projects. The main difference lies in how the page structure is modified. In one case, it can be changed (new project from a project template), while in the other (new project from a basic project), it cannot be changed.

NOTE: From version EPLAN Electric P8 1.9.x, basic projects always have the file extension *zw9. The previous, old, format *epb still exists and may be used as a basis for new projects. However, it is no longer possible (this is also true of version 1.9.x and up) to create basic projects with the *epb extension.

Question: How do I modify the line type of Autoconnecting connections?

Answer: To modify the formatting of Autoconnecting lines, you can set a *potential definition point* on the Autoconnecting line and influence the formatting of the Autoconnecting line in the symbol properties on the *Connection graphic* tab. A potential definition point can be inserted via the INSERT / POTENTIAL DEFINITION POINT menu.

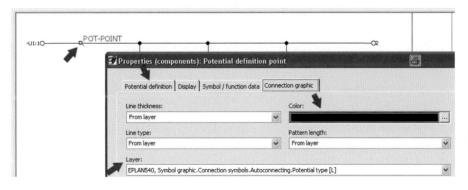

Fig. 13.13.4 Change formatting of Autoconnecting lines globally

Since this involves a connection, the connections should then be updated (via the PROJECT DATA / CONNECTIONS / UPDATE menu). This way, all connections affected (e.g., PE connections) are updated, and they adopt the format of the potential definition point.

If you want to change the line type for a partial connection only, you must use a connection definition point. A connection definition point can be inserted via the **Insert / Connection definition point** menu, which can then be placed on the intended section of the connection.

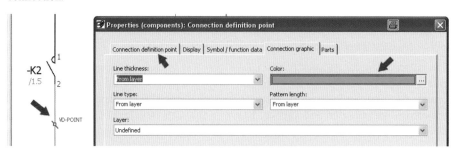

Fig. 13.5 Modify property of a partial connection

Question: Can macros with value sets be used across pages?

Answer: No. Macros with value sets can only be used on, or with, one page. If you wish to use macros with value sets across pages, you will have to rely on the **Project Options** add-on (to be purchased separately).

Question: Why do my interruption points in the project always refer to themselves on each page?

Answer: To ensure that cross-references work in connection with interruption points, you must note the following.

- *First page* -> insert an outbound interruption point
- *Second page* -> insert an inbound and an outbound interruption point
- *Third page* -> insert an inbound and an outbound interruption point
- and so on.
- *Last page* -> insert an inbound interruption point

Interruption points must always be seen in pairs, that is, the top of an interruption point must always come together with the bottom of the next interruption point; then again, top and bottom, and so on. Now, the cross-references (if all other properties are set to default) will work, and the interruption points will no longer result in wrong references.

Fig. 13.6 Default sorting of interruption points

By defining a sort code for interruption points, it is also possible to connect specific interruption points with each other in a targeted manner (which requires identical DTs). Here, too, you must note the following: There must always be a pair with the same sort code.

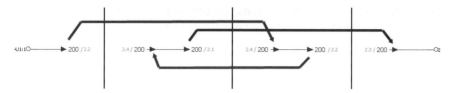

Fig. 13.7 Sort interruption points against default

Question: Why do I sometimes see red exclamation marks on my devices in the device navigator?

Answer: If you see a red exclamation mark on a device, it means that a message has been received for this device via the message navigator.

Fig. 13.8 Symbol properties

Simply open the message navigator, check off *Selection*, select the device in the device navigator, and precisely the message or messages will be displayed that have been generated for this device during the check run.

Fig. 13.9 Message in the message navigator

Question: Why does EPLAN Electric P8 not create connections, and why are no cross-references displayed?

Answer: In this case, the project in question is not of the **schematic project** type, but a project of the **macro project** type. These limitations are typical of macro projects.

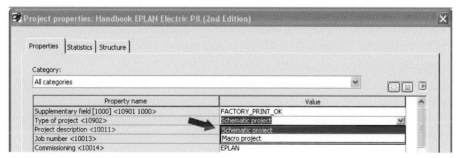

Fig. 13.10 Set project type

If this is not what you want, you must change the **Type of project <10902>** property in the project properties.

Question: Can I change the project structure (page structure) subsequently?

Answer: No, you cannot do that. The underlying structure of the project (page structure) cannot be modified subsequently; under regular circumstances, it can be selected only once (when creating the project).

Exceptions: The page structure cannot be changed either when a new project is created from a basic project. Here, too, it is impossible to change the page structure of the project, because it was defined as such in the basic project. Or a new project is created based on a copy of another project. Here, too, it is impossible to change the page structure of the project, because it was defined as such in the project to be copied.

All other structures, such as devices, cables, etc. can, however, be adjusted subsequently to the requirements of the project.

Question: Why does copy and paste using CRTL + C and CTRL + V not work anymore as before?

Answer: If copy with CTRL + C and paste with CTRL + V do not work in the usual manner anymore (the usual manner would be, for example: select object, copy it with CRTL + C and then paste it using CTRL + V), the DESIGN MODE has been enabled in the OPTIONS menu. This should be disabled again to restore the usual functionality of copy/paste.

Question: What is the difference between Delete and Delete placement?

Answer: *Delete* - the selected object (device) is deleted completely (graphical placement and any device information). From this time, it will no longer be available in the project data or in any of the navigators.

Delete placement - the object is deleted only "graphically". That is, it is no longer visible graphically (e.g., on a schematic page), but the object itself still exists in the project data with all its device information, e.g., in the device navigator, and can be placed "graphically" again later on or be retrieved from the corresponding navigator.

Question: Why can I not enter data into the "Displayed DT" field anymore (it is grayed out)?

Answer: If in the symbol properties the *Displayed DT* field is grayed out (i.e., no data can be entered here), the *Properties (global)* option has been activated. This must be deactivated first to edit the *Displayed DT* field in the usual manner again.

Question: Where can I find a demo version of EPLAN Electric P8?

Answer: There is a P8 version for students, with a trial license of 270 days. It is called Education and can be downloaded from the EPLAN website. Apart from EPLAN Electric P8, it contains EPLAN Fluid, EPLAN PPE and other programs depending on the stage of development. There are also installation instructions for the download.

NOTE: The data of the Education version are not compatible with the data of the industrial (purchased) version! As well, data of the industrial version cannot be edited with the Education version.

Question: How can I edit a single object within a grouping?

Answer: It is possible to edit a single object embedded in a grouping (without having to undo the grouping), by keeping the SHIFT key pressed and double-left-clicking the intended element. EPLAN then opens exactly this one object, and you can modify its data.

Question: How can I reduce five lines simultaneously by a specific length in EPLAN?

Answer: This involves the stretch function. It stems from the CAD area and works as follows (ideally, the OBJECT SNAP mode should be activated in the OPTIONS menu). Call the STRETCH function via the EDIT / GRAPHIC menu.

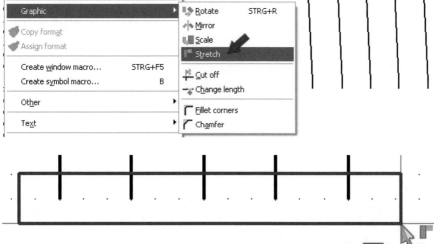

Fig. 13.11 Stretch function

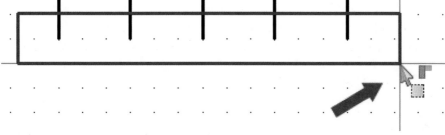

Fig. 13.12 Pull open window

Now, all line ends must be selected by pulling open a window (left mouse button) (the ends are "framed"). EPLAN marks the ends of the lines with a circle.

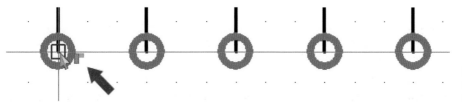

Fig. 13.13 Define starting point of the stretch

Then define the starting point of the stretch. Usually, this is done by clicking on the end of a line. If object snap is enabled, the cursor will "snap" into place there (right rectangle).

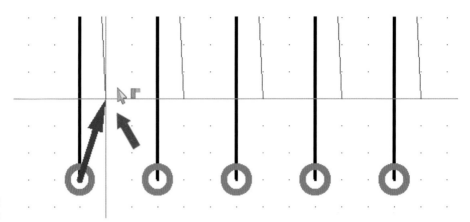

Fig. 13.14 Define end point of the stretch

Then, define the end point of the stretch (can be selected in a variable manner via the left mouse button - EPLAN now displays a cursor with a blue arrow while it has not been placed yet). The end point can be adjusted in any random manner.

Fig. 13.15 Line data have been changed collectively.

If the end point was selected from the placement, confirm it via the left mouse button (click once). Now, the lines are "stretched" toward this end point, i.e., shortened in this case. Of course, it also possible to extend them or to change the angle for all of them.

Question: How can one fill the creator, customer or end customer data in the project properties automatically?

Answer: To avoid having to manually enter each time the creator, customer or end customer data, such as address, name, etc., in the project properties, EPLAN provides an option that allows this data (which, of course, must have been set up in parts management in the **Customer** area first) to be written into the project properties by EPLAN automatically. For this purpose, open the (optional) *Project management* in the PROJECT menu and select the intended project. Now click the EXTRAS button, then select the READ CUSTOMER DATA function. Here, you can select the data for the creator, customer or end customer from the **Select address** dialog. Click on the OK button to import them into the project properties. EPLAN then fills automatically the corresponding project properties with the data stored in parts management.

Question: What is the difference between *Project properties* and *Project settings*?

Answer: *Project properties* are properties of the project, such as settings for the project structure or the types of devices used in it, such as terminal strips or general devices. Also part of project properties are specific properties that apply to the entire project globally, such as properties of the creator or the date of the project start. Project properties also include statistical information, such as the number of specific page types, etc. Project properties are called via the PROJECT menu and the PROPERTIES menu item. Or the project is selected in the **page navigator**, with the PROJECT item in the popup menu and subsequently the PROPERTIES item being executed.

Project settings are settings that also affect the project, but focus more on the "content" of the project, such as the forms to be used for different reports in the project or the online DT numbering scheme to be used, or the type of display to be applied to the project structure in the page navigator. Project settings are called via the OPTIONS menu, followed by the SETTINGS menu item. Then, the correct project must be selected in the **Projects** node (in case of several open projects).

Question: How can I interrupt an Autoconnecting line?

Answer: An Autoconnecting line can be interrupted by means of the ACBP symbol (Autoconnect break point). The symbol can be inserted by pressing simultaneously the CTRL + SHIFT + U shortcut key, or can be selected in the INSERT / CONNECTION SYMBOL / BREAK POINT menu and then placed randomly like any other symbol.

Question: How can I change the position of the contact image?

Answer: The position of the contact image can be modified in different ways: in the page properties of the current page by changing the **Contact image offset <12061>** property. This value applies only to the current page and all symbols affected by it.

The position can be changed globally by adjusting the plot frame in use. To do this, adjust the **Contact image offset <12061>** property in the plot frame editor. This setting will then apply to the entire project.

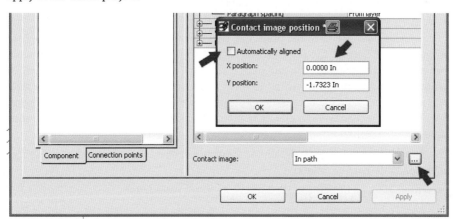

Fig. 13.16 Manually change position of the contact image

A third option involves calling the **Contact image position** dialog by using the More button directly on the affected symbol on the *Display* tab in the *Contact image* area and manually defining the X and Y positions.

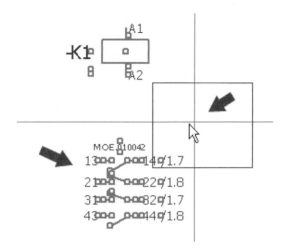

Fig. 13.17 Move

It is also possible to select the symbol, press the CTRL + B shortcut key and click on the contact image, while moving it into any position while keeping the left mouse button pressed.

Question: What is the meaning of the "Superior" option in the project structure in case of different devices?

Answer: If the **Superior** option has been set, EPLAN, in the report, allocates to this device the structure identifier or it does not. This depends on whether the device has been assigned the "-" *preceding sign*.

Fig. 13.18 Superior setting

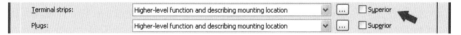

If the device has the preceding sign "-", the device will be allocated the structure identifiers (of the page, etc.) as defined in the project structure.

If the device does not have the preceding sign „-", the device will not be assigned the structure identifiers (of the page, etc.) as defined in the project structure. For example, you can take one terminal strip and have it created globally across several mounting locations by omitting the „-" preceding sign. But at the same time, it is also possible to generate a report on the remaining terminal strips, depending on the project structure defined, based on higher-level function and mounting location by placing the „-" preceding sign.

The **Superior** option is possible for the following device types: Terminal strips, plugs, cables and interruption points.

Question: Where and how can I re-sort my structure identifiers?

Answer: To re-sort structure identifiers such as higher-level function or mounting location, use the structure identifier management tool. It can be found in the PROJECT DATA / STRUCTURE IDENTIFIER MANAGEMENT menu. Here you will find editing options to re-sort or rename identifiers, to create new identifiers or to edit existing or create new descriptions of structure identifiers, and much more.

Question: How can I display for a specific object the message(s) in message management?

Answer: To view the relevant messages from message management for an item in the device navigator, you must take the following steps.

1. Launch PROJECT DATA / MESSAGES / MANAGEMENT message management with the proper check run.
2. In the message management window, put the check mark next to Selection.
3. Launch the device navigator.
4. Click on the desired (defective) item in the device navigator.
5. Now you will see only messages in message management that concern this item.

Question: Where in EPLAN can I enter my own connection point designations or connection point descriptions for symbols?

Answer: For example, if you want to have for a horn (signal device) the existing connection point designations [x1 x2] as well as the connection point designation [L N] (in the symbol properties dialog of the horn), you must take the following steps:

1. Via the UTILITIES / MASTER DATA menu, open the CONNECTION POINT DESIGNATION / DESCRIPTION menu item.
2. On the *Connection point designations* *) tab, locate the corresponding function definition (here: Signal device, acoustic, single) in the function definition column.

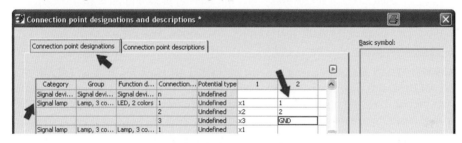

Fig. 13.19 Add connection point designations

3. In column 2, enter the additional connection point designations per connection point.
4. Click the APPLY button and close the dialog with OK.
5. In the UTILITIES / MASTER DATA menu, select the UPDATE CURRENT PROJECT menu item. EPLAN now updates the project master data.
6. Effective immediately, the new connection point designation [L N] is available at the horn for selection.

*) For your own connection point descriptions, follow the same approach. The only difference is that you must select the *Connection point descriptions* tab in the dialog.

Question: What is the difference between [Update current project] and [Synchronize current project]?

Answer: *Update current project* - EPLAN automatically updates the project master data with the system master data. No manual editing is possible at this point. Once this function is launched, there will be no more confirmation prompt (to abort the function, you would have to click on the CANCEL button in the dialog). After the update, EPLAN will display a message.

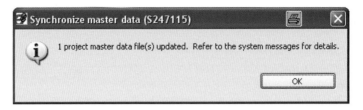

Fig. 13.20 Message regarding synchronization of master data

Synchronize current project - first, the **Synchronize master data** dialog is launched. In this dialog, master data can be exchanged in a purposeful manner (in both directions, i.e., system to project or project to system). Among other things, you can update (synchronize) specific master data manually.

Both menu items are located in the UTILITIES / MASTER DATA menu.

Question: Can EPLAN create a macro from several pages?

Answer: Yes, EPLAN can do that. Such macros are known as page macros in EPLAN.

A page macro of a page can be created via the CTRL + F10 shortcut key. Alternatively, you can select the page in the page navigator and then select in the popup menu (right mouse button) the CREATE PAGE MACRO item. Then, enter the desired information in the **Save as** dialog.

To create a page macro from several pages, the same approach applies accordingly. The pages are selected in the page navigator; then, press the CTRL + F10 shortcut key or call, in the popup menu of the right mouse button, the **Create page macro** menu item.

Question: EPLAN Electric P8 and Windows Vista or Windows 7 - is it possible?

Answer: EPLAN Electric P8 (and the rest of the EPLAN platform, such as EPLAN PPE, EPLAN Fluid, etc.), apart from Windows XP Professional SP3 (32-bit), has also been approved - version 1.9.x and higher - for Windows Vista and - version 2.0 and higher - for Windows 7 regarding the following versions:

The following Windows Vista versions are supported: Windows Vista Enterprise; Windows Vista Business N and Windows Vista Ultimate, in each case SP2 and 32-bit or 64-bit.

The following Windows 7 versions are supported: Windows 7 Enterprise; Windows 7 Business and Windows 7 Ultimate (32-bit or 64-bit).

Question: Why are graphical symbols not shown in the reports, but blue boxes instead?

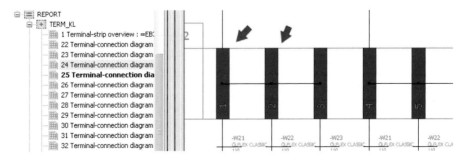

Fig. 13.21 Blue boxes

Answer: For graphical symbols, for example, from the GRAPHICS symbol library, to be displayed in reports, such as terminal or cable diagrams, the following user setting must be enabled.

Under OPTIONS / SETTINGS / USER / GRAPHICAL EDITING / 2D and here, in the selection field for the *Graphical symbol* color setting, you must select the *From symbol* setting.

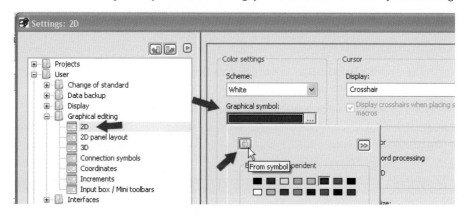

Fig. 13.22 From symbol color setting

This way, the graphical symbols will be displayed completely even in the reports.

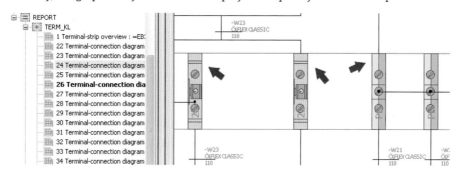

Fig. 13.23 Symbol color settings

Question: What is the difference between a window macro and a symbol macro?

Answer: Generally, they are not vastly different. Both macro types (*window macro* *.ema; *symbol macro* *.ems) can combine and store parts of a page or different objects in a macro, which can then be retrieved as a window or symbol macro.

A window macro is called via the M key or via the INSERT / WINDOW MACRO menu; a symbol macro, via the CTRL + INS shortcut key or via the INSERT / SYMBOL MACRO menu. Either one can then be placed on the page freely.

Question: How can I change the EPLAN installation drive later on - i.e., without reinstalling?

Answer: For example, to change the installation drive from *C:\Program Files\EPLAN Electric P8* to *E:\EPLANP8 V2*, no reinstallation is required. All you need to do is adjust the installation paths in the **Install.xml** file.

Fig. 13.24 Subsequent change to installation paths

The Install.xml file is located in the installation directory *Drive:\EPLAN8 V2\2.0.5\CFG* (depending on your installation, the path may be different).

Question: How can I change the company code later on - i.e., without reinstalling?

Answer: For example, in order to change the *Company code* from CompanyXYZ to CompanyTheBestOfTheWorld, no reinstallation is required. Simply adjust/change the *Company code* in the **Install.xml** file again.

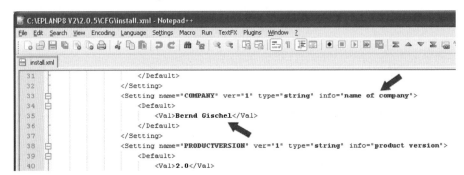

Fig. 13.25 Subsequent change to the company code

The Install.xml file is located in the installation directory *Drive:\EPLAN8 V2\2.0.5\CFG* (depending on your installation, the path may be different).

> **NOTE: Before EPLAN is restarted** (after making changes to the company code in the Install.xml file), the rights database must be located in the new directory (new company code). **Otherwise, EPLAN will not launch!** The rights database is located in the directory (example): *E:\EPLANP8\Management\CompanyCode* (put your own company code here)!

Question: What is a main function?

Answer: A *main function* is the "leading" device with (distributed) devices displayed (e.g., the coil of a contactor) or a single switch. Only the main function, e.g., bears the device definition (parts data) of the device. All other possible functions (e.g., the contacts of the coil) are thus automatically "not main functions", or auxiliary functions.

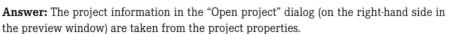

Fig. 13.26 Main function coil

In EPLAN, for each device there must always be **exactly one main function** only of the device used. This is true of all objects, whether they be contactors, motor overload switches or cable definitions.

Question: Where are the page sortings (subpage, characters) of reports stored?

Answer: To arrive at a desired sorting of report pages (subpages yes/no, numeric, letters upper-case/lower-case), launch from the UTILITIES / REPORTS / GENERATE menu the **Reports** dialog. Here, click on the SETTINGS button and then select the OUTPUT TO PAGES menu item. Then, set in the dialog the desired *Sorting* in the *Character* column and whether subpages (subpage column) are to be created at all. Finally close and leave the dialog with OK. Now the reports can/must be generated and/or updated.

Question: What project information appears in the "Open project" dialog on the right-hand side of the dialog?

Answer: The project information in the "Open project" dialog (on the right-hand side in the preview window) are taken from the project properties.

Fig. 13.27 Information in the Open project dialog

The following project properties are displayed (if they have been completed in the project properties): Project description <10011>; Job number <10013>; Commission <10014>; Company name <10015>; Company address 1 <10016>; Company address 2 <10017>; Supplementary field [1] <10901 1>; Last editor: Logon name <10022> and the Modification date <10023>.

Question: How can I restore EPLAN to the values of the default installation?

Answer: To restore the default installation values of EPLAN (when confusing situations arise that cannot be explained or removed in any other way), EPLAN can be launched with the following call-up:

"C:\Program Files\EPLAN Electric P8\2.0\BIN\W3u.exe" /setup

The program directory shown here must be adjusted to reflect your own installation directory.

TIP: With this call-up, all personal settings will be lost. These must be backed up before.

 Question: The object snap for the dimension function does not work anymore.

Answer: If graphical objects on pages - e.g., for the dimension function - are no longer captured by the mouse automatically (the small red rectangle), it means that OBJECT SNAP in the OPTIONS menu has been disabled.

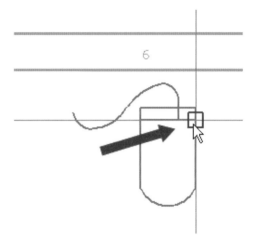

Fig. 13.28 Object snap on

Once object snap is enabled, you will be able to capture objects (points) again.

 Question: How can I create completely new connections or repair connections?

Answer: If connections (or their additional information) exhibit strange behavior, it is possible to recreate all project connections from scratch. To do so, keep the SHIFT and CTRL keys pressed and call from the PROJECT DATA / CONNECTIONS menu the UPDATE menu item. Now, all connections (and/or the connection database) will be recreated. Depending on the number of connections in the project, this may take some time, because EPLAN first deletes all connection data and then recreates them fully from scratch.

 Question: How can I change, later on, the sorting of the structure identifiers, such as higher-level function or mounting location?

Answer: To sort identifiers (structure identifiers), such as higher-level function or mounting location, you must start STRUCTURE IDENTIFIER MANAGEMENT from the PROJECT DATA menu. In the dialog that follows, using the corresponding tabs, you can change the sequence of the various identifiers used in the project (and also those not currently in use).

 Question: What are the default keyboard shortcuts available in EPLAN?

Answer: The following keyboard shortcuts (shortcut keys) are available in P8 by default (e.g., in the graphical editor):

Command [key or shortcut key]

Insert interruption point [SHIFT] + [F4]
Insert shield [SHIFT] + [F6]
Cancel action [Esc]
Select all [CTRL] + [A]
Interrupt connections [CTRL] + [SHIFT] + [U]
Update view ("Redraw") [F5]
Display of insertion points on/off [I]
Call editing mode (in certain table-format representations such as the Multilingual input dialog) [F2]
Base point shift on / off [O]
Move image section left [SHIFT] + [cursor key to the left]
Move image section up [SHIFT] + [cursor key up]
Move image section right [SHIFT] + [cursor key to the right]
Move image section down [SHIFT] + [cursor key down]
Insert arc through center [CTRL] + [G]
Insert jumper (junction) [SHIFT] + [F8]
Move cursor to the left of screen [Home]
Move cursor to the right of screen [End]
Edit properties of objects [CTRL] + [D]
Move property texts [CTRL] + [B]
Insert linear dimension [CTRL] + [SHIFT] + [A]
Paste elements from EPLAN clipboard [CTRL] + [V]
Cut elements and copy to EPLAN clipboard [CTRL] + [X], [SHIFT] + [Del]
Group elements [G]
Copy elements to EPLAN clipboard [CTRL] + [C]
Insert elements into list of search results [CTRL] + [I]
Insert ellipse [E]
Define window (selection area) [SPACE]
Duplicate window (selection area) [D]
Insert window macro [M]
Create window macro [CTRL] + [F5]
Zoom 100% [Alt] + [3]
Go to (graphic) [CTRL] + [J]
Insert device connection point [Alt] + [Ins]
Insert device connection [SHIFT] + [F3]
Draw black box [SHIFT] + [F11]
Graphic: Rotate [CTRL] + [R]
Close graphical editing [CTRL] + [F4]
Delete contents of a window (selection area) [Del]
Insert cable definition [SHIFT] + [F5]
Call context help [F1], [CTRL] + [F1]
Coordinate input [P]
Insert circle [K]
Undo last step [CTRL] + [Z], [Alt] + [Back]
Insert line [CTRL] + [F2]

Open/close message management [CTRL] + [SHIFT] + [E]
Next page [Page down]
Activate/deactivate orthogonal function [SHIFT] + [<]
Activate orthogonal function [X]
Activate orthogonal function [Y]
Activate/deactivate orthogonal function in horizontal/vertical direction, deactivate active orthogonal function [<]
Insert location box [CTRL] + [F11]
Insert path function text [CTRL] + [T]
Insert polyline [L]
Show grid on/off [CTRL] + [SHIFT] + [F6]
Insert rectangle [R]
Enter relative coordinates [SHIFT] + [R]
Open/close panel layout navigator [CTRL] + [SHIFT] + [M]
Set increment [S]
Create page [CTRL] + [N]
Select page [CTRL] + [M]
Print page [CTRL] + [P]
Create page macro [CTRL] + [F10]
Open/close page navigator [F12]
Start search function [CTRL] + [F]
Search function: Jump to counterpiece [F]
Search function: Jump to next entry [CTRL] + [SHIFT] + [F]
Search function: Jump to previous entry [CTRL] + [SHIFT] + [V]
Insert symbol [Ins]
Insert symbol macro [CTRL] + [Ins]
Create symbol macro [B]
Open/close editing in table [CTRL] + [Q]
Insert text [T]
Insert T-node (left) [F10]
Insert T-node (top) [F8]
Insert T-node (right) [F9]
Insert T-node (down) [F7]
Global editing: Edit project data from report [CTRL] + [SHIFT] + [D]
Show/hide invisible elements [U]
Insert connection definition point [SHIFT] + [F7]
Move [V]
Previous page [Page up]
Insert angle (up, left) [F6]
Insert angle (down, left) [F4]
Insert angle (down, right) [F3]
Insert line break [CTRL] + [Enter]
Open zoom [Z]
Jump to element points of selected elements [Tab]
Jump to element points [SHIFT] + [CTRL] + [cursor key]

Jump to element points at same height [SHIFT] + [Alt] + [cursor key]
Jump to left insertion point at same height [Alt] + [cursor key left]
Jump to next insertion point left [CTRL] + [cursor key left]
Jump to next insertion point top [CTRL] + [cursor key up]
Jump to next insertion point right [CTRL] + [cursor key right]
Jump to next insertion point bottom [CTRL] + [cursor key down]
Jump to the next element point, which can also be an end point of an element [Alt] + [Home]
Jump to top of the screen [CTRL] + [Home]
Jump to top insertion point in the same path [Alt] + [cursor key up]
Jump to right insertion point at same height [Alt] + [cursor key right]
Jump to bottom of the screen [CTRL] + [End]
Jump to bottom insertion point in the same path [Alt] + [cursor key down]
Switch between open windows, such as graphical editor, navigators, etc. [CTRL] + [F12]

Question: Can EPLAN also process page names like "Seventy"?

Answer: Yes, this is possible. Usually the page name consists of a number (page number). For example, the name of the page can be "**70**". But EPLAN can also include in page numbers alphanumeric components. Accordingly, a page can also be called "**Seventy**".

> **TIP:** This is why the „number" property of the page is not called „number" but page name, because the „page name" can also include alphanumeric components!

Question: Where can I find a keyboard mask for EPLAN?

Answer: EPLAN does not supply a keyboard mask for EPLAN Electric P8, because users are not expected to stick to the predefined shortcut keys (apart from a few exceptions).

Question: How is it possible, for example, that a device connection point with the potential type PE has the same designation, without EPLAN showing a project check message?

Answer: In this case, the LOGIC (on the **Symbol / function data** tab and, here, the LOGIC button) of the device connection point must be adjusted. In the following **Connection point logic** dialog, there is the *Allow same connection point designations* setting, and this must be activated for the corresponding device connection points. Once this is done, EPLAN will no longer object to the identical designations of the functions.

Question: Does the sequence of the stored symbol libraries affect the representation of the symbols?

Answer: No. The sequence of the stored symbol libraries does not affect EPLAN projects. EPLAN always takes symbols only from the correct symbol library based on the available information, such as the name of the symbol library, the symbol number or name. This requires, of course, that the same symbol libraries are/have been stored in the various projects.

Question: Why, after inserting a macro, do I see a bunch of red crosses instead of the symbol graphics?

Answer: After inserting a macro or similar things, if you see a bunch of red crosses instead of symbol graphics, the cause is often a missing symbol library in the project.

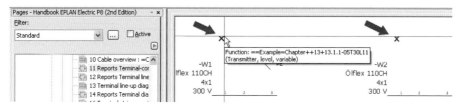

Fig. 13.29 Missing symbols

Since the symbol library or libraries is or are not stored (in the macro, etc.) when creating a macro or similar things (in another project), EPLAN needs the information to display the proper symbol graphic contained in the symbol library that was used in the project (where the macro was created).

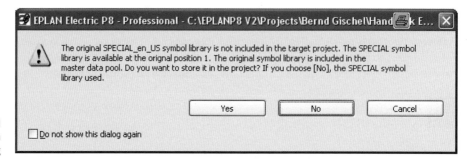

Fig. 13.30 EPLAN warning message when symbols are missing

This can be fixed by placing/storing the corresponding (missing) symbol library in the current project. Then, you must re-insert the macro, or simply select the symbol again.

Question: How can I insert special characters in EPLAN that do not exist in the **Special characters** dialog?

Answer: To insert special characters in texts or similar things in EPLAN, use the CTRL + S shortcut key to call the **Special characters** dialog. But the selection offered there is limited. If you want to insert a special character that is not included in EPLAN by default, you can copy it from another Windows application, for example, WORD, and then paste it into P8.

Question: A project was deleted by mistake - what can I do now?

Answer: If EPLAN was used to delete a project from the server, it is irretrievably lost, because Windows, the operating system, does not have a recycle bin (trash can) on the server. If the project was deleted from the local hard disk, you can restore the project from the Windows recycle bin.

 TIP: To restore projects deleted from the server, the only option - hopefully - is the data backup you created!

Question: Can I change the automatic distance of the contact image on component?

Answer: Yes. To change the automatic distance of the contact image on a component, you must call the *Symbol properties*. Then, you should switch to the *Display* tab and disable the automation in the area of the *contact image* and enter a new, desired, manual value for the X/Y position (shift).

Question: What do the odd frames around my symbols mean that suddenly appear when I move, etc. them?

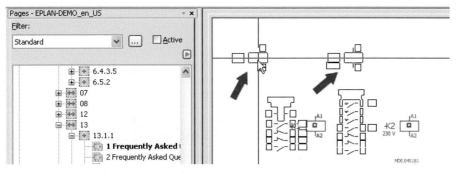

Fig. 13.31 Frames of the terminal server

Answer: If you copy, move, etc. symbols and they show odd frames during those actions, it means that the setting under WORKSTATION / DISPLAY / GENERAL and, here, *Optimize terminal server* has been activated. This setting, for example, makes the cursor change its appearance during actions like copying, moving, etc. To restore the previous behavior/appearance, deactivate this setting.

Question: Where can I find documentation on the EPLAN Electric P8 API interface?

Answer: The documentation on the API interface is currently available only to those who also purchase the API interface from EPLAN.

Question: How do I access the EPLAN Data Portal?

Answer: Anyone who has a software maintenance contract will receive immediate (free) access to the Portal and the data contained therein after installing the EPLAN Data Portal add-on and creating a user account.

Question: What is the easiest way to automatically rename a structure identifier, for example, a mounting location?

Answer: To automatically rename a *structure identifier*, such as a mounting location +OT12 to +U23, globally, i.e., across the entire project, the easiest way to do so involves the *Structure identifier management*. This way, the mounting location used in the project, +OT12, can be found anywhere and renamed to the new structure identifier mounting location +U23, without having to make manual adjustments. The structure identifier management tool can be accessed from the PROJECT DATA menu followed by the STRUCTURE IDENTIFIER MANAGEMENT menu item.

Question: In the "Select scope of menu" start dialog, what do the terms Beginner, Advanced and Expert mean?

Answer: The three settings - *Beginner*, *Advanced* and *Expert* - correspond to three fixed and defined user groups that are selectable if EPLAN is run without rights management.

Beginner refers to users with access to the functionalities that are absolutely necessary to draw schematics as well as work with macros and project data dialogs.

Advanced users can use specific functions on top of the previous ones (e.g., display options such as minimum text size, display of empty text boxes, etc.). They can also use data transfer.

Expert users have access to all functionalities, i.e., they can use the functions they need to prepare work such as system settings, edit master data, work with revisions and project options, back up data, etc.

TIP: Conversely, this means: If one or the other function is missing, but according to the license should exist, check again to see the scope of menu with which EPLAN was launched.

Question: How and where can I change the view of the workspace?

Answer: Via the VIEW / WORKSPACE menu, it is possible to select from default views. In addition, you can create and save new workspace views of your own. As well, it is possible to save anew (overwrite) the current workspace with all its settings to a current scheme.

Question: How do I activate the mode for the direct editing of property texts?

Answer: In EPLAN, you can edit certain properties directly, that is, without having to call the symbol properties dialog.

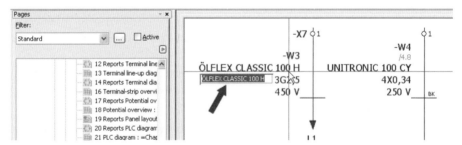

Fig. 13.32 Edit properties directly

Apart from the option of enabling and disabling this edit mode via the OPTIONS menu, you can also activate direct editing temporarily using the mouse or keyboard.

Using the mouse: As soon as you have selected text or a component, keep the left mouse button pressed for a short amount of time. A small window opens where you can edit the data directly.

Using the keyboard: Move close to the text or component and press the [F2] key. Again, a small window opens for direct editing.

TIP: In the case of several properties on symbols, after activating the direct input mode, you can use Tab to „jump" to the individual properties and edit them!

Question: How and where can I change the color scheme (background color) of the graphical editor?

Answer: Under OPTIONS / SETTINGS / USER / GRAPHICAL EDITING / GENERAL, the color scheme of the graphical editor can be changed. The names of the color schemes indicate the color of the screen background: So, you can choose a *white*, *black* or *gray* background. The colors of the various elements are adjusted for the respective background. I recommend that you use the white background.

Question: How can I re-activate suppressed dialogs/messages?

Answer: To re-activate dialogs that have been deactivated before (e.g., Tip of the day), you must again put the following check mark in the settings for the user. Go into the OPTIONS / SETTINGS menu and select in the now open dialog USER / DISPLAY / USER INTERFACE from the tree structure. There, on the right-hand side, you can select the *Reactivate suppressed messages* option. Then, close the dialog again. All deactivated messages/dialogs are available again immediately. For some dialogs/messages, though, it may be necessary to restart EPLAN.

Question: How do I save a project or a page in EPLAN?

Answer: It is not at all necessary to save a project or page in EPLAN. Every change is "written = saved" directly (on the storage medium, such as a hard disk). As a result, additional manual saving is not necessary.

Question: How can the sequence of projects in the page navigator be changed?

Answer: It cannot be changed. The sequence of projects (in the **page navigator**) depends on the sequence in which they were opened.

Question: Can I display the description of structure identifiers in the page navigator?

Answer: No, at this point, neither the tree nor the list view of the **page navigator** allows for that.

Question: Is it possible to install EPLAN Electric P8 (all versions) side by side with an old EPLAN 5 version (also all versions)?

Answer: Yes, absolutely, because both programs operate separately from each other.

Question: Is it possible to increase the maximum number of entries (12) in the clipboard?

Answer: No. The *clipboard* is limited to 12 entries. The clipboard >1 is activated as follows: Via the OPTIONS / SETTINGS menu, open the USER / DISPLAY / GENERAL node and adjust the settings in the right area of the dialog for the clipboard.

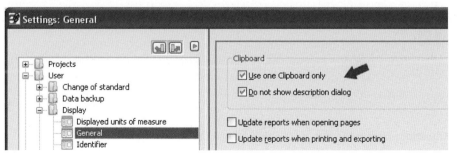

Fig. 13.33 Clipboard settings

Question: Can the "Update connections" function be assigned to a shortcut key?

Fig. 13.34 User-created shortcut key F11

Answer: Yes, you can do that in EPLAN. In fact, it is recommended, because this function is used quite frequently. I recommend that you use the F11 function key. This one is not yet in use, and it is also easy to reach and remember.

Question: How can I suppress the check message "Coil without contacts"?

Answer: If the project uses contactors/relays, etc. as spares (without using contacts), you may receive the check message *P007009: Coil without contact*.

Solution 1: This can be remedied only if you assign this coil a part with a correct function template, i.e., once you have made a device selection. Then, all spare coils in the project will be ignored for the purposes of the check run messages.

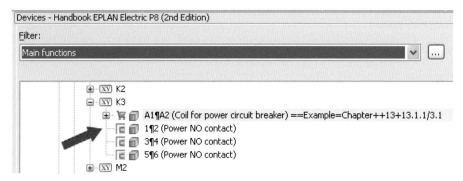

Fig. 13.35 Unplaced functions

Solution 2: Possible contact assignments (i.e., contacts like NO contact, NC contact, etc.) are added as new functions of the coil to the device navigator as unplaced functions.

Question: Can the PDF comment function be used in EPLAN only with Adobe Acrobat Writer (full version)?

Answer: No. Among other things, in EPLAN you can use the Redlining function (reading out to EPLAN comment entries contained in the generated EPLAN-internal PDF) with the freeware PDF-XChange Viewer (last update: September 2010).

NOTE: The write protection for the internal EPLAN PDF output should be disabled (OPTIONS / SETTINGS / USER / INTERFACES / PDF EXPORT followed by the *General* tab). Otherwise, there could be problems when importing the commented PDF (it will not be able to be read)! This also applies to the solution involving the Acrobat software.

Question: What is the fastest way to find a specific structure identifier in the project?

Answer: For this purpose, use the F12 standard shortcut key to launch the **page navigator**. In the navigator, use the right mouse button to select the DETAILED SELECTION menu item. The **Select pages** dialog is displayed. In this dialog, a prefilter is created, which is then used to filter the structure identifiers onto the project.

Question: Why can I not find a specific structure identifier in the project despite an activated prefilter?

Answer: If a structure identifier in the project cannot be located graphically via an activated prefilter in the **page navigator**, it is part of an unplaced object.

Unplaced objects and their structure identifiers can be found, for example, in the **device navigator**. The arrangement/sequence of the structure identifiers is imported from the project settings (DISPLAY / PROJECT STRUCTURE (NAVIGATORS)).

Question: How can lower-case letters be also used and/or entered for a DT designation?

Answer: In order to be able to also enter lower-case letters for a DT, the following setting should be deactivated: OPTIONS / SETTINGS / PROJECTS / [PROJECT NAME] / DEVICES / DT and, here, deactivate the *Conversion to uppercase* setting (remove check mark).

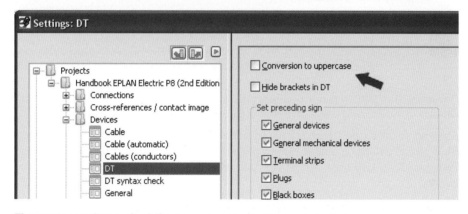

Fig. 13.36 Upper-case letters or lower-case letters

This setting applies to the following properties in the symbols dialog:
- Displayed DT
- Full DT
- Connection point designations
- Terminal designation
- Plug DT, Channel designation, and Address (for PLC connection points)
- Structure identifiers.

 NOTE: If this setting is activated/deactivated, lower-case letters will not be converted to upper-case letters automatically and vice versa. The „conversion" does not take effect until the setting has been activated/deactivated! This way, all „old" DTs entered remain in lower-case even if the setting is activated. Only the next new DT designation entered will be „converted" automatically to uppercase.

Question: How can I insert a mechanical action line between two objects?

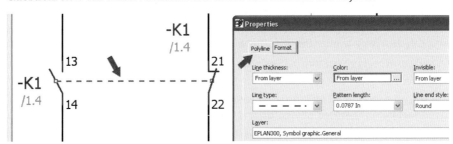

Fig. 13.37 Mechanical action line

Answer: A mechanical *action line* in EPLAN is a purely graphical representation without any logical function. It is represented simply by a graphical line between two objects, or two contacts, which is then formatted according to the properties of the line.

Question: How can I call up a script via a button click?

Answer: Using the "ExecuteScript" command, you can start a script file by clicking the mouse button.

 TIP: The EPLAN online help contains additional useful information under the keyword „Scripts".

Question: How can I open a document/image for a stored part directly from project editing?

Answer: To open a document (or image) stored for a part in parts management directly from project editing, you must call up *properties of the symbol* and then switch to the *Parts* tab. Then, select the desired part and call the popup menu with the right mouse button.

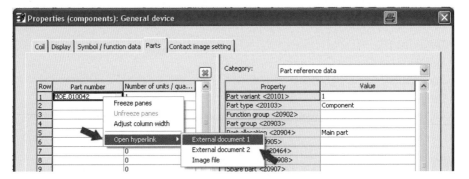

Fig. 13.38 Open hyperlinks

The desired information can be seen in this popup menu under the HYPERLINKS item. Here, you can find the stored external documents 1 to n (depending on the number on the part in parts management) as well as an image of the part.

NOTE: If no such information (such as external documents or an image) is stored on the part in parts management, the popup menu items will be grayed out and cannot be selected.

Question: What is the difference between a main function and a main function of superior device?

Answer: To understand this, you should picture a nested device. Accordingly, a device A (being the *main function*) is located, for example, in a black box B (also a main function). This makes black box B the *superior main function* of device A.

Question: What are the differences between "Save additionally", "File off" and "Archive" in the data backup dialog?

Answer: *Save additionally:* When backing up data, a complete (depending on the options selected) copy of the project is saved to another storage medium. It is also possible to back up several projects. Several projects can be backed up via project management. Only project management allows for a multiple selection.

File off: If a project is to be passed on (e.g., to a customer), you can use the *File off* option. This creates a copy of the project (e.g., on a different storage medium), and the source project is set to read-only. This prevents changes in the meantime to this (filed off) project. If an attempt should still be made to edit such a filed off project, the user doing this will receive a message that the project has been filed off and locked for editing.

Archive: Finished (completed) projects can be archived in order to free up "space" on the hard disk. In the process, a copy of the project is saved to a different storage medium (e.g., the local hard disk), and the (finished) source project is deleted from the hard disk except for an information file.

Question: How can I start, for example, an Excel table via the external programs?

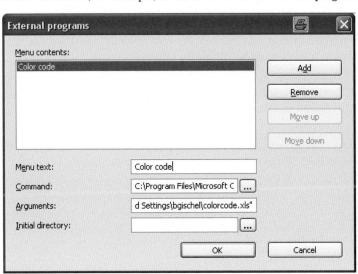

Answer: To integrate, for example, an Excel table via the EXTERNAL PROGRAMS menu and to later call it from the OPTIONS menu, you must take the following steps.

1. Open the external programs via the OPTIONS / EXTERNAL PROGRAMS menu.
2. Create a new menu item using the ADD button.
3. Enter a suitable (subsequently accessible via the OPTIONS menu) *menu text*, e.g., *"Color codes for EPLAN and others..."* or something similar.
4. In the *Command* field, enter the path to Excel.exe, for example: *D:/Program Files/Microsoft Office/OFFICE11/EXCEL.EXE*.
5. Enter the path to the Excel table in the *Arguments* field. In case of blanks in the path, please put the path in quotes, for example: *"E:/EPLANP8 V2/Tools/ExcelColor codes for Eplan E5, P8.xls"*

Fig. 13.39 Using external programs

Fig. 13.40 New menu item

Save all by clicking on OK. In the OPTIONS menu, the "COLOR CODES FOR EPLAN AND OTHERS..." menu item is instantly available and opens the Excel table if it is selected.

 TIP: This menu item can also be assigned its own shortcut key.

 Question: Why is the unit of the cross-section/diameter of a connection definition not displayed?

Answer: If the *unit* of the cross-section/diameter is not displayed, it means that the wrong property arrangement has been applied. The unit is not shown by default. This is why a different property must be set to display the unit.

To display "mm^2" together with the cross-section information, you will require, for example, the *Connection: Cross-section / diameter with unit <31007>* property. This must then be added and formatted accordingly on the *Display* tab of the **Properties** (...) dialog.

 NOTE: The selection of the unit can be set or selected manually in the **Properties** (...) dialog (this approach is not recommended).

But the value of the **Unit** setting can also be left in the default value of the project. Then, the unit is pre-set in the project settings under CONNECTIONS / PROPERTIES on the **Electrical engineering** tab in the **Unit** field (it is then applied globally and should be the preferred approach).

Question: Is it possible to expand the display window on the right in the "Open project" dialog to include further information?

Answer: No. This information cannot be extended and is permanently defined by EPLAN.

Question: Where and how can I define a description for my structure identifiers?

Answer: To define a description for structure identifiers (depending on the setting), you can use the **Place identifiers** dialog (which is opened automatically by EPLAN when entering an unknown structure identifier to the project), or you can add it later directly in STRUCTURE IDENTIFIER MANAGEMENT (PROJECT DATA menu).

 NOTE: The **Place identifiers** dialog, which opens for new and still unknown identifiers, can be activated or deactivated in the settings under USER / DISPLAY / IDENTIFIER.

Question: Can I change the row height (and thus font size of the display), for example, in the **Properties** (components) dialog, or also in the 'Edit in table' mode, etc., in order to be able to read it better?

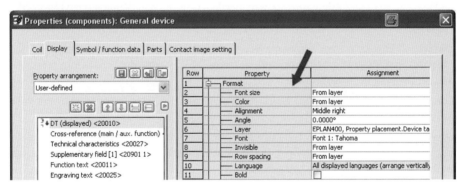

Fig. 13.41 Change size of rows - before

Answer: Yes, this is possible. To increase the *row height* (and thus font size of the display), you must click on the corresponding display, then keep the CTRL key pressed and turn the scroll wheel of the mouse. Depending on the direction, the display will be increased or reduced.

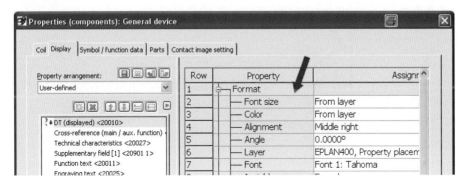

Fig. 13.42 Change size of rows - after

Question: Question: Can I assign the *Find new DT* function in the **Properties** (components) ... dialog in the *Displayed DT* field (right mouse button) to a shortcut key?

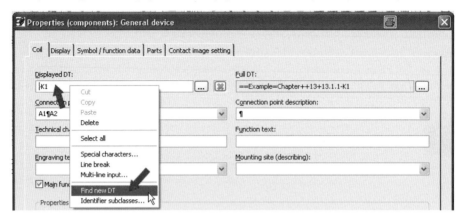

Fig. 13.43 Find new DT

Answer: You cannot use your own shortcut keys, but when you are in the *Displayed DT* field and press the CTRL + N shortcut key (you must keep the CTRL key pressed), EPLAN will execute the exact same function that also exists in the popup menu (FIND NEW DT).

Question: I received a check run message that does not mean anything to me.

Answer: If checks are not necessary, you can disable them in the test scheme. For this purpose, open the test scheme for editing, locate the corresponding check run messages and set the type of check to "No".

Question: Is there a way to fully reset the manual page sorting in the page navigator with a simple click?

Answer: Yes, this is possible. You must take the following steps to fully reset the manual page sorting to default sorting. Open the **page navigator** and in there select *List view* (tab at the bottom end of the page navigator).

6. Click the right mouse button while in the page navigator and select the MANUAL PAGE SORTING menu item.
7. Now click the right mouse button again while in the dialog and click on the DEFAULT SORTING menu item.

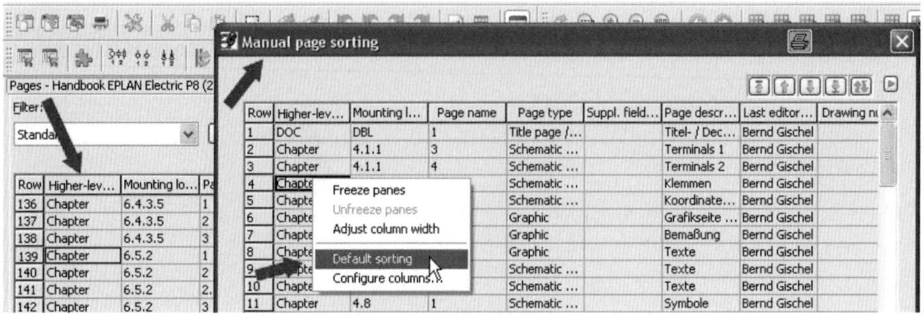

Fig. 13.44 Reset manual page sorting

Now EPLAN resets manual page sorting in the page navigator to default sorting.

Question: Is it possible to let objects placed outside the plot frames be found automatically?

Answer: Yes. EPLAN has a *check run message* for this. This must be activated for the check run so that EPLAN can find objects placed outside the plot frame automatically. It can be found in the check run (check run settings) under ID 010 Cross-references and has the message number "010001" with the message text "The placement lies outside the evaluation range of the plot frame".

TIP: The check run message captures only logical objects. Free objects like texts, etc. are not captured by this check!

13.2 Parts

Question: I saw during device selection a typing error in Designation 1 of my part. Can I remove this typing error directly without having to launch parts management first?

Answer: Yes, this is possible. To do so, you must activate the following setting: USER / MANAGEMENT / PARTS MANAGEMENT and, here, *Modification allowed during selection*.

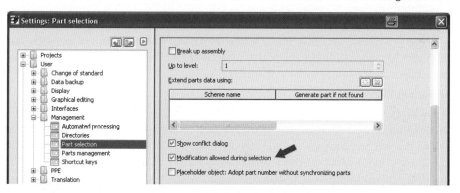

Fig. 13.45 Change parts data setting during the selection

Question: Is it possible to synchronize the stored parts data automatically upon opening the project?

Answer: Yes, this is possible. To do so, you must activate in the project settings the *Synchronize stored parts when opening* setting.

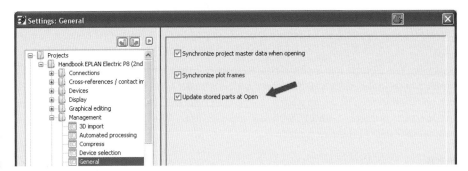

Fig. 13.46 Update stored parts when opening the project

Question: Can I store more than a document on a part in parts management?

Answer: Yes, absolutely. EPLAN allows you to store up to 20 documents on a part in parts management. To do so, they must be entered in parts management for the desired part on the *Documents* tab and/or selected from the directories and imported.

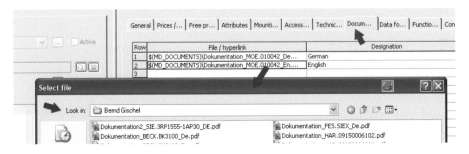

Fig. 13.47 Documents tab

Question: Why can I not enter a part on my function (the Parts tab is missing)?

Answer: Only *main functions* can have parts data. So, a function must be a main function in order to support parts data. If this feature is missing (the **Main function** setting has not been checked, e.g., on the *Symbol / function data* tab), the *Parts* tab will not be available either.

Question: What is the difference between device selection and part selection?

Answer: In the case of *part selection*, EPLAN does not check whether the part fits the device in the project "horizontally" when the part on the device is selected via the *Parts* tab (in the Part number field via the More button [...]). To put it in exaggerated terms, one could assign to the motor overload switch device also a PLC card part.

Fig. 13.48 Simple part selection

In the case of *device selection*, however, there is an immediate check upon clicking the device selection button to determine which device functions already exist. Subsequently, only parts (devices) are displayed that fit the function definitions. That is to say, with device selection, it is not possible to assign to a motor overload switch the PLC card device.

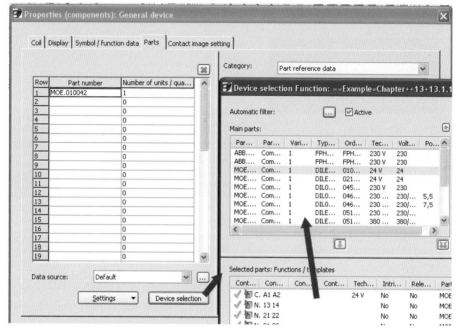

Fig. 13.49 Device selection

Question: How can I export all (or individual) parts in the project to a new parts management?

Answer: To export project parts used (stored) to a *new parts management*, it is first necessary to create a *new database* in parts management.

For this purpose, open parts management and press the EXTRAS button. Here, select the NEW DATABASE function. Then, adopt the new database as your current database and leave parts management.

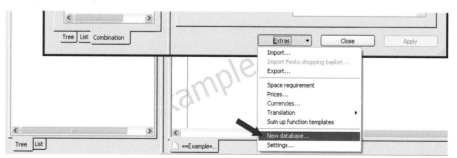

Fig. 13.50 Create a new database

Now, select from the UTILITIES / PARTS menu the SYNCHRONIZE CURRENT PROJECT function. EPLAN opens the **Synchronization of parts** dialog. In this dialog, select the desired or all parts and move them to the new database on the right using the button.

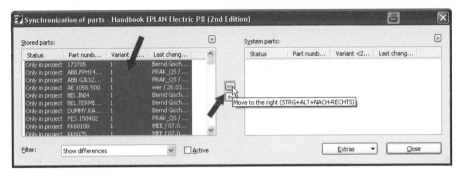

Fig. 13.51 Store project parts in new database

This way, all project parts are stored in a separate parts database.

Question: In parts management, how do I create new product groups, generic product groups or subgroups?

Answer: For reasons related to the data exchange, this is generally impossible. That is, new generic product groups, product groups as well as new product subgroups cannot be created in the EPLAN parts management by the user. These groups are fixed and defined by EPLAN.

Question: In assemblies, how can I enter the function templates of individual devices?

Answer: You do not have to do this manually. EPLAN does this automatically. Simply click on the EXTRAS button in parts management and select the SUM UP FUNCTION TEMPLATES menu item. EPLAN now enters (collects) the function templates of the individual devices and enters them automatically in the *Function template* tab of the assembly.

Question: In parts management, is it possible to copy an existing function template of a part?

Answer: Yes, this is possible. To transfer (copy) a suitable function template of an existing part to another part that does not have a function template, you must do the following:

1. Open parts management.
2. Select a part with a suitable function template in tree or list view.
3. Open the popup menu with the right mouse button and select the COPY function.
4. Now select the desired part (the one without a function template) in the tree or list view.
5. Open the popup menu with the right mouse button and select the PASTE FUNCTION TEMPLATES function.
6. EPLAN now transfers the function template to the part.

NOTE: If the part already has a function template, a confirmation prompt before pasting will ask you whether the existing function template is to be overwritten or not. Make sure to confirm this.

Question: Is it possible to create from stored parts data a device list for an external project?

Answer: Yes, this is possible. To do so, you must export the parts data using a suitable labeling scheme and then import them in the external project as a device list.

Question: What is the meaning of the identifier on the part in parts management (*Technical data* tab), because sometimes it seems it is not transferred when inserting a device?

Answer: The identifier, which can be entered for the part, allows for filtering in parts management on the basis of such identifiers (e.g., for a part selection in graphical editing). It has no other meaning.

Question: How can I generate a translation of the Designation 1 text of parts?

Answer: To translate the Designation 1 text of a part manually, you must do the following:

1. Open parts management.
2. Click on the Designation 1 field of the desired part.
3. Call the popup menu of the right mouse button and select the TRANSLATE entry.
4. EPLAN may open the **Found words** dialog. Here, select the corresponding translation.
5. Adopt the translation by clicking the OK button.

The Designation 1 text has now been translated and can be used in multilingual forms and other reports.

 TO DO: The following condition must be met prior to any translation: The text to be translated must already exist in the dictionary.

Question: How can I make sure that parts to be provided with delivery are not listed in the reports?

Answer: To prevent, for example, parts to be provided with delivery by the customer from appearing in reports, all you need to do is to give this part a characteristic that you can use later on for filtering purposes.

For example, you could use the *Part group* field in the part reference data for this purpose. You could see the entry CS here (CS = customer-supplied; but essentially this is random), and for the purposes of creating reports, these parts are filtered out using a filter scheme and, accordingly, not included in the report output.

 TIP: The *Part group* field can be preset with the entry CS even for parts that are generally customer-supplied in parts management on the *Technical data* tab. This way, the *Part group* is filled automatically with the entry CS on the device when the part is assigned.

Question: Where is a project-specific parts database stored?

Answer: The project-specific parts database is stored in the project directory \...*.edb under the file name part.mdb.

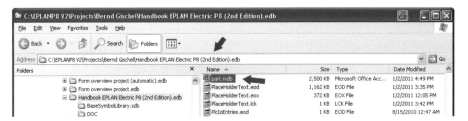

Fig. 13.52 Project-specific parts database

Question: There are many parts missing from my parts management. Where can I obtain the desired parts?

Answer: With the parts database ESS_part001.mdb, EPLAN provides only a limited selection of various parts (as sample parts). To obtain additional parts, you have several options. You can use the EPLAN Data Portal or visit different manufacturers' websites to look for EPLAN parts data. Finally, though, you can and will have to create the missing and/or desired parts in parts management yourself.

■ 13.3 Terminals, plugs

Question: What is a main terminal?

Answer: A main terminal is comparable to a main function in EPLAN. A main function can support further functions, such as a contactor and its contacts. These can then be taken from the navigators and used or placed.

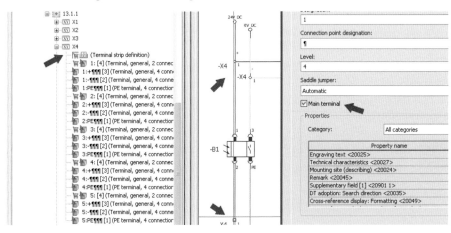

Fig. 13.53 Main terminals and auxiliary terminals in the navigator

A main terminal works the same way. An example of this is the initiator terminal. Apart from the switched connection point, there are also other connection points on the terminal, such as plus, minus or a PE connection point. These functions also exist in the navigators and can be used.

Question: How can I position terminals, i.e., define a sequence?

Answer: Terminals, for example, have properties like *Sorting <20809>* and *Sorting (graphical) <20810>*. On the basis of these two properties, terminals can be positioned freely on a terminal strip independent of their terminal designation. However, one property set excludes the other.

The *Sort code (terminal / pin) <20809>* property (free sorting (input)) also allows for input with decimal points like 0.1. The following applies to the *Sorting (graphical) <20810>* property: If this property has been set, EPLAN will position this terminal in the terminal sequence behind the terminal on the left.

Question: Can one and the same terminal strip be output in two different reports?

Answer: No, this does not work (unless you freeze the first report). Since terminal strips are function-specific reports, there is no way in EPLAN to output one and the same terminal strip twice in a terminal diagram. Function-specific reports may and can occur in a project only once.

Question: Why do the terminal strip parts entered not appear in the parts list?

Fig. 13.54 Include parts

Answer: If articles like *terminal strip parts, terminal parts, cable parts, cable project parts, cable conductor parts, connection parts*, etc. are to be included in reports, for example, in the parts list, you must place the check mark at the corresponding parts (to be included) in the report settings under *Include parts*. The same is true of the desired option to output a parts list with all devices, even if they do not have any parts data.

Question: How can I assign to a terminal strip a "superior" function text?

Answer: To assign a function text to a terminal strip (for example, so that a report can be created on it in a terminal-strip overview), you must enter the function text at the terminal strip definition.

Fig. 13.55 Function text on terminal strip definition

If the terminal strip definitions do not exist, under PROJECT DATA / TERMINAL STRIPS you can call the CORRECT function and have the terminal strip definitions be generated automatically by EPLAN.

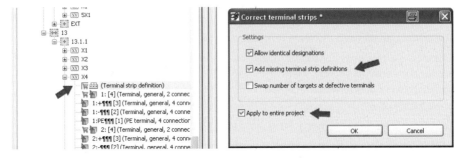

Fig. 13.56 Correct terminal strip

NOTE: For the correction function to work, you must select at least one terminal in the project.

Question: How can I generate plug definitions automatically?

Answer: To create plug definitions automatically, it is possible to call from the PROJECT DATA / PLUGS menu the CORRECT menu item and to select here the *Add missing plug definitions* setting. If the function is to be executed for the entire project, the *Apply to entire project* setting must be selected.

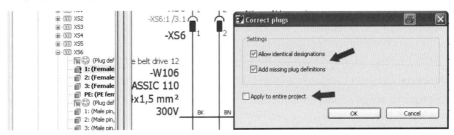

Fig. 13.57 Correct plugs

After you click on the OK button, EPLAN generates the definitions as unplaced functions (visible in the corresponding navigators, and from here they can also be placed on pages if necessary).

Question: How is it possible that terminals have the same terminal designation, without EPLAN bringing up an error message?

Answer: To avoid EPLAN error messages in the check run for terminals sharing the same terminal designation, e.g., in the case of PE terminals and the PE terminal designation, you must activate the *Allow same designations <20811>* property in the symbol properties of the terminal.

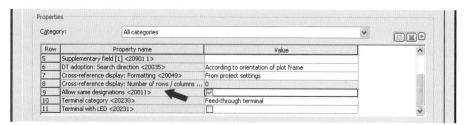

Fig. 13.58 Allow same terminal designations property

Question: How can I remove superfluous terminal strip or also plug definitions from the project automatically?

Answer: To avoid having to delete superfluous terminal strip definitions (or also plug definitions) from the terminal strip or plug navigator manually, you create a scheme in the PROJECT / ORGANIZE / COMPRESS menu that will remove this superfluous data from the project automatically.

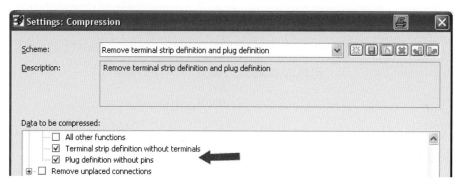

Fig. 13.59 Compression scheme

Question: How can several terminal diagrams be output on a report page in a space-saving manner?

Answer: To output several terminal diagrams (terminal strips) on a single report page, several conditions must be met. The form to be used must be a dynamic form. As well, in the REPORTS dialog (UTILITIES / REPORTS / GENERATE menu) under the SETTING / OUTPUT TO PAGES button, a check mark must be placed for the respective report in the *Combine* column.

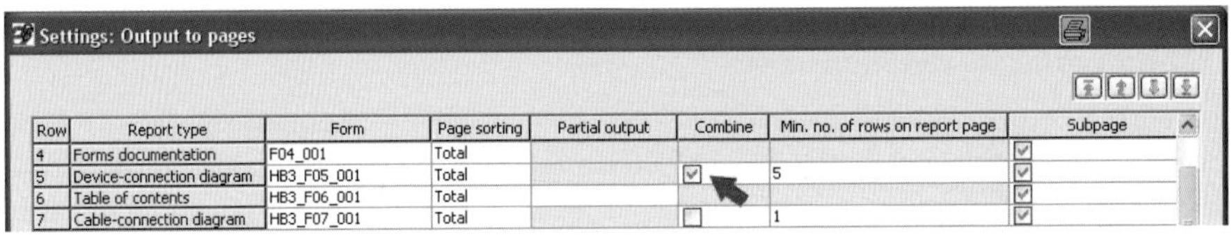

Fig. 13.60 Activate combination for a report

Apart from the terminal diagram report, the *Combine* option is also available for the following reports: Device connection diagram *.f05; Cable diagram *.f09; Cable-connection diagram *.f07; Terminal diagram *.f13; Terminal-connection diagram *.f11; Terminal line-up diagram *.f12; Enclosure legend *.f18; PLC diagram *.f19; Plug diagram *.f22; Pin-connection diagram *.f21 and Symbol overview *.f25

13.4 Cables

Question: Why can EPLAN not execute a correct device selection of cables (incl. assignment of conductor information) on open interruption points?

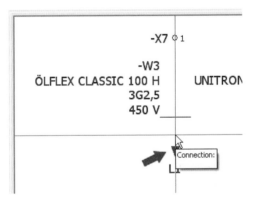

Fig. 13.61 Open connection

Answer: An open interruption point in EPLAN is **not automatically** considered a logical connection. This can be seen very well when you place the cursor on the Autoconnecting line and then view the tool tip. This is why device selection and the automatic placement of conductors can occur only on valid (connected) logical connections.

Question: Which are the identifying properties of a cable definition?

Answer: The following properties of a cable are identifying properties. Accordingly, these properties must match the function definitions of a device selection when the cable and/or its connection data were previously edited manually (e.g., the cable conductors or the pair index may have been inserted manually).

The actual *function definition*, the *shield designation* (Shielded by), the *pair index*, the *potential type*, the (conductor) *color* and/or *number* and the *cross-section* and/or *diameter*.

Fig. 13.62 Identifying properties of a cable part

If one of these properties does not match the cable (in case of a device selection) in parts management, the device selection will not be able to find a suitable cable.

Question: Why are cable details, e.g., conductors of a cable, listed multiple times in the cable navigator?

Answer: If cable details, e.g., *conductors* or the *shopping cart*, are listed in the cable navigator multiple times, it indicates that several cable definitions (or lines) (displayed in a distributed manner) exist in the project and that they are all marked as main functions containing one part each.

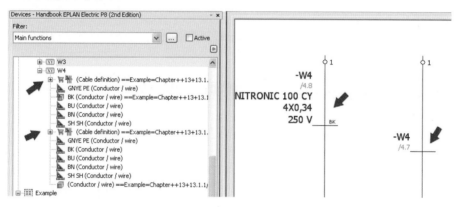

Fig. 13.63 Duplicate cable definition in the navigator

In the case of cables, too, EPLAN only allows for one main function. However, it does not matter to EPLAN where, with which representation type or on which cable definition the main function and the corresponding parts data are located!

Question: How can the conductors of a cable be assigned automatically in EPLAN?

Answer: To let the conductor information like color or cross-section be "filled" by EPLAN automatically for a cable, it is necessary to create the cable as a part complete with all conductors in parts management.

If this cable is then selected and adopted during the input of the cable in the schematic via the device selection, upon exiting of the *Symbol properties cable* dialog, the conductors (crossing the connections) will be labeled by EPLAN automatically based on the parts management data.

Question: How can I place free conductors of a cable on terminals, without drawing them in the project?

Answer: To place free conductors of a cable on terminals, without representing them in the diagram, use the INTERCONNECT DEVICES function in the PROJECT DATA / DEVICES menu.

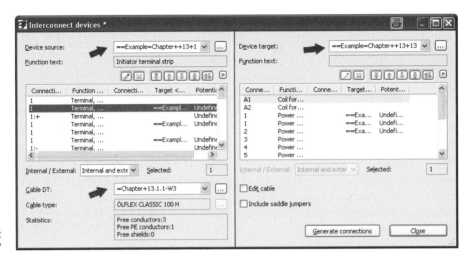

Fig. 13.64 Interconnect devices „virtually"

Here, you can select the desired terminal strips/devices and the desired cable. Then, the corresponding terminals/devices are selected, followed by the GENERATE CONNECTIONS button to establish the connections.

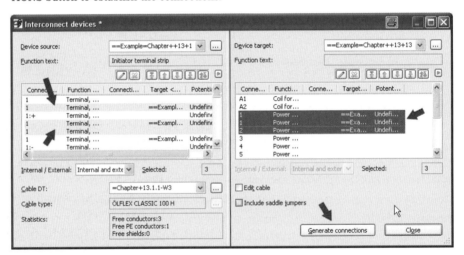

Fig. 13.65 Select devices and generate connections

A "virtual" connection is displayed visually in EPLAN with a small red slash.

Question: How does EPLAN define the source and target of a cable connection?

Answer: When creating reports, EPLAN goes from right to left and from bottom to top and then determines the source and target. This means that when the cable runs from left to right, then the source would be on the left and the target on the right. The same is true of bottom to top. In this case, the source would be at the bottom and the target at the top.

Fig. 13.66 Selection

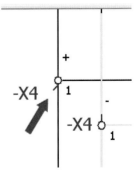

13.5 PLC

Question: How can identical address ranges be used with different PLCs in a project?

Answer: To assign identical address ranges within a project (e.g., when using several CPUs), it is absolutely necessary to "attach" to the individual PLC main functions a CPU name. This way, EPLAN can then differentiate, without a check run message, between several identical addresses on the basis of the CPU (name) having been assigned.

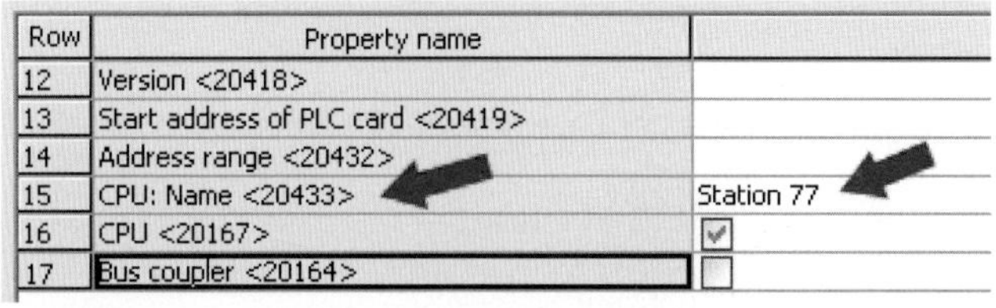

Row	Property name	
12	Version <20418>	
13	Start address of PLC card <20419>	
14	Address range <20432>	
15	CPU: Name <20433>	Station 77
16	CPU <20167>	✓
17	Bus coupler <20164>	

Fig. 13.67 CPU name property

The setting of the CPU name is entered in the *CPU: Name <20433>* property of the PLC box of the main function (for all main functions of the PLC components that are part of this CPU).

Question: Do I have to assign to the PLC connection points (inputs or outputs) or PLC boxes a CPU name?

Answer: It depends on the number of CPUs (PLC stations) in use in the project. If there is only one PLC station, it is not necessary to assign a CPU name (CPU: Name <20433> property). But if there are several PLC stations, a CPU name should always be assigned. This will also allow you to output separate assignment lists.

13.6 Properties, layers

Question: Can I assign in the project properties my own property names?

Answer: Yes, this is possible. To do so, open the project settings (OPTIONS / SETTINGS menu) and open here in the MANAGEMENT branch the SUPPLEMENTARY FIELDS entry. In the *Project* tab, you can now assign your own names to *User supplementary field n <4000n>*.

Question: What information is located in the <10200> project property (Number of pages per page type)?

Answer: This property and/or its value is determined automatically from the project. Nothing is being added here; the values are derived from the project properties on the *Statistics* tab.

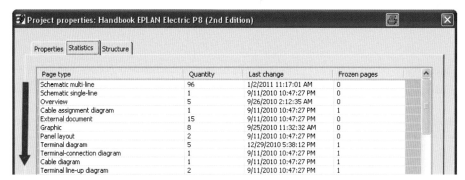

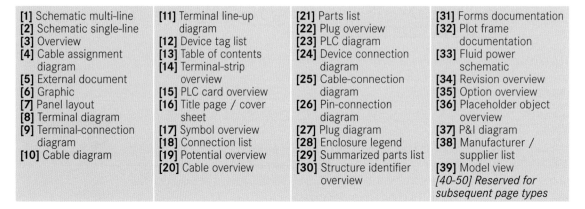

Fig. 13.68 Statistics of the different page types

This is how the Number of pages per page type 10200 [n] property breaks down:

[1] Schematic multi-line
[2] Schematic single-line
[3] Overview
[4] Cable assignment diagram
[5] External document
[6] Graphic
[7] Panel layout
[8] Terminal diagram
[9] Terminal-connection diagram
[10] Cable diagram
[11] Terminal line-up diagram
[12] Device tag list
[13] Table of contents
[14] Terminal-strip overview
[15] PLC card overview
[16] Title page / cover sheet
[17] Symbol overview
[18] Connection list
[19] Potential overview
[20] Cable overview
[21] Parts list
[22] Plug overview
[23] PLC diagram
[24] Device connection diagram
[25] Cable-connection diagram
[26] Pin-connection diagram
[27] Plug diagram
[28] Enclosure legend
[29] Summarized parts list
[30] Structure identifier overview
[31] Forms documentation
[32] Plot frame documentation
[33] Fluid power schematic
[34] Revision overview
[35] Option overview
[36] Placeholder object overview
[37] P&I diagram
[38] Manufacturer / supplier list
[39] Model view
[40-50] Reserved for subsequent page types

Question: How can I carry over the property arrangements quickly from one device to another?

Answer: In order to carry over the painstakingly created arrangements of properties from one device to another, you can use the COPY FORMAT and ASSIGN FORMAT functions in the EDIT menu in EPLAN.

How is a property arrangement carried over? To do so, you should select the (completely) formatted device. Now go to the EDIT menu to select the COPY FORMAT command. Then, select the device to which the previously copied format is to be carried over. Open the EDIT menu again and select the ASSIGN FORMAT command. The format of the property arrangement has been carried over to the device.

TIP: Block functions like multiple selection can be used here. That is, it is possible to select several devices on a page and to carry over to them the format simultaneously. This also works in the navigators. In this case, also select the devices to be modified and assign the format.

NOTE: Formats can be carried over only to similar devices. For example, you could not carry over a cable format to a motor.

Question: What layer is the "anchor" (placeholder object) of a macro with value sets located on?

Answer: The layer of the so-called anchor (placeholder object; symbol 323 / PLHO) of a macro with value sets is located on *EPLAN322 Symbol graphic.Macro.Placeholder object*. Here, for example, you can make settings whether the symbol is to be printed or whether it is to be visible in the diagram.

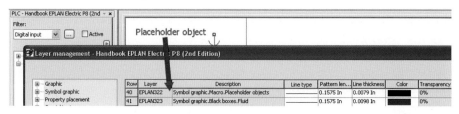

Fig. 13.69 Properties of the EPLAN322 layer

The symbol with the symbol number 323 (symbol name PLHO) originates from the SPECIAL symbol library and cannot be modified, because the SPECIAL symbol library is protected against modifications by EPLAN.

Question: What layer is the part definition point symbol (symbol number 80, symbol name PDP) located on?

Answer: The part definition point symbol (Insert - Part definition point menu) is located on the layer *EPLAN321 Symbol graphic.Part definition points*.

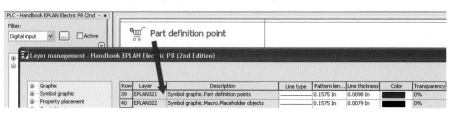

Fig. 13.70 Properties of LAYER321

The symbol with the symbol number 80 (symbol name PDP) also originates from the SPECIAL symbol library and cannot be modified, because the SPECIAL symbol library is protected against modifications by EPLAN.

Question: How can I change the thickness (stroke width) of a dimension line (dimensioning functions)?

Answer: To change the thickness of a dimension line of different dimensioning functions, you can/must edit in the layer management the *EPLAN107 (Graphic.Dimensions)* layer accordingly.

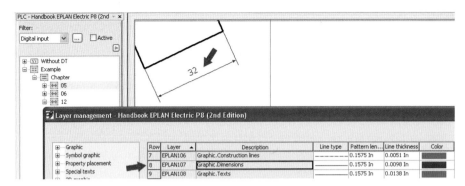

Fig. 13.71 Properties of the EPLAN107 layer

Question: What is the meaning of the three different options of the cross-reference display <20021> property?

Answer: There are three options for the *Cross-reference display <20021>* property. They regulate the display of automatic cross-references between the main function and its auxiliary functions. They have the following meaning.

Fig. 13.72 Differences in cross-reference display <20021>

Automatic display - a cross-reference is displayed only if the *Displayed DT* field is filled with a correct DT, i.e., when it is not empty.

Never display - a cross-reference is never displayed; existing cross-references are suppressed in the display.

Always display - cross-references are generally displayed.

Independent of the displays, there is always a logical connection between the main function and the auxiliary functions. All navigation commands, such as GO TO..., work as usual!

Question: Where can I change the text size of the part number and/or type number above/below the contact image?

Answer: This information (formatting options) is located on the layer *EPLAN480 Property placement.Part number* (for the part number) and on the layer *EPLAN481 Property placement.Type number* (for type designation).

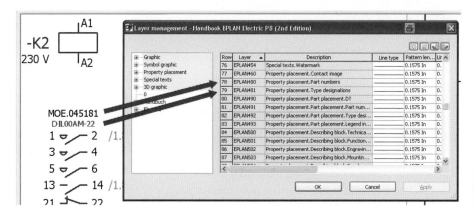

Fig. 13.73 Properties of the EPLAN480 and EPLAN481 layers

Question: Do the settings of layer management generally apply globally, or are they project-specific?

Answer: The layer management settings (called via the OPTIONS / LAYER MANAGEMENT menu) always refer to the current project.

Fig. 13.74 Layer management and the project

Question: What layer are the color settings for the alignment box of texts located on, for example?

Answer: The alignment box, which, for example, can be adjusted on the *Format* tab in the text dialog, does not have a separate layer of its own. All settings (color, invisible, etc.) have so far been controlled via the *EPLAN106 Graphic.Texts* or *EPLAN110 Graphic.Path function texts* layers.

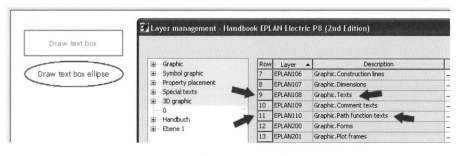

Fig. 13.75 Properties of the EPLAN108 and EPLAN110 layers

Question: How can I display the description of structure identifiers on devices?

Answer: If you wish to display the descriptions of structure identifiers on devices, you simply have to show on the desired device the corresponding property in the display of properties (*Display - Property arrangements - Components* tab). The following properties allow for descriptions of structure identifiers to be shown:

- Description: Higher-level function <1130>
- Description: Higher-level function number <1730>

- Description: Installation site <1430>
- Description: User-defined <1630>
- Description: Document type <1530>
- Description: Mounting location <1230>
- Description: Functional assignment <1330>

13.6.1 Master data

Question: How can I include row numbers in the plot frame?

Answer: You must adjust the following plot frame settings in the plot frame editor. First, you must open the plot frame via the UTILITIES / MASTER DATA / PLOT FRAME / OPEN menu to edit it.

- Insert the *row texts* in the graphical editor with the desired number (INSERT / SPECIAL TEXT menu).

Fig. 13.76 Insert row texts

- Use CTRL + M followed by CTRL + D to call the *Plot frame properties*.
- Enter the *number of rows*, define the *row height*, the *strings* as well as the *numbering format* for the rows and exit the plot frame properties by clicking on OK.

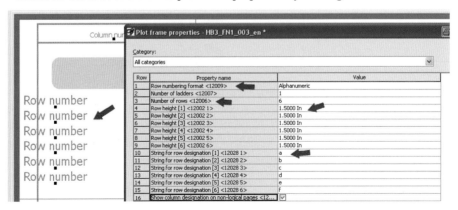

Fig. 13.77 Enter properties for the rows

- Then, select from the menu UTILITIES / PLACE COLUMN AND ROW TEXT AGAIN. EPLAN automatically places the *column texts* and *row texts* anew on the basis of the widths and heights taken from the plot frame properties. These distances may have to be corrected manually and a graphical line may have to be entered, if so desired.

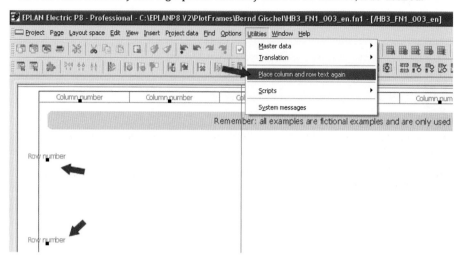

Fig. 13.78 Letting EPLAN determine the distances automatically

Thus, the editing of the plot frame is complete. It can now be closed and used.

Question: What is the difference between the symbols XBD and XBD2?

Answer: In connection with the correct "countersymbols", one combination does not generate a connection (XBD), while the other combination does generate a connection (XBD2). This way, as well, the counterpiece of the plug (female pin) is found and cross-referenced.

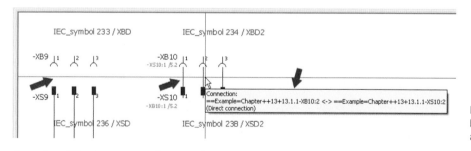

Fig. 13.79 Differences between XBD/XBD2 and XSD/XSD2 symbols

Question: Why is my report form, such as a terminal diagram, larger than my plot frame?

Answer: Here is how this works. It involves an imported EPLAN 5 project. The plot frame (a holdover from the old EPLAN 5 version) was left in the project settings, but reports are now created on the basis of the EPLAN Electric P8 report forms. Since plot frames imported from the old EPLAN 5 version are smaller in size than the EPLAN Electric P8 report forms, the report forms are "written out", or placed, via the plot frame. This can be remedied only by a new plot frame from the EPLAN system master data, such as FN1_001.FN1, for example. Select it from the project settings or modify an existing plot frame according to the old EPLAN 5 plot frame in order to add this new plot frame to the project settings.

The settings of the plot frame are adjusted in the project settings as follows: OPTIONS / SETTINGS / PROJECTS / [PROJECT NAME] / MANAGEMENT / PAGES.

Question: Where can I exchange the global standard plot frame for another?

Answer: The global standard plot frame (considered a default setting for all pages) can be modified or replaced in the project settings under MANAGEMENT / PAGES.

 NOTE: If a page has been assigned a plot frame directly (via the page properties, plot frame name 11016 property), the global standard plot frame will not apply. In this case, it is always the plot frame assigned to the page that is displayed!

Question: Where can I assign a different legend form to a specific mounting panel (in the 2D panel layout)?

Answer: To select for a specific mounting panel a form other than the one globally set in the reports, select the mounting panel, open the properties of the mounting panel and then insert here the *Legend form <20440>* property (unless available).

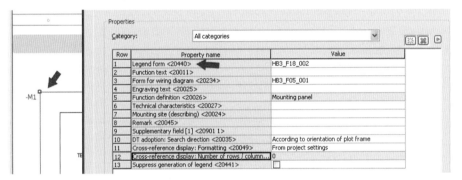

Fig. 13.80 Legend form entry

In the Value column, you can now enter a different form for this mounting panel.

Question: How can I assign to a form a separate plot frame as default, so that the standard plot frame is not imported from the project settings?

Answer: To assign to a report (e.g., a terminal diagram) generally a plot frame different from the standard plot frame of the project, you must do the following.

- Open the desired report (form) via the UTILITIES / MASTER DATA / FORM / OPEN menu item.
- Select and open the desired form in the OPEN FORM dialog.
- In the *Form properties* (accessible by pressing the CTRL + M shortcut key followed by Ctrl + D), set the *Use 'Plot frame to edit form' property for reports <13055>* property and then set the desired plot frame in the *Plot frame to edit form <13001>* property that is to be set automatically in the future for this report (form).

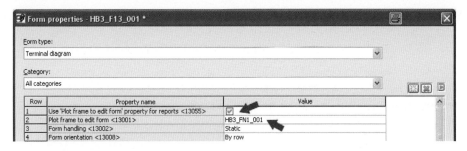

Fig. 13.81 Customize form properties

- Close the form properties and then the form.

From now on, this form will always be automatically assigned the plot frame defined in the form properties for the purposes of graphical output.

Question: Which form information is shown in the "Open form" dialog on the right-hand side, and is it possible to extend this information?

Answer: The following *form information* appears in the right-hand field of the **Open** dialog.

- Description: Description (form, plot frame, outline) <18011>
- Company code:
- Creator: Bernd Gischel
- Creation date: 9/18/2010
- Last editor: Bernd Gischel
- Modification date: 9/18/2010
- Form handling: Static form

Fig. 13.82 Form properties in the Open dialog

This information shown cannot be extended.

Question: Which plot frame information is shown in the "Open plot frame" dialog on the right-hand side, and is it possible to extend this information?

Answer: The following *plot frame information* appears in the right-hand field of the **Open** dialog.

- Description: Description (form, plot frame, outline) <18011>
- Company code:
- Creator: Bernd Gischel
- Creation date: 9/18/2010
- Last editor: Bernd Gischel
- Modification date: 9/18/2010

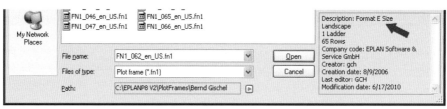

Fig. 13.83 Plot frame properties in the Open dialog

Here, too, it is not possible to add further information to this display.

Question: How can I edit a form that is located only in the project received from a supplier and not in the system?

Answer: To edit a form (report) that is located only in the project master data and not in the system master data - e.g., a project sent by an external supplier who uses proprietary forms in the project - you must do the following:

6. Open the project that contains the form.
7. In the UTILITIES menu, call the MASTER DATA / SYNCHRONIZE CURRENT PROJECT item. The **Synchronize master data** dialog opens.
8. Select the form not stored in the system master data in the left area of the dialog (the project master data) and copy it via the button in the center to the right to the system master data of EPLAN.
9. Close the dialog again. Now, the form (report) can be edited as usual in EPLAN (master data editor).

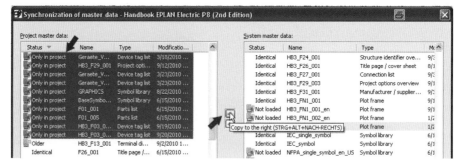

Fig. 13.84 Synchronize master data when working with external master data

TIP: This approach also applies to the remaining master data only stored in the project, such as plot frames, etc.

Question: How can I change the "fixed" project properties, e.g., on the cover/title page?

Answer: You can open the project properties directly via the PROJECT / PROPERTIES menu path and change the desired entries in the *Properties* tab.

Or you could select the project in the **page navigator** (you only need to select one page), call up the popup menu via the right mouse button and then also select PROJECT / PROPERTIES.

Question: When generating a report, I receive the message "S079005 form does not exist". What should I do?

Answer: If the message is received while generating a report of a project or a specific report of EPLAN, you are missing the corresponding form that has been entered under Output to pages or the report template.

Open in the **Reports** dialog under the SETTINGS button the OUTPUT TO PAGES menu item. Here, select an existing form in the Report row, e.g., terminal diagram, from the *Form* column. Or select and adopt an existing form in the corresponding (report) template (in the **Reports** dialog; *Templates* tab) from the *Form* row.

Question: How can I set a symbol (or its symbol graphic) to invisible?

Answer: It is not possible to set a symbol (or the actual symbol graphic) to invisible. To set a symbol (or the actual symbol graphic) to invisible, the existing symbol must be exchanged for a symbol from the symbol libraries that precisely matches this symbol, with the exception that its symbol graphic properties are set to "invisible".

If there is no "invisible" (counter-)symbol in the symbol libraries, the symbol cannot be set to invisible.

Question: Is it possible to "convert" symbols to graphic, for example, to reduce or enlarge them for internal connections?

Answer: Yes, this is possible, and it is quite easy to do. Select the corresponding symbol and then click on it via the EDIT / OTHER / CONVERT COMPONENT INTO GRAPHIC menu. EPLAN then "dissolves" the symbol and generates for each element a separate handle.

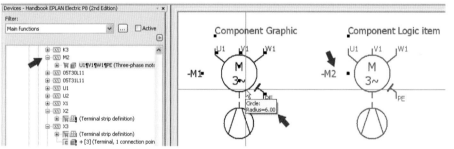

Fig. 13.85 Component converted into graphic

 NOTE: This action can be reversed only immediately (i.e., as soon after the conversion as possible). If the project has been closed, or the UNDO list has been cleared for any other reason, you will not be able to undo the conversion (Graphic -> Component / Symbol) anymore.

Question: How do I remove a symbol library from the project?

Answer: Generally, this works only when all symbols in use from this symbol library have been removed from the project completely. Otherwise, EPLAN will not allow the removal of a symbol library.

13.7 Data exchange

Question: Does EPLAN Electric P8 have an EXF interface like the previous EPLAN 5?

Answer: No, EPLAN Electric P8 does not have such an EXF interface like EPLAN 5.

Question: How do I convert EPLAN Electric P8 projects to the old EPLAN 5 or EPLAN 21 versions?

Answer: Logically (i.e., with evaluable information), this is not possible at all. EPLAN Electric P8 projects are not backward compatible with EPLAN 5 or EPLAN 21. In purely graphical terms, however, e.g., in the DXF/DWG exchange format, this can be done. Of course, this may be subject to certain limitations regarding representation, line thickness or font size.

Question: Is there a viewer for EPLAN Electric P8?

Answer: Yes. There is a viewer for EPLAN, which can be purchased separately.

Question: Should imported EPLAN 5 or EPLAN 21 projects be used as a basis for new projects in EPLAN Electric P8?

Answer: Even though the import function works fairly well (leaving aside special or fully exotic, "distorted" legacy projects), personally I would not use imported EPLAN 5 or EPLAN 21 projects as a basis of future EPLAN Electric P8 projects. Surely, imported EPLAN 5 or EPLAN 21 projects could be enhanced to include the new options of EPLAN Electric P8, and the effort going into this at first does not even appear to be as substantial as creating a new EPLAN Electric P8 project from scratch. But converted data can be buggy or have not really been optimized for the new system (in this case, EPLAN Electric P8).

An example would be the additional auxiliary contacts of a motor overload switch, which in EPLAN 5 had to be drawn on the actual motor overload switch to obtain a cross-reference from the pair cross-reference.

This approach is also possible in EPLAN Electric P8, but not necessary. It is not necessary, because a motor overload switch with auxiliary contacts can be created as a part including the function template, thus displaying the "contact image" on the motor overload switch automatically following a device selection. This means that you do not have to set auxiliary contacts again, only to display the cross-reference to the auxiliary contact used on the motor overload switch (and its auxiliary contacts).

There are a number of examples of where EPLAN Electric P8 has substantially more elegant and more convenient solutions. This is why starting from scratch in EPLAN Electric P8 is the better approach.

Question: How can I export certain information of devices to Excel?

Answer: To export information (properties) of devices easily to Excel, in EPLAN you should use the **Labeling** module. The **Labeling** module is located in the UTILITIES / REPORTS menu. Generally, there are several file formats to choose from: the Excel format *.xls, the text file *.txt and an XML file *.xml.

EPLAN provides a series of complete *schemes* that can be used to import specific properties of certain devices, such as cables, terminals or general devices, to Excel. These schemes can be copied; the properties to be exported can then be modified on the *Label* tab. But it is also possible to create your own, new schemes on the basis of the various *report types* like terminal diagram, device tag list, etc.

Inside the scheme (for exporting to Excel), there are many format output options on the various tabs (*Header*, *Footer* and *Label*). As well, you can set up filter and sorting options in the schemes (**Settings** tab).

Question: Can I use programs other than Microsoft Excel to export and import EPLAN Electric P8 data?

Answer: No, this is not possible. Other program packages, such as OpenOffice, etc., cannot be used for functions such as the export and import of labeling or for external edit functions. This data exchange generally requires Microsoft Excel.

13.8 Reports

Question: Is it possible to generate a report for a device tag list with the graphical overview of complex devices, for example, as demanded by energy supply companies?

Answer: Yes, EPLAN can do that. For this purpose, for example, you can use the conditional forms.

You must do the following in EPLAN.

1. Generate a graphical overview of the device. Ideally, you should create a separate symbol library for this where such overviews can be created and saved.

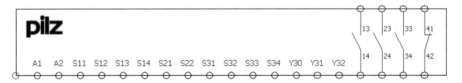

Fig. 13.86 Graphical overview

2. The part for which a report is to be generated in the device tag list with this graphical overview can be assigned this symbol on the *Data for reports* tab in parts management.

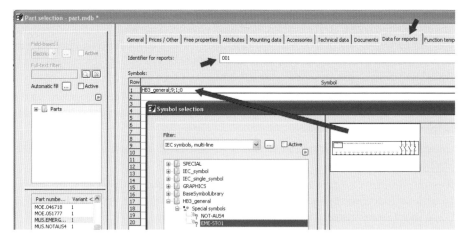

Fig. 13.87 Assign symbols to part

3. Then, you must create corresponding device tag list forms, so that the representation can be displayed on them as well later on. You can start with a basic form that displays all other devices without detailed representation as before.

Fig. 13.88 Basic form editing

Into this standard form you place a conditional form (via the INSERT / CONDITIONAL FORMS menu) and use the [※] button in the **Conditional forms** dialog to select the subform for the graphical detailed representation.

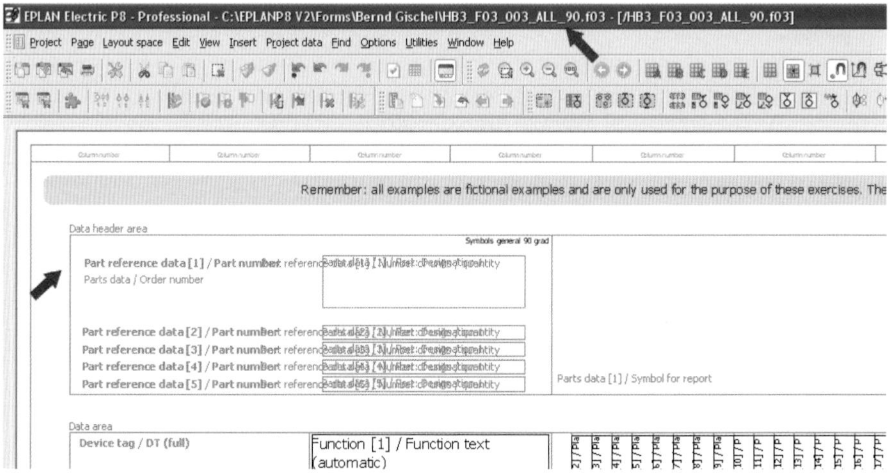

Fig. 13.89 Subform

4. The subform can be called conveniently and directly from the basic form for editing. Once the subform is closed again, EPLAN returns to the **Conditional forms** dialog. In this dialog, it is important to note that a *code* is now defined for the report, so that EPLAN, when generating a report, can assign the part (with its code on the *Data for reports* tab) and the *code* of the conditional form (subform) correctly.
5. If all these preconditions have been met, a report of the device tag list can now be generated. EPLAN generates the report and displays the devices accordingly.

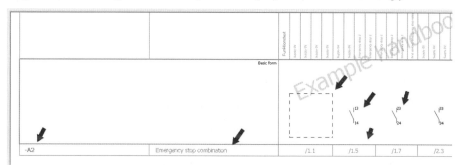

Fig. 13.90 Standard report without detailed representation

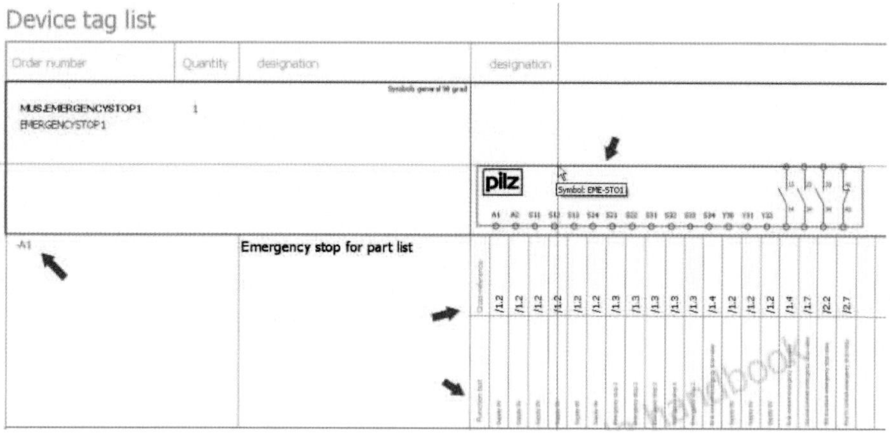

Fig. 13.91 Detailed report with symbol and special report

This "emergency stop combination", of course, is a relatively easy example. It is also possible and conceivable to think of examples of complex devices with internal schematics or plug-in design, etc.

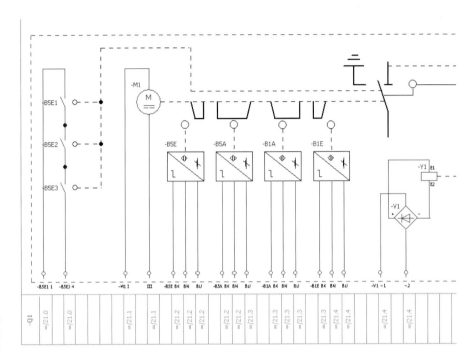

Fig. 13.92 Sample device in the device tag list

Index

Symbols

284
#H# 284
#ID# 284
#RO# 284
#RW# 284
<<...>> string 100

A

Actions 387
Added 132
Address 218
Addresses 216
Adjust page structure 90
Adoption parameters 46
Alignment 58, 126
Alphanumeric 299
Angle variant 135
API 46
Apply 301
Archive 377
Area of application 233
Assigning 164
Assign main function 165
Assignment 201
Assignment lists 216
Assign pages 346
Asterisk 103
AutoComplete 47
AutoCorrect 47
Automated processing 37, 47, 387
Automatic cable generation 204
Automatic cable selection 206
Automatic generation 204
Automatic page description 411
Auxiliary contact 138
Auxiliary function 164
Auxiliary functions 176
Available 277
Available actions 388

B

Back 76
Backed up 373
Back up 375
Back up data 388
Backup directories 384
Backup directory 382
Backup drive 44
Back up entire project 379
Back up external documents 379
Backup file name 391
Backup function 375
Back up image files 379
Back up master data 390
Backup process 394
Backup scope 375
Basic project 73
BIN 27
Black and white 350
Black triangle 58
Blank pages 254
Block editing 191
Block properties 324
bmp 352
Border 128, 134
Brackets 36
Break up assemblies 270
Browse 251
By column or row 36
By position 299

C

Cable assignment diagram 237
Cable conductor 200
Cable connection diagram 236
Cable data 205
Cable definition 209
Cable definition line 204
Cable definition lines 32
Cable diagram 238
Cable editing functions 203
Cable navigator 199
Cable numbering 32, 202
Cable overview 238
Cable parts 250
Cable properties 209, 238
Cables 209
Cable selection 206
Cable type 204
Cancel 76, 202
Categories 304
Category 278, 306
CD / DVD burner 377
CFG 27
Changed pages only 354
Change of standard 45
Change positions 55
Characteristic 329
Characters 254
Check form 409
Check routines 303
Check type no 306
Check type Offline 306
Check type Online / Offline 307
Check usage 299
Circle 110
Clipboard 42
Close 98, 412, 416
Closed 98
Coil contact image 149
Collapse 265
Color codes 36
Color depth 349
Color designation 249
Color information 350
Color settings 45, 367
Column 36
Column configuration 65, 312
Columns 415
Combine 253
Combine main identifiers 249

COM interface *46*
Company fonts *48*
Comparison criteria *329*
Comparison project *70*
Completed messages *307*
Complete option *19*
Component data *329*
Components *132*
Compress *37*, *385*
Compression *349*, *386*
Conductor *200*
Conductor designations *208*
Configure *51*, *55*, *312*
Connection data *199*, *206*, *232*, *248*
Connection definition point *32*
Connection dialog *45*
Connection junction *45*
Connection junctions *34*
Connection levels *236*
Connection list *245*
Connection point description *132*
Connection point designation *132*, *212*, *329*
Connections *36*, *103*, *168*, *200*, *209*, *282*
Consecutively numbered *80*
Contact arrangement *36*
Contact image *36*, *138*, *140*
Contact image margin *141*
Contactor coil *140*
Contacts *140*
Continued dimension *120*
Conversion *419*
Conversion to uppercase *32*
Coordinate input *104*, *107*
Coordinates *102*, *107*, *353*
Coordinate system *103*, *108*
Coordinate tables *45*
Copied *90*, *405*
Copy *91*, *93*, *118*, *298*, *352*
Copy from / to *92*
Copyright notice *17*
Copy to project *380*
Correct *185*, *195*
Correct terminal strips *185*
Create form *413*
Create plot frame *417*
Create report template *265*
Creating *67*
Cross-reference *36*
Cross-reference DTs *167*
Cross-reference representation *36*, *137*
Cross-references *137*
Currency *324*
Cursor position *102*
Cursor settings *45*

D

Data area *410*
Data backup *44*, *373*
Data backups *390*
Database *299*
Data entry fields *100*
Data Portal *468*
Data records *249*, *411*
Data rows *411*
Data to be compressed *386*
Data type *215*
Data volume *30*, *385*
Default cable type *206*
Default directories *28*
Default setting *57*
Default settings *47*, *251*, *384*, *385*
Default values *142*, *203*
Definition *135*
Delete *96*, *118*, *297*, *298*
Deleted *97*
Delete report block *265*
Description *315*, *353*, *376*, *391*
Description dialogs *42*
Designations *322*
Device connection diagram *235*
Device cross-references *36*
Device dialog *54*
Device groups *77*
Device navigator *162*
Devices *278*
Device selection *37*, *207*, *328*
Device structure *67*
Device tag *78*, *162*, *202*
Device tag list *234*
Device tag structure *33*
Device types *158*
Dialog elements *53*
Dialog language *43*
Dictionary *47*, *298*
Digital start address *219*
Dimensioning *119*
Dimension line termination *35*
Direct entry *129*
Directory *68*, *284*, *391*
Directory structure *27*, *41*
Display *31*, *33*, *132*
Displayed DT *132*, *199*
Displayed languages *127*
Distortion *353*
Distributed view *212*
DOC *366*
Document type *295*
Dongle *21*
Drawing directory *236*
DRX *109*
DRY *109*
DT numbering *80*
DWG *340*
DXF *340*
DXF/DWG dialog *340*
DXF/DWG directory *344*
DXF / DWG files *46*, *341*
Dynamic *243*, *244*, *253*, *411*
Dynamic forms *410*

E

edc *281*
Edit *182*, *200*
Edit externally *281*, *286*
Editing in table *65*
Editor *275*. Siehe V5-Editor
Edit properties *99*
Edit terminal strips *183*
elk *69*
ell *69*
elp *69*
elr *70*
els *69*
elt *70*
elx *69*
E-mail *373*, *377*, *392*
E-mail message split size *44*
Embedded reports *233*
Enclosure legend *241*
Engraving text *132*
Enlarge *117*
Enlarged *103*
Enlargement *342*
Entire project *206*, *350*, *354*, *366*
epj *357*
Equals sign *249*
Error *25*, *306*
Excel *231*, *274*, *276*, *279*
Excel template *283*
Exchanging *281*
Exclusive *43*
Expand *265*
Export *339*, *348*, *357*
Exported *288*
Extension *128*
Extent *187*
External application *286*
External files *274*
External format *231*
External PPC system *46*
External project *95*

F

Female pins *195*, *197*, *244*
Filed-off *69*
File off *377*
File type *285*, *342*
File types *281*
Filter *59*, *130*, *234*, *309*, *387*, *391*
Filter dialogs *59*
Filter setting *270*
Find *300*
Finish *76*
Font *48*
Fonts *48*
Font size *126*
Font style *126*
Footer *277*
Form *270*
Format *58*, *126*, *176*
Format elements *277*, *278*, *285*
Format of the subpage *254*
Formatting *57*, *58*
Form editor *405*, *408*
Form errors *410*
Form handling *410*
Form insertion point *256*
Form name *407*
Form overview project *400*
Form project *68*
Form properties *50*, *233*, *410*, *413*
Forms *397*, *404*
Forms documentation *235*
Form settings *248*
Form structure *405*
Found words *298*
Free properties *324*
Full DT *132*
Full name *323*
Full page name *87*
Functional assignment *295*
Function data *66*
Function definition *135*, *139*, *178*, *190*, *196*, *329*
Function definitions *37*, *220*
Function group *327*
Functions *282*
Function template *328*
Function text *132*
Function texts *123*
Function text synchronization *167*

G

General *285*
Generate connections *170*
Generated *85*
Generate file names *342*
Generate functions *220*
Generate pin *198*
Generate plugs *197*
Generate project reports *272*
Generate report *257*
Graphic *85*, *347*
Graphical *85*
Graphical approach *26*
Graphical buttons *53*
Graphical coordinate system *103*
Graphical elements *117*
Graphical functions *110*, *408*
Graphical output *251*, *253*
Graphical output pages *252*, *254*
Graphical outputs *231*, *258*
Graphical reporting *249*
Graphical reports *231*, *257*
Graphics card *15*
Grid *99*, *102*, *104*
Grid sizes *44*
Group *118*
Group number *327*

H

Handle parts when outputting *250*
Handles *58*
Hardware *15*
Hardware protection *21*
Hash *103*
Header *277*, *296*
Higher-level function *295*
Higher-level function number *295*
Highlighted *281*, *405*

I

Icon. Siehe Schaltfläche
Identical function text *249*
Identifier *43*, *80*, *87*, *284*, *291*
Identifier block *295*
Identifiers *33*, *129*, *176*
Identifier set *33*
Identifying property *212*
IEC identifier structure *72*
Illogical values *284*
Image files *46*, *347*
Image formats *347*
IMAGES *352*
Import *55*, *77*, *271*, *286*, *339*, *351*, *357*, *419*
Increment *106*
Increment table *45*
Indirectly *132*
Information *102*
Insert *408*
Inserted *351*
Insertion mode *91*
Insertion point *104*
Installation *15*
Installation language *16*
Installation site *295*
Integer *254*
Intelligence *123*, *364*
Interactive *85*
Interconnect devices *168*
Interconnected *169*
Interfaces *45*
Internal *46*
Internal outputs *231*
Item numbers *299*

J

jpg *352*
Jump functions *364*
Jump point *302*

K

Keep aspect ratio *353*, *355*
Keep source directory *352*
Key combination *61*

L

Label *277*
Labeling *46*, *231*, *274*
Labeling data *275*
Labeling dialog *278*
Language *127*
Language display *134*
Language output *275*
Layer *58*, *126*, *313*
Layer configuration file *314*
Layer designations *313*
Layer export *314*
Layer import *314*
Layer management *313*
Layers *313*
Layer settings *314*
Letters *38*
Level *250*
Levels *237*, *243*
License number *21*

Line *110*
Line thickness *38*
Links *380*
List view *64*, *83*, *92*
Local *27*
Logic 1\
– 1 *102*
Logical *85*, *102*, *213*
Logical coordinate system *104*
Logic elements *105*
Long-term archiving *347*
Lost *98*

M

Macro project *68*
Main form *252*, *256*
Main function *164*, *207*
Main report *256*
Male pins *198*
Managing *68*
Manual *142*, *205*, *232*, *269*, *302*, *313*
Manual placement *242*, *262*
Manual selection *262*, *263*
Manufacturer *322*
Master data *44*, *67*, *390*, *393*
Master data directories *60*
Maximum *410*
Message class *305*
Message description *302*
Message management *25*, *38*, *301*
Message number *306*
Message priority *302*
Message text *306*
Message type *302*
Method *376*
Minimum number *254*
Min. no. of report rows on report page *253*
Motor overload switch contact image *138*
Motor overload switches *36*
Mounting data *325*
Mounting location *252*, *295*
Mounting site *132*
Move *141*, *200*, *297*, *388*
Move base point *108*
Moved *133*
Multi-line *35*, *53*, *85*, *209*
Multiple designations *185*
Multiple selection *393*
Multiple starts *48*
Multi-user operation *43*

N

Name *43*, *269*
Naming conventions *67*
Navigation elements *364*
Navigator *157*
Negative values *104*
Nesting *79*
Network drive *42*
New *72*, *178*, *190*, *196*, *209*, *220*, *265*, *277*, *413*
New device *179*, *222*
New form *407*
New functions *190*, *197*, *220*
New layer *315*
New page *86*
New page if property is changed *270*
New project *72*
Next *76*
Next forms *256*
No output *185*
Normal texts *123*
Note *306*
Number cable DT *210*
Numbered *80*
Numbering *32*, *36*, *80*, *81*, *166*, *175*, *176*, *182*, *201*, *210*, *346*
Numbering format *176*
Numbering function *176*
Numbering (offline) *32*
Numbering (online) *33*
Numbering options *201*
Numbering pattern *220*
Number / Name *135*
Number (offline) *176*
Number plugs *196*

Number terminals *186*
Numeric *187*

O

Object-oriented approach *27*
ODBC *46*
Offline *32*, *176*, *291*
Offline check *303*
Offline numbering *32*, *175*
One language (variable) *127*
Online *26*, *80*, *232*, *248*, *291*, *302*
Online assignment *33*
Online check *303*
Online system *26*
Open *89*, *94*, *251*, *352*, *412*
Open in new window *89*
Operation *68*
Options *275*

Order number *322*
Origin *103*, *104*
Original project *377*, *419*
Output *31*, *280*, *343*
Output directory *342*, *350*
Output file *275*, *286*
Output format *260*, *266*
Output settings *349*
Output sorting *354*
Output to pages *251*
Output type *275*
Overview *47*, *401*
Overview cross-references *35*
Overviews *35*
Overwrite with prompt *352*
Overwriting *70*
Overwritten *406*
Own layers *315*

P

Page *36*, *86*
Page break *249*, *253*, *411*
Page contents *90*
Page description *99*, *270*
Page editing *86*
Page handling *38*
Page name *36*, *87*
Page navigator *83*
Page number *90*, *96*, *254*
Page-oriented *80*
Page properties *50*, *92*, *94*, *99*, *124*, *245*, *281*, *408*
Page range *354*
Pages *282*, *339*
Page sorting *252*, *270*
Page structure *70*, *74*, *88*, *95*, *252*, *346*, *400*
Page type *41*
Page types *35*, *85*
Part assemblies *250*
Partial output *252*, *270*
Part number *137*, *322*
Part quantity list *234*
Parts *137*
Parts data *321*
Parts database *37*
Part selection *37*, *46*, *137*
Parts list *234*
Parts management *162*, *222*, *318*
Parts master data *358*
Paste *92*, *118*
Path areas *53*
Path function text *216*
Path function texts *123*
Path numbering *38*, *80*

Path-oriented *80*
PDF export *365*
PDF file *366*
PDF format *364*
Percent *353*
Performance *42*
PE terminals *187*
Phone *43*
P_ID *27*
Pin connection diagram *243*
PIN number *212*
Pins *195*
Place *194*, *199*, *209*
Placed *168*, *233*, *353*
Placeholder text *408*
Placeholder texts *415*
Place identifiers *88*, *261*
Place multi-line *180*, *194*
Placing *253*
Plain text *123*
PLC address *216*
PLC addresses *213*
PLC assignment lists *33*
PLC box *220*
PLC card *212*, *221*
PLC card overview *242*
PLC cards *158*
PLC components *242*
PLC data *223*
PLC diagram *242*
PLC navigator *211*
PLC numbering *32*, *81*
PLC overview *212*, *214*
PLC-specific settings *219*
PLC terminal *212*
PLC terminals *215*
Plot frame *38*, *415*
Plot frame documentation *240*
Plot frame editor *414*
Plot frame properties *50*, *141*, *415*
Plot frames *397*, *414*
Plug definitions *195*
Plug diagram *243*
Plug navigator *195*
Plug overview *244*
Plugs *195*, *197*, *244*
Plug / Socket strips *197*
png *352*
Point wiring *34*
Polyline *110*
Positioning *296*
Positions *141*
Post-processing *401*
Potential overview *241*
Preceding sign *32*
Preselection *37*
Preview *93*, *344*

Preview dialog *166*
Preview of result *177*, *202*
Previous versions *419*
Prices / Other *323*
Print *354*
Print command *353*
Printer *354*
Print margins *355*
Print position *355*
Print preview *356*
Print settings *354*
Print size *355*
Print to scale *355*
Program directory *27*
Program language *16*
Program start *29*
Program variant *16*
Project *67*, *76*
Project cables *237*
Project checks *38*, *303*
Project data *157*, *195*, *232*, *249*, *251*, *254*, *262*, *279*, *404*, *408*
Project database *48*
Project documentation *231*, *236*, *404*
Project editing *38*, *70*, *72*
Project file *374*
ProjectInfo.xml *373*
Project languages *127*, *275*
Project management *29*
Project master data *31*, *37*, *45*, *60*, *68*, *251*, *255*
Project messages *291*
Project name *76*, *98*, *382*
project options *477*
Project-oriented *80*
Project path *382*
Project properties *30*, *50*, *74*, *81*, *374*
Project property texts *124*
Project selection *93*
Project setting *256*
Project-specific *313*
Project structures *33*
Project template *70*
Project type *68*
Project types *68*, *69*
Project wizards *75*
Properties *49*, *74*, *81*, *99*, *180*, *269*
Property *132*
Property arrangement *56*
Property dialog *104*
Property name *132*
Property names *49*
Property number *43*, *49*
Property selection *50*, *53*, *99*, *132*
Purchase price *323*

Q

Quantity unit *323*

R

Range *41*
Read back *286*
read-only *284*
read write *284*
Rectangle *110*
Reduce *117*
Reduction *342*
Relative coordinate input *108*
Remove project data *386*
Rename *96*
Renaming *281*
Repetition *249*
Repetitions *275*
Replace *297*, *300*
Report *53*
Report pages *233*, *254*
Report runs *248*
Reports *31*, *42*, *123*, *231*, *257*
Report type *53*, *233*, *251*, *259*, *266*
Representation type *135*
Responsible for project *81*
Restore *381*, *382*
Restore back *381*
Restored *373*
Restore master data *392*
Restriction *266*
Review *303*
Revised *70*
Revision control *38*
Revision overview *241*
Revision tracking *69*
Round *255*
Row texts *415*
RX *102*
RY *102*

S

Save *367*
Save additionally *377*
Saved *289*
Scale *99*, *117*
Scaling factor *342*
Schematic-oriented view *222*
Schematic project *68*
Scheme *78*, *219*, *276*, *285*, *302*, *341*, *345*
Schemes *175*, *202*, *304*
Script *37*, *388*

Script name 388
Scrolling 45
Select 96, 168
Select devices 169
Selected 26, 204, 206, 214, 263, 277, 281, 285, 343
Select form 251
Selection 90, 354
Selection list 81, 132
Select plot frame 416
Separators 36, 46, 79
Sequence 53, 187, 312
Sequential numbering structure 72
Service time 327
Set data types 215
Settings 30, 144, 275
Settings – Company 48
Settings – Project 30
Settings – User 41
Settings – Workstation 47
Shortcut keys 47
Single-line 35, 85
Single-screen solution 15
Size 353
Socket strip 196
Sort 183, 296, 299
Sorting 85, 299
Sorting into 346
Sorting sequence 88
Sort setting 270
Source 169, 200, 210, 340, 366
Special characters 32
Special text 50, 52
Special texts 124, 408
Split size 378
Standard 183
Standard base point 108
Standard layers 313
Standard plot frame 80
Start application 278, 286
Start page 269
Static 108, 253, 411
Static forms 410
Status 388
Status bar 102
Storage location 67, 76
Storage medium 377, 392
Stored 38, 49, 60, 68, 80, 251, 385, 412
Storing 30
Structure 74, 130
Structure identifier management 85, 291
Structure identifier overview 244
Structure identifiers 32, 33, 38, 43, 87, 261, 267, 291
Style version 342

Sub-entries 313
Subpages 38, 254
Subsequent data 256
Subsequent editing 281
Summarized 252
Summarized parts list 234, 256
Superior 78
Supplementary field
 – Sheet no. 269
Supplementary fields 38
Supplier 323
Switched on 105
Symbol data 135
Symbol / Function data 135
Symbol libraries 31, 38
Symbol library 130
Symbol library assignment 46
Symbol list 130
Symbol names 129
Symbol overview 245
Symbol preview 130
Symbol project 68
Symbol properties 36, 50
Symbol representation type 130
Symbols 35, 129
Symbol selection 129, 131, 160
Synchronization 68, 398
Synchronize 37, 167
Synchronize function texts 168
Synchronize master data 407
Synchronize parts 358
Synchronizing master data 233
System master data 31, 37, 60, 68, 255, 397, 398
System messages 43, 389

T

TAB 281
Tabbed page list 89
Table of contents 236
Tabs 75
Target 169, 200, 210
Target and point wiring 34
Target designations 346
Target directory 382
Target file 278
Target project 73
Target system 342
Target wiring 34
Technical characteristic 132
Technical data 327
Template 76, 279
Templates 256, 257, 266
Temporary 68, 70, 398, 415, 417
Temporary page 407, 412

Terminal connection diagram 239
Terminal editing 158
Terminal line-up diagram 239
Terminal number 212
Terminal parts 250
Terminal schemes 188
Terminal strip definitions 185
Terminal strip navigator 182
Terminal strip overview 240
Textdatei. Siehe Datei; Siehe Datei; Siehe Datei
Text dialog 57
Text expansion 128
Text file 276
Text input field 125
Texts 122
Title page / cover sheet 245
T-node 34, 45
Total 252
Total + higher-level function 252
Transferred 168
Translation 40, 123, 125, 298
Translation run 41
Translations 128
Tree structure 84
Tree view 63, 92, 130
Two-screen solution 15
Types of check 301

U

UNDO command 98
Undo function 414
UNDO function 281
Ungroup 118
Unit 36
Unplaced 168, 180, 194, 199, 220, 234, 385
Unzipped 373
Update pages 273
Updating 265
Upper-case letters 32
Use available function data 144
Used 199
Use existing 159
Use graphical coordinates 106
Use original DT 205
User-defined 57, 295
User interface 41, 62, 101
User settings 44
User supplementary field 38
Using the mouse 62
Utilities 231

V

Value *132*
Variant *135*
Variants *130*
Viewpoints *157*

W

Warning *306*
Watermark *38*
Wearing part *327*
Which cables *206*
Window legend *242*
Window macros *68*
Wire *200*
Wizard *75*

Workbook *89*
Working area *101*
Workspace *42*, *43*
Write back *213*, *232*, *282*
Write protection *377*
Writing back *203*

X

XML file *276*, *281*, *358*
X,Y *103*

Y

Y positions *107*

Z

Zipped *69*, *373*
zw1 *390*
zw2 *390*
zw3 *390*
zw4 *390*
zw5 *390*
zw6 *390*
zw7 *390*
zw8 *390*